AF572635

SATELLITE COMMUNICATIONS: ADVANCED TECHNOLOGIES

Edited by
David Jarett
TRW, Inc.
Redondo Beach, California

Volume 55
PROGRESS IN
ASTRONAUTICS AND AERONAUTICS

Martin Summerfield, Series Editor-in-Chief
Princeton University, Princeton, New Jersey

Technical papers mainly selected from AIAA/CASI 6th Communications Satellite Systems Conference, April 1976, subsequently revised for this volume.

Published by the American Institute of Aeronautics and Astronautics

American Institute of Aeronautics and Astronautics
New York, New York

Library of Congress Cataloging in Publication Data

AIAA Communications Satellte Systems Conference, 6th, Montreal, 1976.
Satellite communications.

(Progress in astronautics and aeronautics; v. 55)
"Papers . . . selected from the 1976 AIAA Communication Satellite Systems Conference co-sponsored by the Canadian Aeronautics and Space Institute."
Includes bibilographies and index.
1. Articificial satellites in telecommunication—Congresses. I. Jarett, David. II. American Institute of Aeronautics and Astronautics. III. Canadian Aeronautics and Space Institute. IV. Title. V. Series.

TL507.P75 vol. 55 [TK5104] 629.1'08 77-8524
ISBN 0-915928-19-1 [621.38'0422]

Table of Contents

Progress in Astronautics and Aeronautics	Martin Summerfield, Series Editor PRINCETON UNIVERSITY
VOLUMES	EDITORS
1. Solid Propellant Rocket Research. 1960	Martin Summerfield PRINCETON UNIVERSITY
2. Liquid Rockets and Propellants. 1960	Loren E. Bollinger THE OHIO STATE UNIVERSITY Martin Goldsmith THE RAND CORPORATION Alexis W. Lemmon Jr. BATTELLE MEMORIAL INSTITUTE
3. Energy Conversion for Space Power. 1961	Nathan W. Snyder INSTITUTE FOR DEFENSE ANALYSES
4. Space Power Systems. 1961	Nathan W. Snyder INSTITUTE FOR DEFENSE ANALYSES
5. Electrostatic Propulsion. 1961	David B. Langmuir SPACE TECHNOLOGY LABORATORIES, INC. Ernst Stuhlinger NASA GEORGE C. MARSHALL SPACE FLIGHT CENTER J.M. Sellen Jr. SPACE TECHNOLOGY LABORATORIES
6. Detonation and Two-Phase Flow. 1962	S.S. Penner CALIFORNIA INSTITUTE OF TECHNOLOGY F.A. Williams HARVARD UNIVERSITY
7. Hypersonic Flow Research. 1962	Frederick R. Riddell AVCO CORPORATION
8. Guidance and Control. 1962	Robert E. Roberson CONSULTANT James S. Farrior LOCKHEED MISSILES AND SPACE COMPANY
9. Electric Propulsion Development. 1963	Ernst Stuhlinger NASA GEORGE C. MARSHALL SPACE FLIGHT CENTER

10. Technology of Lunar Exploration. 1963	Clifford I. Cummings and Harold R. Lawrence JET PROPULSION LABORATORY
11. Power Systems for Space Flight. 1963	Morris A. Zipkin and Russell N. Edwards GENERAL ELECTRIC COMPANY
12. Ionization in High-Temperature Gases. 1963	Kurt E. Shuler, Editor NATIONAL BUREAU OF STANDARDS John B. Fenn, Associate Editor PRINCETON UNIVERSITY
13. Guidance ano Control - II. 1964	Robert C. Langford GENERAL PRECISION INC. Charles J. Mundo INSTITUTE OF NAVAL STUDIES
14. Celestial Mechanics and Astrodynamics. 1964	Victor G. Szebehely YALE UNIVERSITY OBSERVATORY
15. Heterogeneous Combustion. 1964	Hans G. Wolfhard INSTITUTE FOR DEFENSE ANALYSES Irvin Glassman PRINCETON UNIVERSITY Leon Green Jr. AIR FORCE SYSTEMS COMMAND
16. Space Power Systems Engineering. 1966	George C. Szego INSTITUTE FOR DEFENSE ANALYSES J. Edward Taylor TRW INC.
17. Methods in Astrodynamics and Celestial Mechanics. 1966	Raynor L. Duncombe U.S. NAVAL OBSERVATORY Victor G. Szebehely YALE UNIVERSITY OBSERVATORY
18. Thermophysics and Temperature Control of Spacecraft and Entry Vehicles. 1966	Gerhard B. Heller NASA GEORGE C. MARSHALL SPACE FLIGHT CENTER
19. Communication Satellite Systems Technology. 1966	Richard B. Marsten RADIO CORPORATION OF AMERICA
20. Thermophysics of Spacecraft and Planetary Bodies Radiation Properties of Solids and the Electromagnetic Radiation Environment in Space. 1967	Gerhard B. Heller NASA GEORGE C. MARSHALL SPACE FLIGHT CENTER

21. Thermal Design Principles of Spacecraft and Entry Bodies. 1969	Jerry T. Bevans TRW SYSTEMS
22. Stratospheric Circulation. 1969	Willis L. Webb ATMOSPHERIC SCIENCES LABORATORY, WHITE SANDS, AND UNIVERSITY OF TEXAS AT EL PASO
23. Thermophysics: Applications to Thermal Design of Spacecraft. 1970	Jerry T. Bevans TRW SYSTEMS
24. Heat Transfer and Spacecraft Thermal Control. 1971	John W. Lucas JET PROPULSION LABORATORY
25. Communication Satellites for the 70's: Technology. 1971	Nathaniel E. Feldman THE RAND CORPORATION Charles M. Kelly THE AEROSPACE CORPORATION
26. Communications Satellites for the 70's: Systems. 1971	Nathaniel E. Feldman THE RAND CORPORATION Charles M. Kelly THE AEROSPACE CORPORATION
27. Thermospheric Circulation. 1972	Willis L. Webb ATMOSPHERIC SCIENCES LABORATORY, WHITE SANDS, AND UNIVERSITY OF TEXAS AT EL PASO
28. Thermal Characteristics of the Moon. 1972	John W. Lucas JET PROPULSION LABORATORY
29. Fundamentals of Spacecraft Thermal Design. 1972	John W. Lucas JET PROPULSION LABORATORY
30. Solar Activity Observations and Predictions. 1972	Patrick S. McIntosh and Murray Dryer ENVIRONMENTAL RESEARCH LABORATORIES, NATIONAL OCEANIC AND ATMOSPHERIC ADMINISTRATION
31. Thermal Control and Radiation. 1973	Chang-Lin Tien UNIVERSITY OF CALIFORNIA, BERKLEY
32. Communications Satellite Systems. 1974	P.L. Bargellini COMSTAT LABORATORIES

33. Communications Satellite Technology. 1974	P.L. Bargellini COMSTAT LABORATORIES
34. Instrumentation for Airbreathing Propulsion. 1974	Allen E. Fuhs NAVAL POSTGRADUATE SCHOOL Marshall Kingery ARNOLD ENGINEERING DEVELOPMENT CENTER
35. Thermophysics and Spacecraft Thermal Control. 1974	Robert G. Hering UNIVERSITY OF IOWA
36. Thermal Pollution Analysis. 1975	Joseph A. Schetz VIRGINIA POLYTECHNIC INSTITUTE
37. Aeroacoustics: Jet and Combustion Noise; Duct Acoustics. 1975	Henry T. Nagamatsu, Editor GENERAL ELECTRIC RESEARCH AND DEVELOPMENT CENTER Jack V. O'Keefe, Associate Editor THE BOEING COMPANY Ira R. Schwartz, Associate Editor NASA AMES RESEARCH CENTER
38. Aeroacoustics: Fan, STOL, and Boundary Layer Noise; Sonic Boom; Aeroacoustic Instrumentation. 1975	Henry T. Nagamatsu, Editor GENERAL ELECTRIC RESEARCH AND DEVELOPMENT CENTER Jack V. O'Keefe, Associate Editor THE BOEING COMPANY Ira R. Schwartz, Associate Editor NASA AMES RESEARCH CENTER
39. Heat Transfer with Thermal Control Applications. 1975	M. Michael Yovanovich UNIVERSITY OF WATERLOO
40. Aerodynamics of Base Combustion. 1976	S.N.B. Murthy PURDUE UNIVERSITY **J. R. Osborn, Associate Editor** **PURDUE UNIVERSITY** **A. W. Barrows and J. R. Ward, Associate Editors** **BALLISTICS RESEARCH LABORATORIES**
41. Communications Satellite Developments: Technology. 1976	Gilbert E. LaVean DEFENSE COMMUNICATIONS ENGINEERING CENTER William G. Schmidt CML SATELLITE CORPORATION

42. Communications Satellite Developments: Technology. 1976	William G. Schmidt CML SATELLITE CORPORATION Gilbert E. LaVean DEFENSE COMMUNICATIONS ENGINEERING CENTER
43. Aeroacoustics: Jet Noise, Combustion and Core Engine Noise. 1976	**Ira. R. Schwartz, Editor** NASA AMES RESEARCH CENTER **Henry T. Nagamatsu, Associate Editor** GENERAL ELECTRIC RESEARCH AND DEVELOPMENT CENTER **Warren C. Strahle, Associate Editor** GEORGIA INSTITUTE OF TECHNOLOGY
44. Aeroacoustics: Fan Noise and Control; Duct Acoustics: Rotor Noise. 1976	**Ira R. Schwartz, Editor** NASA AMES RESEARCH CENTER **Henry T. Nagamatsu, Associate Editor** GENERAL ELECTRIC RESEARCH AND DEVELOPMENT CENTER **Warren C. Strahle, Associate Editor** GEORGIA INSTITUTE OF TECHNOLOGY
45. Aeroacoustics: STOL Noise; Airframe and Airfoil Noise. 1976	**Ira R. Schwartz, Editor** NASA AMES RESEARCH CENTER **Henry T. Nagamatsu, Associate Editor** GENERAL ELECTRIC RESEARCH AND DEVELOPMENT CENTER **Warren C. Strahle, Associate Editor** GEORGIA INSTITUTE OF TECHNOLOGY
46. Aeroacoustics: Acoustic Wave Propagation; Aircraft Noise Prediction; Aeroacoustic Instrumentation. 1976	**Ira R. Schwartz, Editor** NASA AMES RESEARCH CENTER **Henry T. Nagamatsu, Associate Editor** GENERAL ELECTRIC RESEARCH AND DEVELOPMENT CENTER **Warren C. Strahle, Associate Editor** GEORGIA INSTITUTE OF TECHNOLOGY
47. Spacecraft Charging by Magnetospheric Plasmas. 1976	Alan Rosen TRW INC.
48. Scientific Investigations on the Skylab Satellite. 1976	Marion I. Kent and Ernst Stuhlinger NASA GEORGE C. MARSHALL SPACE FLIGHT CENTER Shi-Tsan Wu THE UNIVERSITY OF ALABAMA

49. Radiative Transfer and Thermal Control. 1976	Allie M. Smith ARO INC.
50. Exploration of the Outer Solar System. 1977	Eugene W. Greenstadt TRW INC. Murray Dryer NATIONAL OCEANIC AND ATMOSPHERIC ADMINISTRATION Devrie S. Intriligator UNIVERSITY OF SOUTHERN CALIFORNIA
51. Rarefied Gas Dynamics Parts I and II. (two volumes) 1977	J. Leith Potter ARO INC.
52. Materials Sciences in Space with Application to Space Processing. 1977	**Leo Steg** **GENERAL ELECTRIC COMPANY**
53. Experimental Diagnostics in Gas Phase Combustion Systems. 1977	**Ben T. Zinn, Editor** **GEORGIA INSTITUTE OF TECHNOLOGY** **Craig T. Bowman, Associate Editor** **STANFORD UNIVERSITY** **Daniel L. Hartley, Associate Editor** **SANDIA LABORATORIES** **Edward W. Price, Associate Editor** **GEORGIA INSTITUTE OF TECHNOLOGY** **James G. Skifstad, Associate Editor** **PURDUE UNIVERSITY**
54. Satellite Communications: Future Systems	**David Jarett** **TRW Inc.**
55. Satellite Communications: Advanced Technologies	**David Jarett** **TRW Inc.**

(Other volumes are planned.)

PREFACE

The theme of these companion volumes is Satellite Communications Systems for the 1980's, the third decade, and the advances in technology needed to make these new systems possible. In this preface, a perspective view is offered of the relationship between demand and improved cost effectiveness, and approaches are suggested in which new systems and technologies can be brought to bear to accommodate the required dramatic increases in capacity while at the same time conserving our natural resources of frequency spectrum and orbital space.

At the time of this writing, we are just coming to the end of the second decade of the use of communications satellites. Today, we take the availability of high quality and reliable communications "via satellite" for granted. In addition to the Intelsat Global Network, we now have several operational national domestic systems, and regional systems are being planned. The total communications capacity of today's space segment exceeds that of the 1965 Intelsat 1 by a factor exceeding 500! Intelsat continues to grow in capacity at 15-18% per year, and this growth is expected to continue into the 1980's; a number of domestic carriers project growth rates of 12-15%. Some predict that this growth rate will flatten out, citing such examples as "What happens when each individual in the world has a telephone?" My own subjective answer to this question is that it will not be the ability of any individual in the world to talk to any other individual at any time that will limit growth, but rather his ability to pay for it. This telephony example was used here since it is used by all of us and since it is the largest user today of communication satellite systems capacity. Similar growth arguments can be made for newer services such as: high-rate data transmission, specialized television services, and eventually videotelephony. It also is important to remember that the communication needs of the underdeveloped nations are just beginning to be met. It is very probable that dramatic improvements in cost effectiveness of communication satellite systems will exert an equally dramatic effect on growth and demand for these services.

It is important to consider, in this connection, the availability of frequency bands and the availability of geosynchronous orbit positions to meet this growing demand. It is well known that the number of frequency bands available for satellite telecommunications is limited. In addition to defense systems bands, three commercial bands have been allocated by international agreements in the region between 4 and 31 GHz. These are the 4/6-GHz, 12/14-GHz, and 20/30-GHz bands. Next, it is clear that satellites operating within the same frequency band must be separated adequately in orbit to assure an acceptably low level of radio interference between them. It is also necessary that satellite earth stations be located so that they do not interfere with terrestrial microwave radio relay stations operating in the same frequency bands. These factors, together with a limited spectrum, limit the number of orbit positions available for satellite telecommunication. For example, the region of orbital space most useful to North America is limited to about 60°. The number of orbital positions available in the 4/6-GHz bands is about 15, and of these only six permit full United States coverage – i.e., all 50 states and Puerto Rico. Twelve of the fifteen slots already are occupied or soon will be by the United States or Canada. These positions do not include any allocations of orbital space to Central or South American needs. In the 12/14-GHz bands, which are planned for high power television broadcast or telecommunications to relatively small earth terminals, there are perhaps 10 to 15 positions available (depending on the mix between them), of which 10 already are committed to North American use. We see, therefore, that before the needs of most of the developing world (in this case, Central America and the Western States of South America) are met, we already will have nearly consumed the available spectral and orbital space in the 4/6-GHz and 12/14-GHz frequency bands. This same trend will soon follow around the world.

It seems inevitable that the 4/6-GHz and 12/14-GHz bands cannot satisfy all potential users and that it will be necessary to move to higher frequency bands to satisfy our demands for communication. Not only do the higher frequency bands, 20/30-GHz, provide greater bandwidth, but they permit closer spacing of satellites operating in these bands as well. These advantages come at the price of overcoming the increased attentuation in rain. New concepts and technologies can be used here with great effect. In addition to the concepts of power control (both in the uplinks and downlinks) and of earth station space diversity, new technologies appropriate for the 1980's will lead to significantly new trends in system design.

We should take full advantage of the new technologies and the increased orbital weight the shuttle will afford in the 1980's to place larger payloads into the limited number of geosynchronous orbital positions. New technologies, such as high-speed satellite switched time division multiple access (SSTDMA), higher frequency components, and better spacecraft pointing accuracy associated with narrower antenna beams, can further increase the telecommunication capacity of future geosynchronous satellites. It will be possible to build large communication satellites operating at digital rates of 1 to 2 Gbits/sec in the 20/30-GHz frequency band. The full frequency band can be fully reused in each of many spot beams. These satellites will handle tens of thousands of voice circuits, television, and high-speed computer data, all simultaneously. The number of equivalent voice circuits of high-speed digital satellites operating in 20/30-GHz bands will be five to ten times greater than at the 4/6-GHz bands predominantly used today, and these new satellites can be spaced more closely in orbit by a factor of four, since the satellites can be spaced less than 1° apart instead of the 4° spacing required at 4/6 GHz. This capacity can be employed by using the available spectral and orbital space more efficiently and by dedicating satellites to single high-capacity users, or by simultaneously serving several users in one satellite, using an on-orbit sharing approach. All this is not to say that the 4/6-GHz and the 12/14-GHz frequency bands will not have their place in the 1980's – quite the contrary – but their use should be properly balanced with those services planned for the higher frequency bands. A single large shuttle-launched spacecraft can accommodate communications capability in all three frequency bands simultaneously at a single orbital location although, of course, not every 1° of arc can be so equipped.

To summarize, this writer's view is that the next decade will present a dramatic increase in demands for telecommunications on a worldwide basis. These demands can be accommodated by new technologies, at the same time conserving our precious natural resources of radio-frequency spectrum and geosynchronous orbital space.

Volumes 54 and 55 of *Progress in Astronautics and Aeronautics* are organized to follow a twin-thread theme. Volume 54 describes satellite communication systems for the 1980's. The first four chapters cover specific classes of satellite systems: North American Domestic, Intelsat, National and Regional, and Defense Systems. Chapter 5 provides a significant and coherent overview of launch options for the satellite systems engineer. Chapter 6 presents several outstanding advanced systems concepts.

Companion Volume 55 describes the new technologies which will make the systems in Volume 54 possible. A key principle used in the preparation and organization of Volume 55 has been to present more papers in the "other than communications" disciplines. Although we recently have gone through a period when the major emphasis in communications satellites was in communications technologies, significant and major advances have been made in other technologies equally important to the future of these satellites.

The papers presented in these volumes were selected (with one exception) from the 1976 AIAA Communication Satellite Systems Conference co-sponsored by the Canadian Aeronautics and Space Institute, for which I was privileged to serve as Conference General Chairman. The selected papers were then revised and edited for these AIAA archive volumes. Acknowledgments are due the many contributors to that Conference, the Conference Committee, and the Editorial Committee. I owe special thanks to Mr. Richard S. Davies, who aided immensely as a key member of both Committees. Acknowledgment is also due Dr. Martin Summerfield, Series Editor-in-Chief; Miss Ruth F. Bryans, Director of Scientific Publications at AIAA Headquarters; her secretary, Mrs. Jeanne Godette; and my secretary, Mrs. Helen Miller, who provided most of the coordination and all of the secretarial assistance.

David Jarett
April 1977

Chapter I – Advanced Spacecraft Engineering Mechanics

ADVANCED COMPOSITES IN SPACECRAFT DESIGN

J. S. Archer* and L. D. Berman+
TRW Defense and Space Systems Group, Redondo Beach, Calif.

Abstract

Advanced composite materials rapidly are assuming a key role in the design and fabrication of critical components for high-performance spacecraft. Boron, graphite, and Kevlar fiber-reinforced epoxy structures have been flight-qualified and utilized for major spacecraft assemblies. The advantages of advanced composite materials compared to conventional fiberglass and aluminum spacecraft construction are most pronounced for applications requiring low thermal expansion, lightweight construction, high structural rigidity, rf transparency, or thermal isolation. Cost tradeoffs for commercial communication satellite systems favor the extensive utilization of advanced composite materials to minimize the weight of the spacecraft structure and other components that are noncritical to the operation in orbit, so that maximum weight can be made available to improve the payload performance.

Introduction

All spacecraft are designed to satisfy an absolute overall weight constraint. Increased weight allowances for components deemed to be most critical can be achieved only with compensating weight reductions elsewhere in the spacecraft. With more weight available, the orbit operation critical communication and attitude control components can be made more reliable by increased ruggedness and redundancy, made longer lasting with increased fuel capacity for attitude control, or have greater capacity by the addition of more communication channels. Fairchild Space and Electronics Company performed a study[1] addressing the economic value

---

Presented as Paper 76-237 at the AIAA/CASI 6th Communications Satellite Systems Conference, April 5-8, 1976, Montreal, Canada.

*Special Assistant for Engineering Development.
+Assistant Manager, Materials Engineering Department.

of this weight tradeoff for commercial communication satellites. They concluded that the revenue resulting from a communications satellite is proportional to the operational life of a critical component whose life is a function of its weight. This revenue potential was translated to as much as 0.5 to 1.5 \$M/lb available for additional payload in orbit. This tradeoff is a major motivation for minimizing the weight of any spacecraft component that does not contribute critically to the spacecraft operation in orbit, such as the structural subsystem.

Weight savings in the structural subsystem for reallocation to other components can be achieved by configuration improvements and by use of more efficient materials. Basic structure configuration concepts have been well developed over the years, and hence the most fruitful area of improvement is in the application of new materials such as advanced composites to structures because material improvements are being developed at a rapid pace. Aluminum alloy, because of its light weight, low cost, easy fabricability, and high thermal conductivity, is used widely in spacecraft structure, but its stiffness-to-weight ratio is less than one-sixth that of some advanced composites. Moreover, some advanced composites have up to five times the strength-to-weight ratio of aluminum alloys. However, the use of advanced composites, particularly those that are orthotropic or which may be highly notch-sensitive, in spacecraft structures presents a problem. Confidence must be gained by proving the material properties and by developing satisfactory hardware designs that are proven by test. Depending upon the specific materials, this generally implies material characterization and development of designs uniquely suited to the intrinsic constituent materials properties.

The designation "advanced composites" is intended to apply to materials of construction such as boron fiber-reinforced epoxies, graphite fiber-reinforced epoxies, Kevlar fiber-reinforced epoxies, graphite fiber-reinforced aluminum, and boron and borsic fiber-reinforced metal matrix materials. These materials are analogous to the more conventional composites such as glass fiber-reinforced epoxy and steel-reinforced concrete construction. The ultrahigh-tensile-strength boron and graphite fibers are bonded together and stabilized in compression by the low-stiffness epoxy structural matrix in which the fibers are imbedded. Many of the advanced composite material advantages, as well as their disadvantages, arise from the unique orthotropic material properties that they exhibit, in contrast to the more homogeneous and isotropic material

Table 1 Spacecraft material characteristics: metal

Material	Type	Physical and mechanical properties at room temperature (RT)							Thermal properties			
		Density	Ultimate tensile strength	Tensile yield	Young's modulus	Shear modulus	Specific strength	Specific stiffness	Thermal conductivity	Specific heat	Thermal expansion α, in./in./°Fx10^{-6}	
		ρ, lb/in.^{3}	F^{tu}, psix10^3	F^{ty}, psix10^3	E, psix10^6	G, psix10^6	F/ρ, in.x10^3	E/ρ, in/x10^6	K, Btu/hr-ft-°F	σ, Btu/lb-°F	-300 to RT	RT to 212
Aluminum alloy												
Sheet	2014-T6	0.101	67A	59	10.5	4.0	663	104	90.0	0.23	10.4	12.5
Sheet	2024-T3	0.100	65A	47	10.5	4.0	650	105	70.0	0.21	9.5	12.8
Sheet	6061-T6	0.098	42A	36	9.9	3.8	429	101	96.0	0.23	11.0	13.0
Sheet	7075-T6	0.101	78A	70	10.3	3.9	772	102	7[illegible].0	0.23	11.6	12.9
Beryllium												
Extrusion	...	0.067	70-90	40	42.5	20.0	1044-1343	634	104.0	0.5	4.8	6.2
Sheet	Cross-rolled	0.067	65A	42	42.5	20.0	970	634	104.0	0.5	4.8	6.2
Block	Hot-pressed	0.067	40A	26	42.5	20.0	597	634	104.0	0.5	4.8	6.2
Invar 36												
Strip	Annealed	0.291	65	40	20.5	8.1	223	70	6.1	0.123	1.2	0.67
Lockalloy												
Sheet	Be-38%Al	0.076	55	40	27.0	...	723	355	120.0	0.39	...	9.4
Magnesium												
Sheet	AZ31B-H24	0.0639	39A	29	6.5	2.4	610	102	47.5	0.25	...	13.6
Extruded tube	AZ31B	0.0639	32S	16	6.5	2.4	501	102	47.5	0.25	...	13.6
Steel												
Sheet-TH 1050	PH15-7Mo	0.277	190S	170	29.0	11.0	686	105	9.25	0.114	4.6	6.1
Sheet	4130	0.283	95S	75	29.0	11.0	336	102	22.0	0.114	4.8	6.3
Sheet	4340	0.283	260S	215	29.0	11.0	919	102	22.0	0.114	4.8	6.3
Titanium												
Sheet-STA	6Al-4V	0.160	160A	145	16.0	6.2	1000	100	4.3	0.120	4.0	4.9
Forgings-STA	6Al-4V	0.160	150A	140	16.0	6.2	938	100	4.3	0.120	4.0	4.9

Table 2 Spacecraft material characteristics: fiber-reinforced plastics

Material	Layup	Physical properties: Cured resin content, wt %	Density ρ, lb/in.^{3}	Cured ply thickness, mils	Mechanical properties, Longitudinal: Tensile ultimate F_x^{tu}, psix10^3	Tensile modulus E_x^t, psix10^6	Compr. ultimate F_x^{cu}, psix10^3	Compr. modulus E_x^c, psix10^6	Specific strength F_x^t/ρ, in.x10^3	Specific stiffness E_x/ρ, in.x10^6	Interlam. shear F^{isu}, psix10^3	Shear modulus G, psix10^6	Thermal properties: Thermal conductivity K, btu/hr-ft-°F	Specific heat σ, btu/lb-°F	Thermal expansion (-300° to 212°F) α_x, in./in./°Fx10^{-6}	Thermal Expansion (-300° to 212°F) α_y, in./in./°Fx10^{-6}
G/E																
GY70/934	0	...	...	...	90	43.0	70	44.0	1574	705	8.6	0.75	30.0	0.26	-0.58	17.9
	±45	...	...	...	13	2.5	13	2.5	213	41	...	...	...	...	-0.28	- 0.28
	0,±60	...	...	...	33	15.0	37	17.0	541	246	...	...	...	...	-0.16	...
	0,±60,90,∓30	...	...	...	23	13.0	23	13.0	377	213	...	2.9	...	...	...	...
	0,90,90,0	...	0.061	5.2	43	21.0	43	21.0	705	344	...	...	...	...	...	...
	0,±40,0	30	...	...	48	25.0	47	20.0	787	410	...	3.0	26.0	...	-0.83	...
	0,40,0,-40,0	30	...	...	58	27.0	53	28.0	951	443	...	3.0	...	...	-0.76	...
	0,±25,90,∓25	30	...	...	48	24.0	47	20.0	787	393	...	3.0	...	...	-0.48	...
HM-S/934	0	...	...	...	125	27.0	80	28.0	1810	466	9.0	...	6.3	0.26	-0.27	16.6
or	±45	...	...	...	16	2.5	18	2.5	276	431	24.0	...	...	...	0.28	0.28
I/5208	0,±60	...	...	...	37	10.0	39	11.0	638	172	...	3.6	...	...	0.4	0.6
	0,±60,90,∓30	...	0.058	4.0	26	9.0	25	8.0	448	155	...	...	...	...	0.12	0.18
	0,90,90,0	...	...	...	48	16.5	41	15.5	828	284	...	...	3.7	...	0.5	0.5
	0,±25,90,∓25	...	...	...	60	14.0	51	12.0	1034	241	...	...	...	...	0.01	...
HT-S/3002	0	...	...	...	188	22.0	224	22.0	3481	407	13.0	...	...	...	-0.2	16.0
or	±45	...	...	...	30	2.4	25	2.4	556	44	...	...	...	...	1.7	1.7
5206-II	0,90	...	0.054	6.5	80	10.0	80	10.4	1481	185	...	...	...	...	...	...
	0_2,±45	...	...	...	96	12.0	119	11.6	1778	222	...	...	...	...	-0.25	5.5
	0,40,0,-40,0	...	...	...	106	13.0	19	5.5	1963	241	...	...	...	...	...	...
HM-S/934 or I/5208 (thin ply)	...	...	0.057	3.0	...	...	...	...	...	...	...	...	...	...	...	...
Boron/epoxy																
5505/4	0	...	...	...	188	30.0	362	32.0	2611	416	6.0	0.93	1.1	0.22	2.2	16.0
	±45	...	...	...	19	3.0	37	4.5	264	42	...	8.8	...	...	...	...
	0,±60	27	0.072	5.2	68	11.1	226	13.2	944	154	...	...	...	...	3.3	3.4
	0,90	...	...	...	87	18.4	166	16.0	1208	256	...	0.93	...	...	...	...
	0_2±45	...	...	...	104	16.8	238	18.5	1444	233	...	...	...	...	2.6	6.1
SP296-5.6	0	...	...	...	215	30.5	...	...	3028	429	15.9	...	...	...	...	...
	±45	...	...	...	...	...	...	...	...	...	...	...	...	...	...	...
	0±60	34	0.071	7.0	...	...	...	...	...	...	...	...	...	...	...	...
SP296-8.0	0	...	...	...	220	30.3	...	...	...	...	15.3	...	...	...	...	...
	±45	...	...	...	...	...	...	...	...	...	...	...	...	...	...	...
	0±60	...	0.070	7.8	...	...	...	...	...	...	...	...	...	...	...	...
	0,90	...	...	...	...	...	...	...	...	...	...	...	...	...	...	...
Kevlar																
K49/8517	0	...	0.050	...	185	11.0	40	11.5	3700	...	8.0	0.3	1.0	0.4	-2.2	34.0
K49/8517	0,90	...	...	...	44	3.2	...	...	880	...	...	0.23	1.0	0.4	0.8	0.8
(120 clth)	±45	...	0.050	3.5	...	...	...	...	...	...	...	...	...	...	-0.6	- 0.6
Thornel																
HMF-330C (T300/934)	0.90	30	0.060	13.0	90	10.5	...	...	...	...	10.0	...	...	...	1.16	1.24
(wvn clth)	±45	...	...	...	24	2.7	...	...	...	...	...	0.765	...	...	2.14	2.14

Table 3 Spacecraft material characteristics: metal matrix composites

		Physical properties			Mechanical properties								Thermal properties			
					Longitudinal											
Material	Layup	Fiber volume V_f, %	Density ρ, lb/in.^{3}	Ply thickness, mils	Tensile ultimate F_x^{tu}, psix10^3	Tensile modulus E_x^t, psix10^6	Compr. ultimate F_x^{cu}, psix10^3	Compr. modulus E_x^c, psix10^6	Specific strength F_x^t/ρ, in.x10^3	Specific stiffness E_x/ρ, in.x10^6	Interlam. shear F^{isu}, psix10^3	Shear modulus G, psix10^6	Thermal conductivity K, Btu/hr-ft-°F	Specific heat σ, Btu/lb-°F	Thermal Expansion (-300° to 212°F) α_x, in./in./°Fx10^{-6}	Thermal expansion (-300° to 212°F) α_y, in./in./°Fx10^{-6}
B_4/6061-F	0	...	...	...	161	32.0	215	30.0	1660	330	15.0	...	...	...	2.8	9.5
	±45	50.0	0.097	5.2	23	20.0	...	...	237	206	8.4	13.6	...	...	...	...
	0,90	...	...	...	95	19.3	...	...	979	199	...	...	...	...	...	...
$B_{5.6}$/6061-F	0	50.0	0.095	6.5	215	33.0	623	33.0	2263	347	22.0	7.0	...	...	3.2	10.5
$B_{8.0}$/6061-F	0	50.6	0.093	9.5	235	32.3	...	...	2527	347	...	...	...	...	...	...
$BSC_{4.2}$/6061-F	0	...	0.100	5.3	163	30.0	209	34.9	1630	300	13.0	7.0	...	0.28	2.9	9.5
	0,±60	50.0	...	...	65	...	...	...	650	...	...	...	...	...	...	...
	0,90	50.0	...	...	80	...	...	...	800	...	...	...	...	...	...	...
	±45	...	...	...	17	...	...	...	170	...	...	...	...	...	...	...
$BSC_{5.7}$/6061-F	0	...	...	6.6	190	33.0	...	...	2000	347	16.0	7.0	...	...	...	...
	0,±60	50.0	0.095	...	76	...	...	...	800	...	...	...	...	...	...	...
	0,90	50.0	0.095	...	95	...	...	...	1000	...	...	...	...	...	...	...
	±45	...	...	...	23	...	...	...	242	...	...	...	...	...	...	...

properties exhibited by the conventional noncomposite metallic structural materials such as aluminum, titanium, beryllium, Invar, and steel (see Tables 1-3).

Background

A distinction must be made between the fiber-reinforced plastic composites (graphite/epoxy, boron/epoxy, and Kevlar/epoxy) and the metal matrix composites. The advantages of metal matrix composites are centered around good mechanical properties and the fact that they are electrically and thermally conductive. In addition, metal matrix materials are not subject to any deterioration from moisture. However, the density of metal matrix composites is much higher than fiber-reinforced plastics, thereby decreasing specific strength and stiffness. Processing of these materials is also much more difficult, since it involves coated frames and spacers, metal envelopes, and high temperatures and pressures for bonding. Thus metal matrix composites have not been considered seriously for spacecraft applications.

The commonly available advanced composite materials are graphite fiber, boron fiber, and Kevlar fiber-reinforced epoxies. The major advantage of the graphite/epoxies lies in their diversity and cheapness. Fibers having a wide range of strengths (200,000 to 450,000 psi) and moduli (30 to 88×10^6 psi) have been produced from a variety of precursors and are readily available. The fiber diameters are extremely small, resulting in complete flexibility in spite of the very high modulus. Products are available as prepregged tapes, broad goods, or woven cloth. The wide range of products makes it possible to tailor materials and configuration selectively to suit almost any application. Tow materials, such as HM-S or type I, can be spread out to produce thin prepreg tapes or broad goods. Tapes as thin as 0.0005 in. actually have been produced, but cannot be considered to be reliable. Tapes and broad goods that will result in a cured ply thickness of 0.0025 in. can be produced reliably, but costs increase by a factor of 4 to 5, and resin contents are high, usually about 50%. All graphite fibers have a negative coefficient of thermal expansion, some more negative than others. By varying the type of fiber, resin, and layup configuration, laminates or shapes can be produced which will exhibit an α ranging from -0.8×10^{-6} to about 6.0×10^{-6} in./in./°F. Thus dimensionally stable components can be produced readily, as well as components with similar expansivities to adjacent materials. An important element in determining the

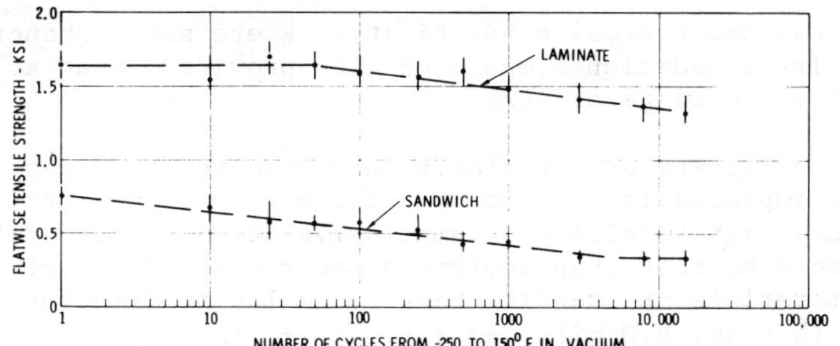

Fig. 1 Effect of thermal cycling on flatwise tensile properties of GY70/934 laminates and cocured sandwich.

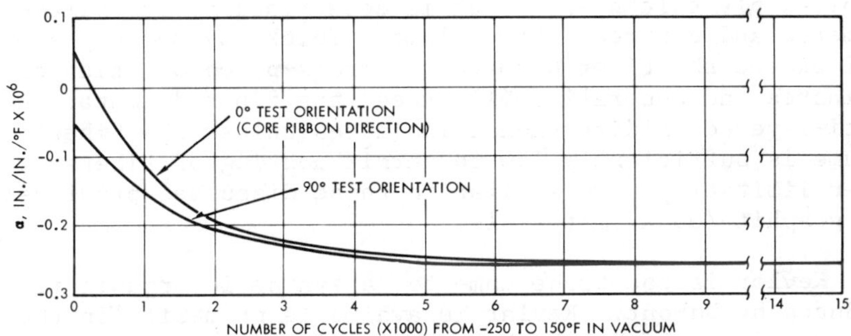

Fig. 2 Effect of thermal cycling on thermal expansion of GY70/934 cocured sandwich.

material behavior is the composition of the epoxy which binds the fibers together. The epoxy composition will determine the cure cycle and affect such properties as creep, shear strength, moisture sensitivity, outgassing, and ultraviolet sensitivity, which affect its long-term stability in the space environment. Figures 1 and 2[2] illustrate the relative changes in strength and thermal expansion which may be expected in a typical graphite/epoxy system as the result of extended exposure to the space environment with continual thermal cycling between temperature extremes. These data also indicate the asymptotic nature of the material changes that provide for selection of practical design values for analysis of performance at end of life. Graphite/epoxy construction has the disadvantage of brittleness, which requires care in the design of mechanical joints and the avoidance of holes when possible, and has poor electrical and thermal conductivity. Where good conduction is required, additional means must be provided, such as conductive coatings or straps.

Boron fibers are available in collimated tape configurations impregnated with epoxy. The boron/epoxy materials have very high tensile and compressive strength and stiffness comparable to most graphite/epoxy materials. Users of these materials are tending toward the larger-diameter fibers (5.6 and 8.0 mils) as a means of increasing strength and lowering density. However, data on laminates made from the large-diameter fibers still are quite limited. The major drawbacks to the use of boron/epoxy lie in its handling characteristics and cost. The fibers are extremely stiff and brittle, difficult to work with, and limited in the radius around which they can be wrapped. Another drawback involves ply thickness, which is determined by the filament diameter and desired fiber volume. Thickness per ply always will exceed the fiber diameter in cross-plied or oriented laminates and generally will exceed the fiber diameter in multi-layered unidirectional laminates unless the fiber volume is sufficiently low to permit nesting of fibers. Other limitations are similar to those discussed previously for graphite/epoxy materials.

Kevlar is the trade name for a synthetic organic fiber produced by DuPont. Kevlar is available primarily in the form of woven cloth, which subsequently is impregnated with epoxy for structural fabrication. The fiber itself has about 450,000-psi tensile strength, a modulus of about $19 \mathrm{x} 10^{-6}$ psi, and a density of 0.052 $\mathrm{lb/in.}^3$. These properties give Kevlar a specific strength higher than either boron or graphite and a reasonably high specific stiffness. In addition, Kevlar composites have a very low dielectric constant and loss tangent, making them very suitable for use in rf communication systems. Kevlar laminates have a near-zero coefficient of thermal expansion, making them viable candidates for thermally stable structures. The major limitation to the use of Kevlar composites in lightweight structures is its modulus. Stiffness cannot compete with either boron or graphite composites. Although it has near-zero thermal expansion, careful processing is required. Test results developed at TRW indicate that fairly large changes in α may occur as the result of very small changes in resin content. Therefore, if a given range of α is required, extremely careful monitoring of raw material resin content and process variables must be exercised.

The relative efficiency of the advanced composite materials is a strong function of the specific structural

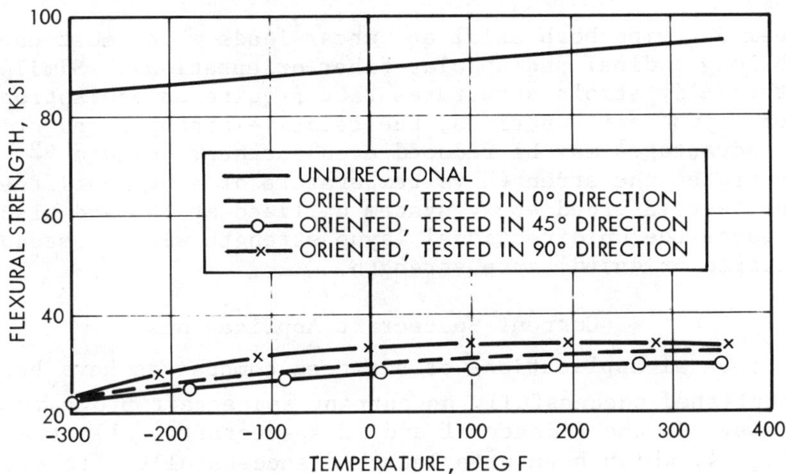

Fig. 3 Flexural strength of GY70/934 laminates vs. temperature.

Fig. 4 Pioneer spacecraft in Jupiter flyby.

configuration in which they are made. For example, when utilized in a strut configuration carrying axial loads only, the required unidirectional fiber orientation theoretically could result in a weight savings as high as 75% for graphite/epoxy compared to an aluminum structure. This compares to a potential savings of approximately 50% when utilized in a beam carrying both axial and shear loads which must use both longitudinal and angular fiber orientations. Similarly, in thermally stable structures that require pseudoisotropic fiber layup configurations, the relative strength and stiffness advantages may be reduced even further. Figure 3[2] illustrates the strength vs temperature of a high-stiffness graphite-reinforced epoxy system utilized at TRW and also contrasts the unidirectional layup strength with a pseudo-isotropic oriented layup strength.

Current Spacecraft Applications

Several applications of advanced composites have been accomplished successfully on current spacecraft programs as follows. On the Pioneer 11 and 12 spacecraft[3], illustrated in Fig. 4, which have been launched successfully into orbits extending beyond Jupiter, structural elements fabricated of boron fiber-reinforced epoxy construction were utilized in 10 critical locations. These consist of six platform support struts, three antenna feed support struts, and the outer segment of the deployable magnetometer boom. A net savings of about 3 lb/spacecraft was realized by this usage, representing a 43% component weight savings compared to conventional construction. In addition to the weight savings that were realized, the high thermal impedance of the boron/epoxy-reinforced construction in the platform struts provided added thermal isolation for the equipment compartment, which minimized heat leakage to space. The boron fiber-reinforced epoxy construction in the magnetometer boom segment provided a high bending stiffness with minimum creep which satisfied a preload retention requirement in the stowed configuration.

NASA's Applications Technology Satellite[4], models F and G, employs graphite fiber-reinforced epoxy for the reflector feed support truss (see Fig. 5). This 14-ft-high truss has critical requirements for light weight and thermal stability to maintain accurate feed/reflector positioning in the antenna system. The HM-S graphite composite construction weighs approximately 50% as much as a comparable aluminum truss design. This design eliminated the thermal distortions that were predicted for an aluminum feed support truss, which otherwise would have caused a 1-dB signal loss due to antenna defocusing.

The unique thermal stability characteristics of graphite fiber-reinforced epoxy have resulted in the use of this construction for a number of antenna parabolic reflector applications. TRW, Hughes Aircraft, General

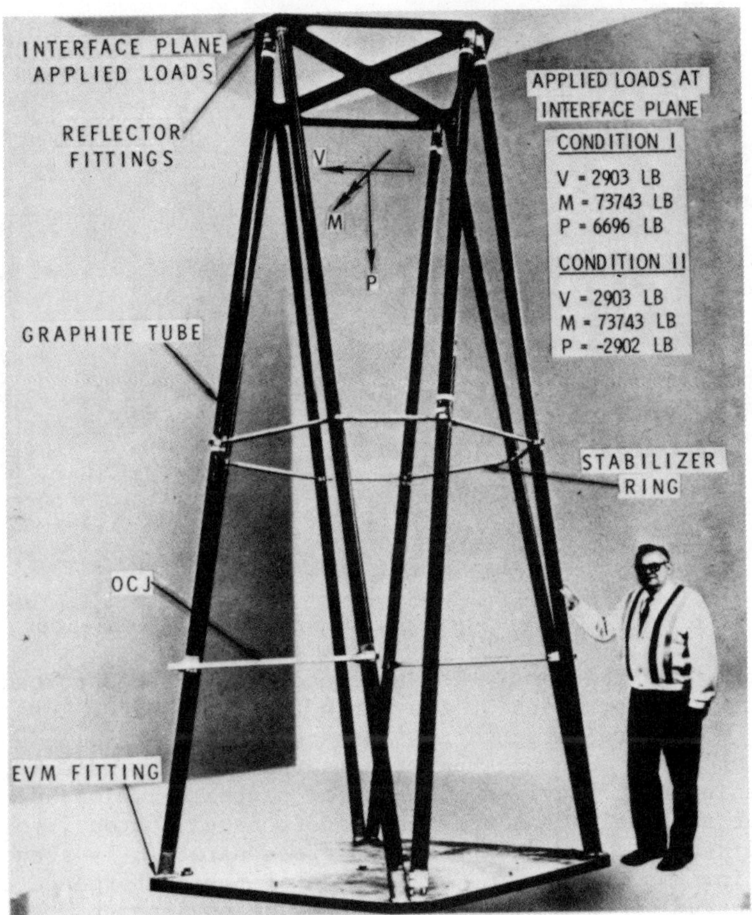

Fig. 5 NASA ATS F/G reflector feed support truss.

Dynamics, and Ford Aerospace and Communications Corporation all have developed and fabricated graphite fiber-reinforced epoxy parabolic reflectors of various configurations for flight applications. TRW's design features the utilization of graphite facesheet sandwich construction with aluminum honeycomb core in a monocoque shell configuration. In contrast, General Dynamics' approach features a truss-supported shell-type construction, and Hughes Aircraft[5] has utilized graphite-reinforced epoxy rib-supported reflecting surfaces successfully.

The TRW reflector is illustrated in Fig. 6. This configuration utilizes GY70 ultrahigh modulus fibers in a 6-ft-dia sandwich construction that was developed for 60-GHz

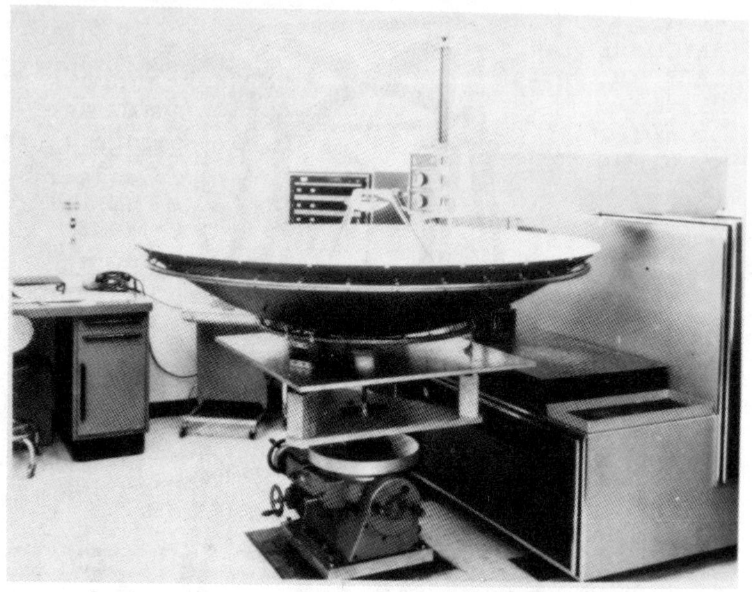

Fig. 6 TRW 6-ft graphite epoxy 60-GHz reflector.

frequency applications. The reflector shell is supported and stiffened by means of two rings fabricated from Invar tubing which are attached to the back of the parabolic shell near the outer edge and near the midradius. The outer stiffening ring provides a means for postfabrication adjustment of the shell contour. The manufactured contour, following adjustment, deviates from a perfect parabola by about 0.001 in. rms. Real-time evaluation of the performance to be expected of this reflector in a space environment has been accomplished by performing full-scale measurements under thermal vacuum exposure utilizing laser holography interferometry. The demonstrated gain and pointing performance of the TRW reflector at 60-GHz frequencies suggests that it would be fully acceptable for use at rf frequencies higher than 100 GHz.

The Hughes Aircraft[5] reflector configuration utilizes HM-S fibers in a 6-ft off-axis reflector for the Canadian Domestic Satellite (Anik I). This design utilizes a sandwich rib construction. Hughes Aircraft also has utilized HM-S fibers as composite construction to stiffen the horizontal cross-arm of the Intelsat IV mast assembly (see Fig. 7). Boron fiber composite also was used to stiffen the elliptical-shaped waveguides that support the reflector feeds.

RCA Limited[6] has made extensive use of advanced composites in the communication spacecraft Satcom utilized for

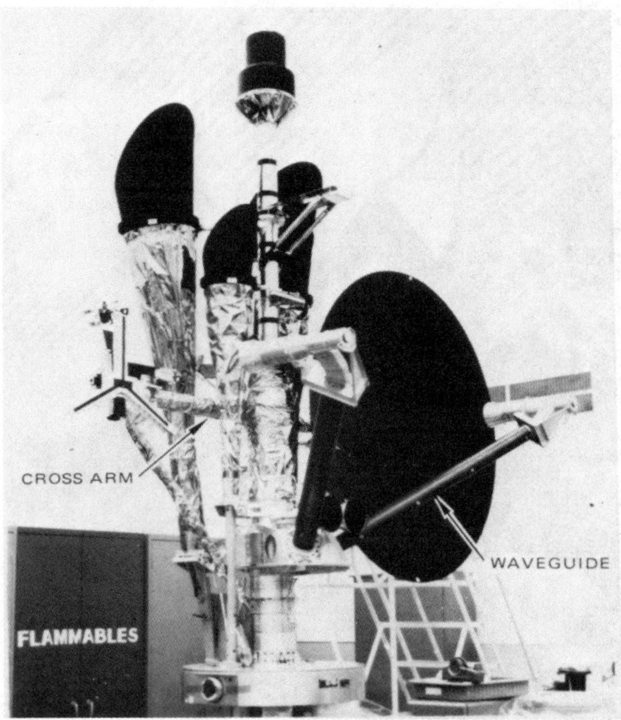

Fig. 7 Hughes Intelsat IV mast assembly.

the RCA communication system. The feed horns, waveguide runs, feed support structure, and rf bandpass filters all are fabricated of graphite fiber-reinforced epoxy construction. The primary requirements were light weight and thermal stability in the space environment. The antenna reflectors, illustrated in Fig. 8, were fabricated of all-Kevlar sandwich construction supported by graphite fiber-reinforced epoxy tubes under subcontract to TRW. The requirements for this application were light weight, thermal stability, and rf transparency, which were satisfied uniquely by the Kevlar construction.

Ford Aerospace and Communications Corporation Western Development Laboratories (WDL)[7] has built graphite/epoxy antenna reflectors for the Mars Viking orbiter. These reflectors were made of thin graphite/epoxy facesheets on aluminum honeycomb core. WDL has utilized graphite/epoxy for three horn antennas and Kevlar for the antenna support structure on the Nato 3 spacecraft. WDL also is utilizing graphite/epoxy and Kevlar for the antennas on the Japanese ETS-2 and the medium-capacity communications satellites being built under contract to Mitsubishi.

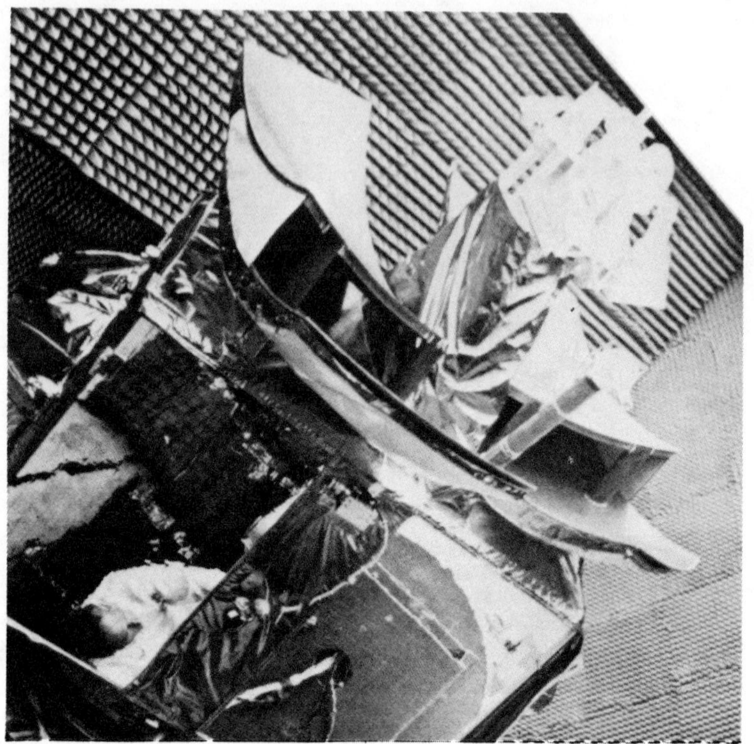

Fig. 8 RCA Satcom spacecraft antenna assembly.

Fig. 9 TRW deployable helical antenna structure.

A final illustration of advanced composites applied to a current spacecraft program at TRW is illustrated in Fig. 9. A deployable off-axis mounted helical antenna boom and supporting structure is fabricated of ultrahigh modulus GY70 graphite fiber-reinforced epoxy. The combined requirements of light weight, high rigidity, and thermal stability could have been satisfied only by this material. In addition, the reinforcing spider that supports the base cup on the helical antenna is fabricated of Kevlar fiber-reinforced epoxy construction. The primary requirements here were lightweight construction, high strength, and thermal stability.

Future Spacecraft Applications

Specific tradeoff studies at TRW have shown that weight savings in the structure subsystem of 25% over existing conventional designs can be accomplished in a spacecraft primary structure by the greater utilization of graphite/epoxy construction. Current work has demonstrated the feasibility of fabricating major structural elements such as integral structural frames, equipment panels, and the central structural cylinder or thrust cone of graphite/epoxy construction.

Technical factors other than weight and thermal geometric stability assume significant importance when evaluating the use of advanced composites for primary structure applications. Conventional spacecraft structures, in addition to carrying loads and providing rigidity, also provide thermal conduction paths, dc grounding, rf shielding, and electrostatic grounding. It must be noted that fiber-reinforced plastics are poor thermal conductors and may not be able to handle severe thermal requirements without auxiliary aids such as heat pipes. Typical spacecraft electrical requirements imposed on conventional construction call for a maximum resistance of 1 mΩ from one end of the spacecraft to the other. If the primary structure is a nonconductor, then a continuous foil or plating applied to one surface of the nonconducting structure may be necessary to satisfy the electrical requirements.

The advantages of advanced composite construction are not limited to lightweight structure subsystems and thermally stable antenna applications. The value of weight savings applies to the construction of components for all spacecraft subsystems. Thus we find potential graphite/epoxy usages in lightweight solar array support structure, solar collection devices, thermally stable housings for critical attitude control sensors, waveguides, and rf bandpass filters.

Fig. 10 TRW graphite composite bandpass filters.

The utilization of graphite/epoxy bandpass filters in large-capacity multichannel communication satellites has become crucial to minimize the payload weight and allow increases in the channel capacity. These transponder components require extreme dimensional stability in the thermal/vacuum space environment which can be provided only by conventional Invar construction or by graphite/epoxy construction. The graphite/epoxy construction (see Fig. 10) that has been developed for this application results in components that weigh approximately one-fourth those of lightweight Invar construction. It is only a matter of time until graphite/epoxy construction becomes commonplace in all high-performance communication satellites, including Intelsat V, TDRSS, and Aerosat.

Two final major applications for graphite/epoxy in space structures must be mentioned. These are the large space telescope (LST) and the Space Shuttle Orbiter[7]. The LST metering truss has the function of maintaining the relative positions of the primary and secondary mirrors to an extremely tight tolerance. This 24-ft-long structural assembly therefore must be strong, stiff, lightweight, and dimensionally stable: the perfect role for graphite/epoxy construction. The Space Shuttle Orbiter payload bay doors now being fabricated of graphite/epoxy construction by Rockwell International/Tulsa are 60 ft long by 15 ft wide. They will provide a weight savings of more than 900 lb (23%) per ship set compared to the aluminum baseline design.

References

[1]Raab, B. and Friedrich; S., "Design Optimization for Profit in Commercial Communications Satellites," Contract AT(49-15)-3063.

[2]Berman, L.D., "Reliability of Composite Zero-Expansion Structures for Use in Orbital Environment," Composite Reliability, STP 580, 1975, American Society for Testing and Materials, pp. 288-297.

[3]"Pioneer Probe Set for Launch Toward Jupiter," Aviation Week and Space Technology, Vol. 96, Feb. 14, 1972, p. 17.

[4]Randolph, R.E., Jensen, L.C., and Martin W., "Graphite Epoxy Reflector Support Truss for Applications Technology Satellite (ATS)," National SAMPE Technical Conference Series, Vol. 4, Oct. 1972, pp. 209-224.

[5]Switz, R.J., "High Modulus Composites Versus Beryllium for Achieving Stiffness in Spacecraft Structural Applications," AIAA Paper 73-374, March 20-22, 1973, Williamsburg, Va.

[6]Stein, K.J., "Synchronous Satellite Offers 24 Channels," Aviation Week and Space Technology, Vol. 103, Dec. 8, 1975, pp. 39-43.

[7]"Orbiter Composite Structures Readied" and "Primary Spacecraft Structures Studies," Aviation Week and Space Technology, Vol. 104, Jan. 26, 1976, pp. 126, 128-129.

APPLICATION OF GRAPHITE COMPOSITES TO FUTURE SPACECRAFT ANTENNAS

John A. Fager*
General Dynamics Corporation, San Diego, Calif.

Abstract

There is a continuing trend toward higher frequencies and larger antenna systems for the next generation of communication satellites. Lightweight graphite composite with high rigidity ($E = 14 \times 10^{-6}$) and near zero coefficient of thermal expansion (4×10^{-8} in./in./°F) appears to be an ideal material to fabricate future close-tolerance systems. The performance characteristics of three types of composite antennas are considered: a solid reflector for millimeter wave application up to 600 GHz; an erectable antenna for 15- to 300-ft applications where packaging is limited by booster envelope; and a phased array system where thermal stability is important in the waveguide network. By increasing a reflector's thermal stability, in-space contour and beam-pointing tolerance can be improved substantially, with corresponding gain improvement. In addition, a 40 to 60% weight reduction of existing metal systems can be achieved with graphite. RF components such as filters, cavities, and oscillators can be built at one-fifth the weight of equivalent Invar components.

Introduction

In the next 20 years, communication satellites will have an ever-increasing problem of rf spectrum availability and minimization of interference. The impact of these problems on antenna systems will drive us toward larger antennas operating at higher frequencies with rigorous control on pointing accuracy and sidelobes. Thermally stable graphite with high rigidity will be used effectively in reflectors, arrays, and lens systems to meet these future requirements.

The ideal antenna material would have a zero thermal coefficient of expansion, high rigidity, and light weight.

Presented as Paper 76-238 at the AIAA/CASI 6th Communications Satellite Systems Conference, April 5-8, 1976, Montreal, Canada.

*Program Manager, Convair Division.

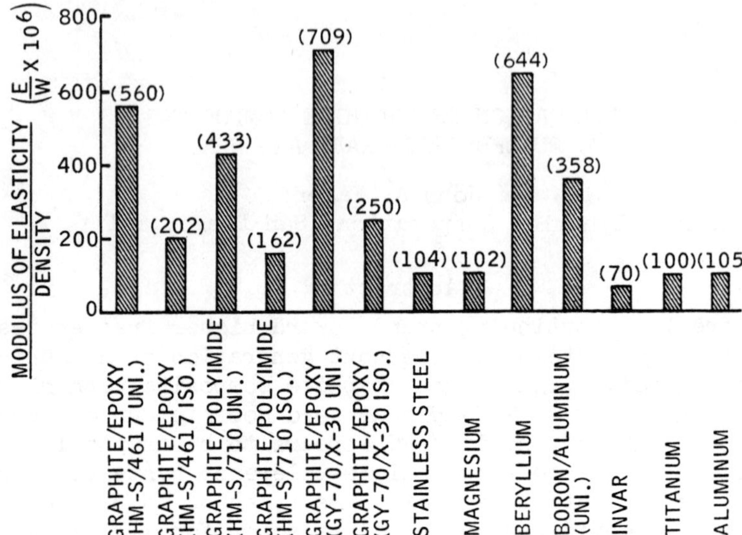

Fig. 1 Relative merits of candidate materials as a function of stiffness and density.

Figure 1 shows the stiffness-to-density ratio of the graphite composite materials to metals. Only high-priced beryllium is

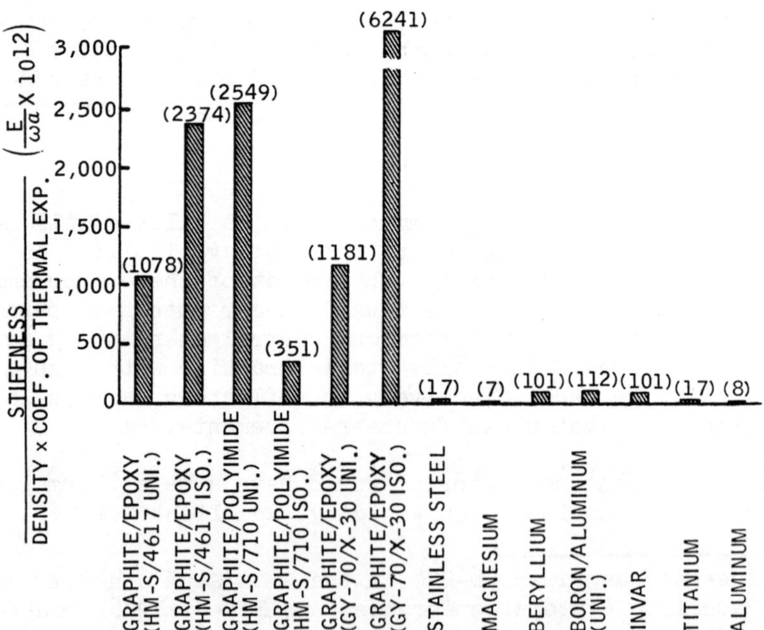

Fig. 2 Relative merits of candidate materials as a function of stiffness, weight, and thermal stability.

Table 1 Candidate material systems

	E, psi x 10^6	F_{TU}, ksi	a, in./in./°F x 10^{-6}	ω, lb/in.^{3}	K, Btu/ft-hr/°F
Graphite/epoxy (HM-S/4617 uni.)	33.62	111.4	-0.52	0.06	0.78, 5.8
Graphite/epoxy (HM-S/4617 iso.)	12.11	36.9	0.085	0.06	0.78, 5.0
Graphite/polyimide (HM-S/710 uni.)	26.00	131.0	-0.17	0.06	0.78, 5.8
Graphite/polyimide (HM-S/710 iso.)	9.71	43.4	0.46	0.06	0.78, 5.0
Graphite/epoxy (GY-70/X-30 uni.)	42.53	79.8	-0.60	0.06	0.78, 5.8
Graphite/epoxy (GY-70/X-30 iso.)	14.98	28.1	0.04	0.06	0.78, 5.0
Stainless steel	29.00	180.0	6.20	0.28	10.0
Magnesium	6.50	32.0	14.00	0.064	56.0
Beryllium	42.50	44.0	6.40	0.066	87.0
Boron/aluminum	34.00	175.0	3.20	0.095	. . .
Invar	20.50	65.0	0.70	0.291	6.05
Titanium	16.00	160.0	5.80	0.160	4.2
Aluminum	10.5	70.0	13.00	0.100	76.0

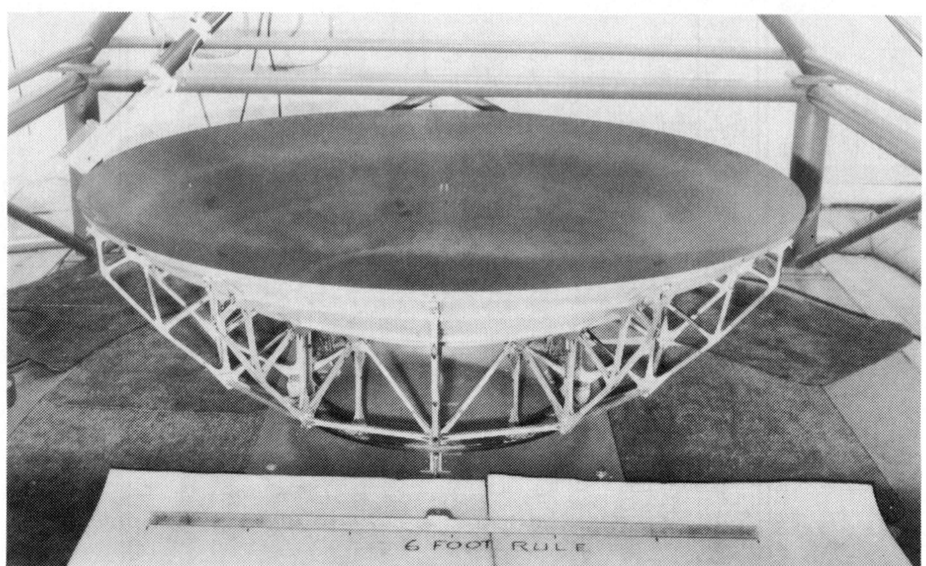

Fig. 3 Eight-ft, 200-GHz graphite reflector.

competitive with the unidirectional graphite in stiffness. In general, beryllium will cost four to five times that of graphite structure. The true figure of merit in material selection is the relationship of stiffness (E) divided by density (μ) and the coefficient of thermal expansion (CTE). Figure 2 shows the dominating aspects of graphite on this basis. Table 1 provides detailed physical properties of composites compared to typical metals. A poor characteristic is its low thermal conductance properties, which are similar to Invar. In a large space antenna, conductance is not critical if the local cross section can be minimized.

In rf reflectance, uncoated graphite will perform similar to copper up to K-band range. Above K band, a vapor deposition of aluminum or other high-conductance metallic film is needed to prevent gain degradation. Fortunately, as the frequency increases, the required metallic thickness decreases.

Moisture absorption in the epoxy binder of the graphite material must be evaluated for highly dimensional stable components. Coatings have been developed to seal out moisture completely if this minor dimensional change is critical. Above 100 GHz, moisture will be critical and could result in misalignment or gain loss. A review of potential performance characteristics of a graphite/epoxy reflector, erectable antenna, and waveguide power divider elements applicable to a phased array or lens system follows.

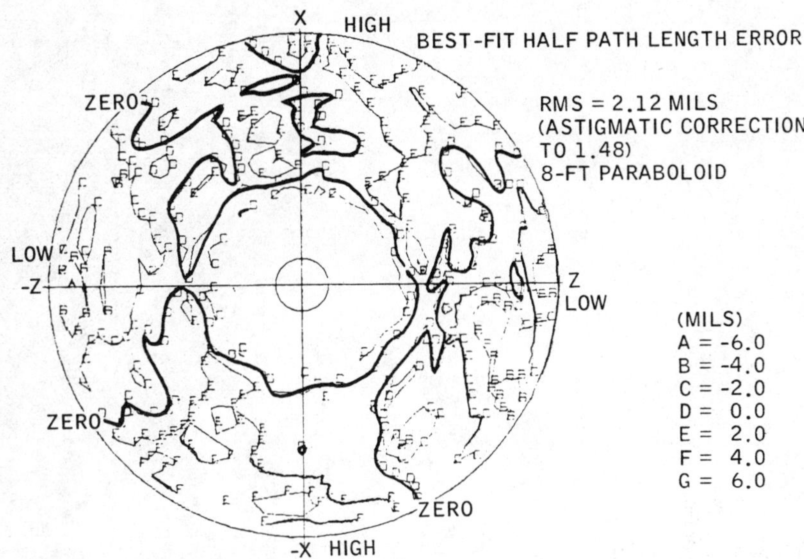

Fig. 4 Eight-ft graphite reflector contour based on 460 points measured over the surface.

Parabolic Reflector

An in-house program developed an 8-ft graphite reflector with capability of operating in the 200-GHz range. Figure 3 shows the reflector fabricated completely of graphite composite material. The lower truss elements, rings, and radial stiffeners all are unidirectional graphite material with E = 44 mpsi with a density of 0.06 $lb/in.^3$ (Note: steel E = 30 mpsi, with a density of 0.30 $lb/in.^3$) Figure 4 shows the 2.12-mil rms over the reflector surface with the ability to achieve a 1.48-mil rms.

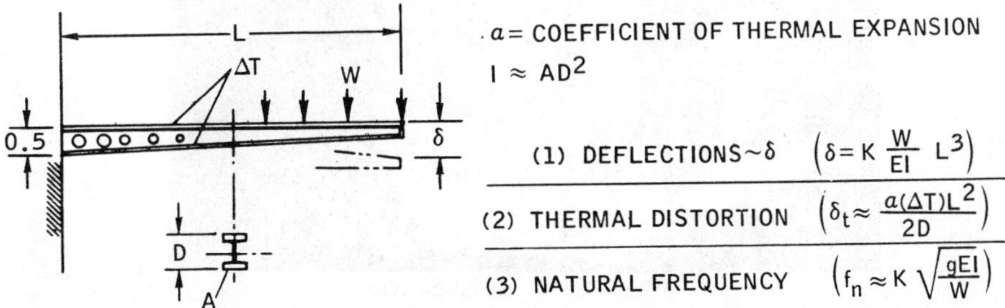

Fig. 5 Relationship of elastic modulus, cross section, and coefficient of thermal expansion on the performance of boom-type antennas.

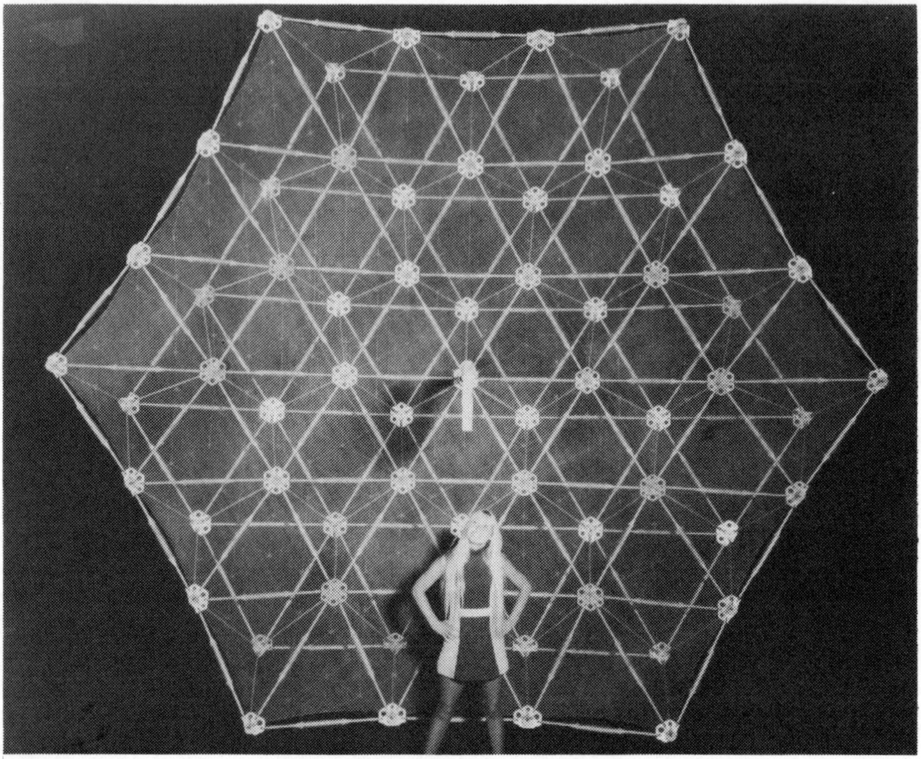

Deployed geotruss antenna

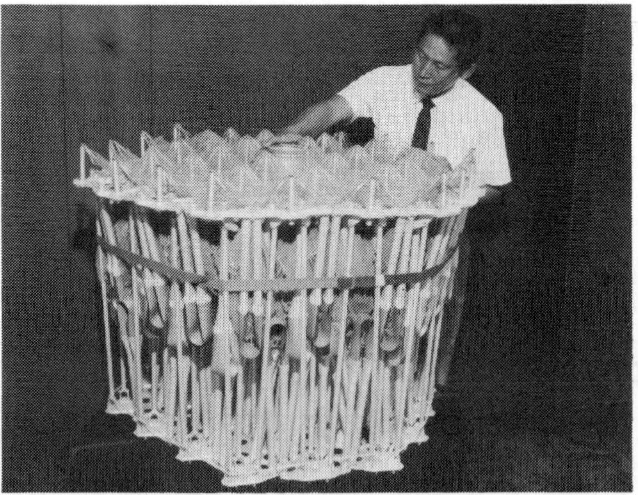

Packaged reflector

Fig. 6 Fifteen-ft erectable geotruss antenna.

Under a thermal load of +200°F on one tip of the reflector and -200°F on the opposite diagonal, the reflector's rms is increased by 0.5 mil. The ability of the material to demonstrate the low-expansion characteristics in an assembled configuration where high-expansion adhesives are included substantiates the application of composites to spaceborne complex rf structures. Load tests at 30 g and 147 dB showed no loss of contour. The completed reflector weighed 40 lb and had a 55-cps structural natural frequency.

A problem area that has occurred with some users of graphite on reflector shells is the loss of isotropicity in the paraboloid shell. With proper control of ply layup, uniformity can be obtained. Tradeoffs have been made between thin-shell reflectors and honeycomb with graphite faces. Because of the difference in stiffness in honeycomb (ribbon direction), the isotropicity required for rigorous thermal

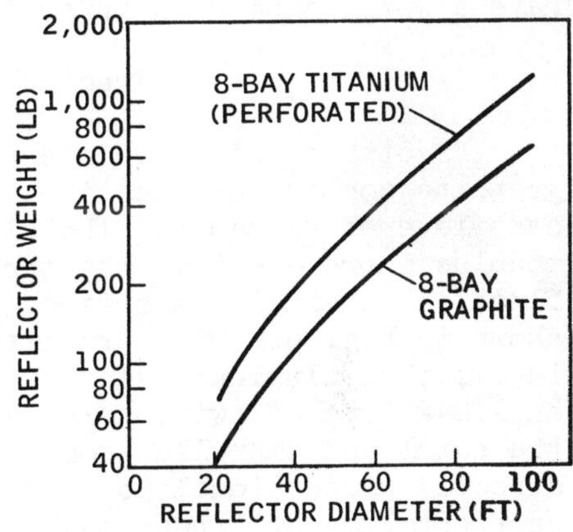

Fig. 7 Geodetic truss antenna weight comparison.

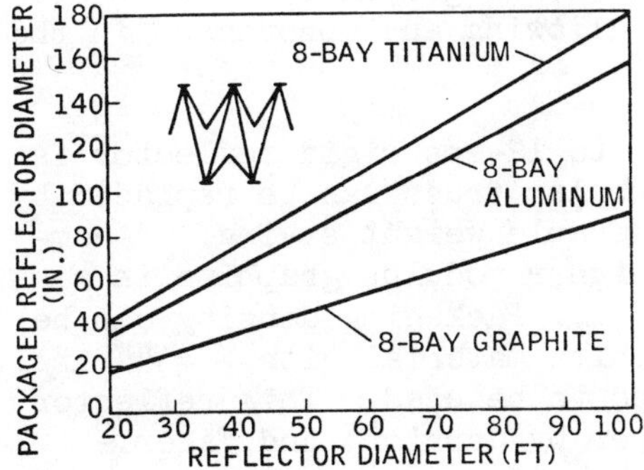

Fig. 8 Packaging diameter is related to material modulus.

control is lost. Thermal gradients between face sheets of the sandwich structure also contribute to distortion. The ring, truss-stiffened shell in Fig. 3 had only a 0.032-in. thick isotropic shell.

Surface deviations for the antenna are random errors that combine in a complicated fashion to reduce the antenna gain. The overall effect of surface deviation on the antenna gain can be estimated by the Ruze equation [1]

$$G = \eta(\pi D/\lambda)^2 \exp\left[-(4\pi\sigma/\lambda)^2\right]$$

where η is the aperture efficiency, D the antenna diameter, and σ/λ the rms surface tolerance in wavelengths. The equation is based on an assumption of random, uniformly distributed aperture errors with small correlation intervals. With a potential 1.48-mil rms, this reflector would have a gain of 71.0 dB at 200 GHz. Without thermal control, completely exposed to space environment, it would degrade only 0.60 dB or have a worst-case gain in orbit of 70.4 dB.

Erectable Antenna

Erectable antenna concepts that employ booms can use graphite composite effectively to reduce the motion of the cantilevered elements. The simple relationship of elastic modulus improvement in the three critical areas of load deflection, thermal distortion, and structural frequency is shown in Fig. 5. With graphite approximately 320 times lower in CTE than aluminum, significant improvements can be made in an ATS-F type of swirl antenna or the umbrella boom type. In the swirl and umbrella boom types, the performance gain is directly proportional to E and CTE, whereas the geotruss concept (Fig. 6) stiffness and thermal stability also can be improved by increasing the depth of the structure. With the significant weight stiffness ratios of graphite, the diameter of the tubes can be decreased, allowing an improvement in the packaging ratio.

Reflector weight for an 8- to 12-cps rigid reflector is shown in Fig. 7. A titanium tubular truss can be replaced by graphite, resulting in a considerable weight saving. A comparison of the effects of the higher modulus graphite in a tubular design is shown in Fig. 8. Packaging density can be improved by unidirectional graphite material with E = 44 mpsi, allowing a smaller-diameter tube to be used. This reflector was presented in an earlier paper by Garriott and Fager.[2] Cabling used to control the mesh contour also employs graphite.

With a precision yield of 20 ksi, both the tube and cable return to contour after repeated deployments.

Figure 9 illustrates the advantages of graphite compared to a heavier titanium reflector for a 70-ft system under space solar heating. Synchronous orbit is used, since most communication satellites employ this orbit. The noon position assumes an antenna looking directly at the sun. Six a.m. and six p.m. are the side-on-sun conditions. Side-on-sun conditions are typically the worst for a reflector. Because of a slight unsymmetry in the geotruss, there is a minor shift off the 6:00-p.m. case from one sun angle.

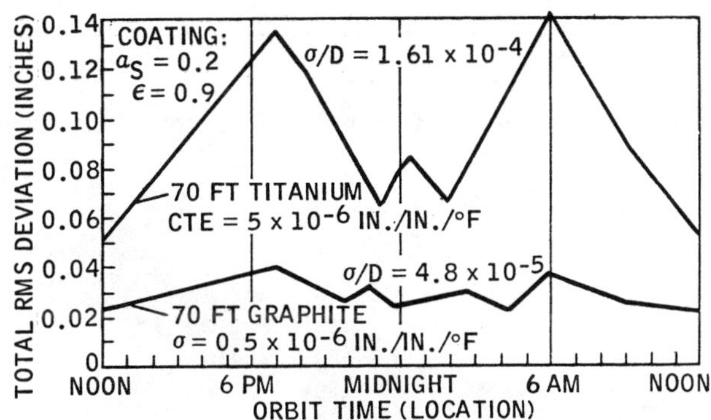

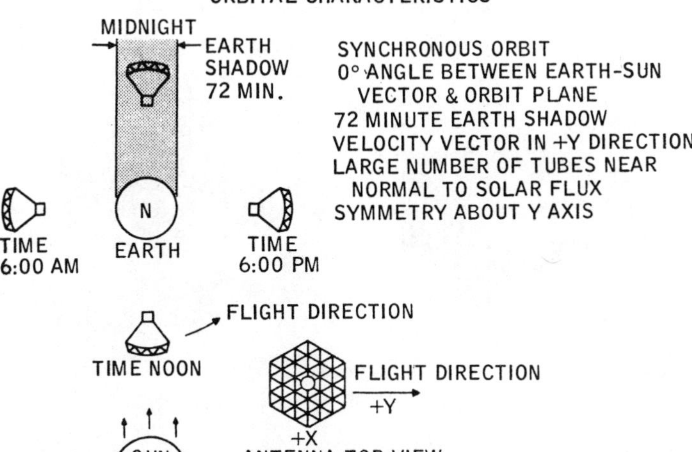

Fig. 9 A graphite composite truss tube design minimizes thermal distortion and yields an extremely good 10^{-5}-rms surface.

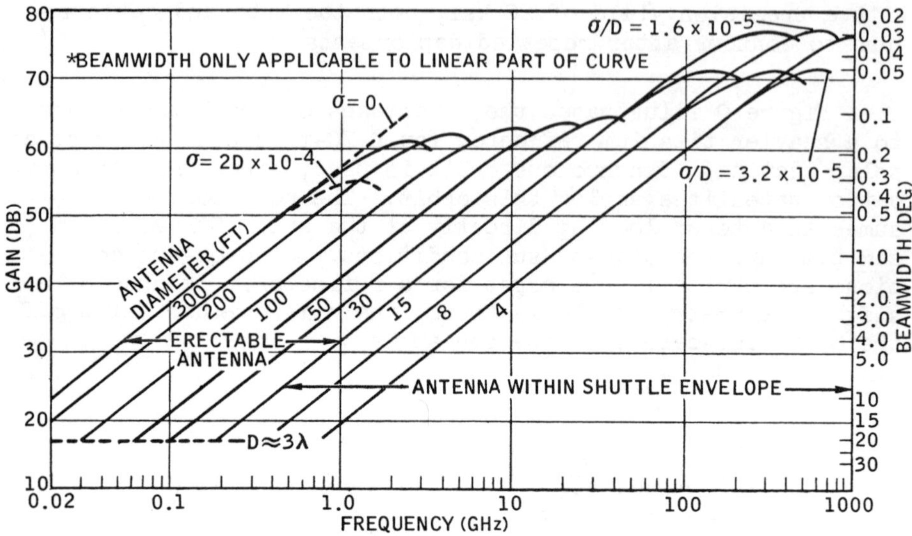

Fig. 10 Projected gain of graphite antenna system.

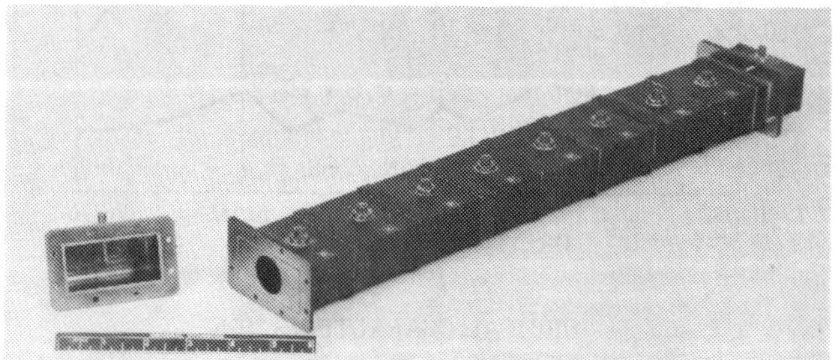

Fig. 11 Graphite WR-229 eight-element graphite filter.

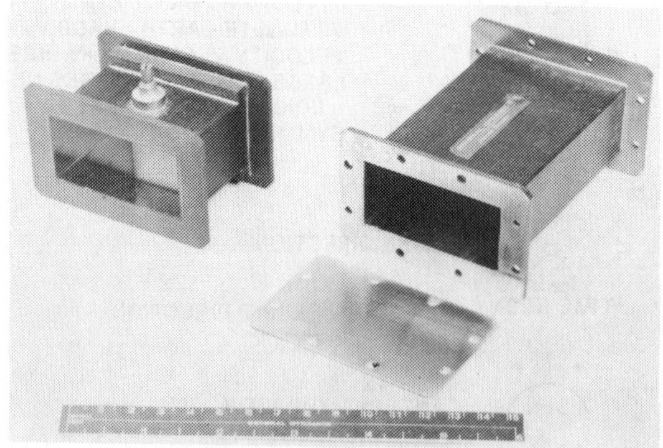

Fig. 12 Graphite cavity with graphite tuning screw.

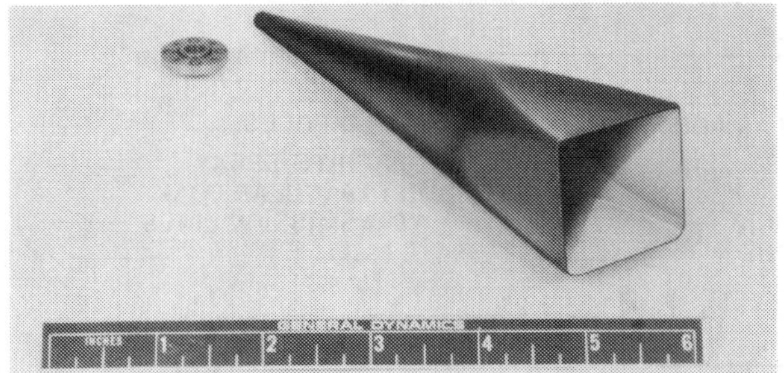

Fig. 13 Graphite W-band feed horn.

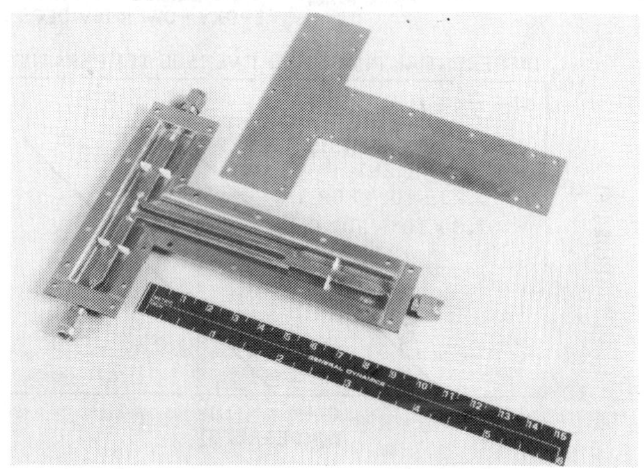

Fig. 14 Graphite L-band, 3-dB power divider.

Conversion of the rms using the Ruze equation for the geotruss erectable antenna is shown in Fig. 10. The impact of graphite is significant if you consider the higher ranges of frequencies that now are available for communications. It is reasonable to consider that a 150-ft erectable reflector could be packaged in the shuttle payload envelope (15 ft diam x 60 ft long) with a standard spacecraft and Interium Upper Stage.

Phased Arrays and Lens Systems

Phased arrays and lenses may be made up of waveguide tubing sections, slotted sections, or dipole power dividers and transmission lines. Figures 11-14 show a group of metallized graphite components that have the basic properties of graphite and the electrical characteristics of metal elements. In rf performance, they are equivalent to copper components. These rf components have been cycled thermally over ± 300°F without breakdown of the metallic/graphite bond.

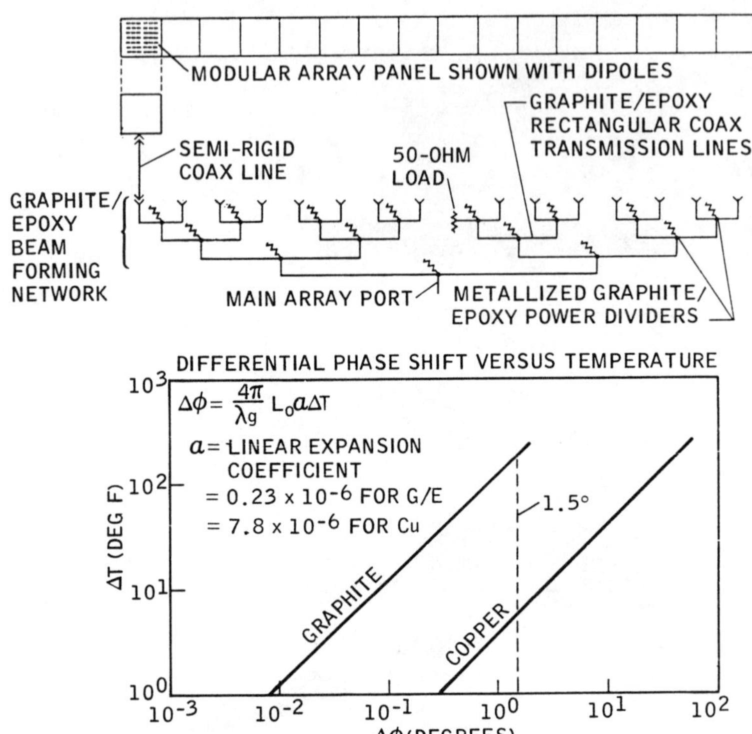

Fig. 15 Graphite/epoxy rf distribution system.

The high degree of thermal stability achievable with an L-band beam-forming network is shown in Fig. 15. Over a 100°F thermal differential could exist in a graphite system before it would suffer the phase shift of an equivalent (5°C) copper system. The network installed on a graphite support structure allows the implementation of very large, low-weight apertures.

Summary

Graphite composite has the potential of significantly improving the next generation of commercial communication satellite antennas. Its stability will allow high-gain, spot-beam systems to use millimeter wavelengths, thus opening up broad-bandwidth areas to meet the ever-increasing spectrum requirements. Graphite materials under development for 10 years are ready to be integrated into satellite hardware.

Acknowledgments

I wish to thank the following personnel, who contributed to the development of composite rf components at General Dynamics/Convair: E. Lipscomb, R. Garriott, G. Howell, D. Vaughan, F. Fujimoto, J. Strachan, R. Ankeny, and J. Fischer.

References

[1]Ruze, J., "Antenna Tolerance Theory - A Review," Proceedings of the Institute of Electrical and Electronics Engineers, Vol. 54, April 1966, pp. 633-640.

[2]Fager, J. A. and Garriott, R., "Large-Aperture Expandable Truss Microwave Antenna, "IEEE Transactions on Antenna & Propagation, Vol. AP-17, July 1969, pp. 452-458.

MAGNETIC BEARING MOMENTUM WHEEL

C. J. Pentlicki*
COMSAT Laboratories, Clarksburg, Md.
and
P. Poubeau†
Aerospatiale, Les Mureaux, France

Abstract

This paper describes a 100-N-m-sec momentum wheel that operates at 24,000 rpm and uses a magnetic bearing suspension and a composite fiber-resin rotor. The device characteristics include the following: 1) passive radial centering by permanent magnetic rings; 2) axial active control by a servoloop utilizing electromagnets supplemented by permanent magnets; 3) damping of the radial and transverse oscillations of the rotor in the magnetic suspension by a passive electrodynamic damping system; 4) acceleration and deceleration of the wheel by a brushless ironless dc motor with electronic computation; 5) emergency bearings to support the rotor in the event of a momentary interruption in the power supply or axial control (there is no contact between these bearings and the rotor during normal operation); 6) a composite fiber-resin rotor centered on a central hub by using a special technique called cycloprofile; and 7) a ventilated housing providing caging for rotor launch restraint. The engineering model that has been constructed is very close to a flight model concept. The test results given herein demonstrate the feasibility of using the design principles over a wide range of rotation speeds for magnetic bearing momentum and reaction wheels. Compared to conventional bearings and rotors, this device concept permits considerable improvement in the tradeoffs relating to mass, power, and reliability.

Presented as Paper 76-239 at the AIAA/CASI 6th Communications Satellite Systems Conference, April 5-8, 1976, Montreal, Canada.

*Manager, Mechanical Design Department.
†Manager, Department of Special Studies.

Introduction

Long-life spacecraft hardware is an objective of many researchers. One particular class of satellites requires angular momentum devices for attitude stabilization. These devices, whether reaction wheels or momentum wheels, rely upon ball bearings for rotor suspension. Hence, the life of the devices is limited inherently by the lifetime of the bearing system. The life-limiting wear processes of ball bearings can be eliminated by using magnetic bearings. This rotor suspension technique requires no physical contact between the stator

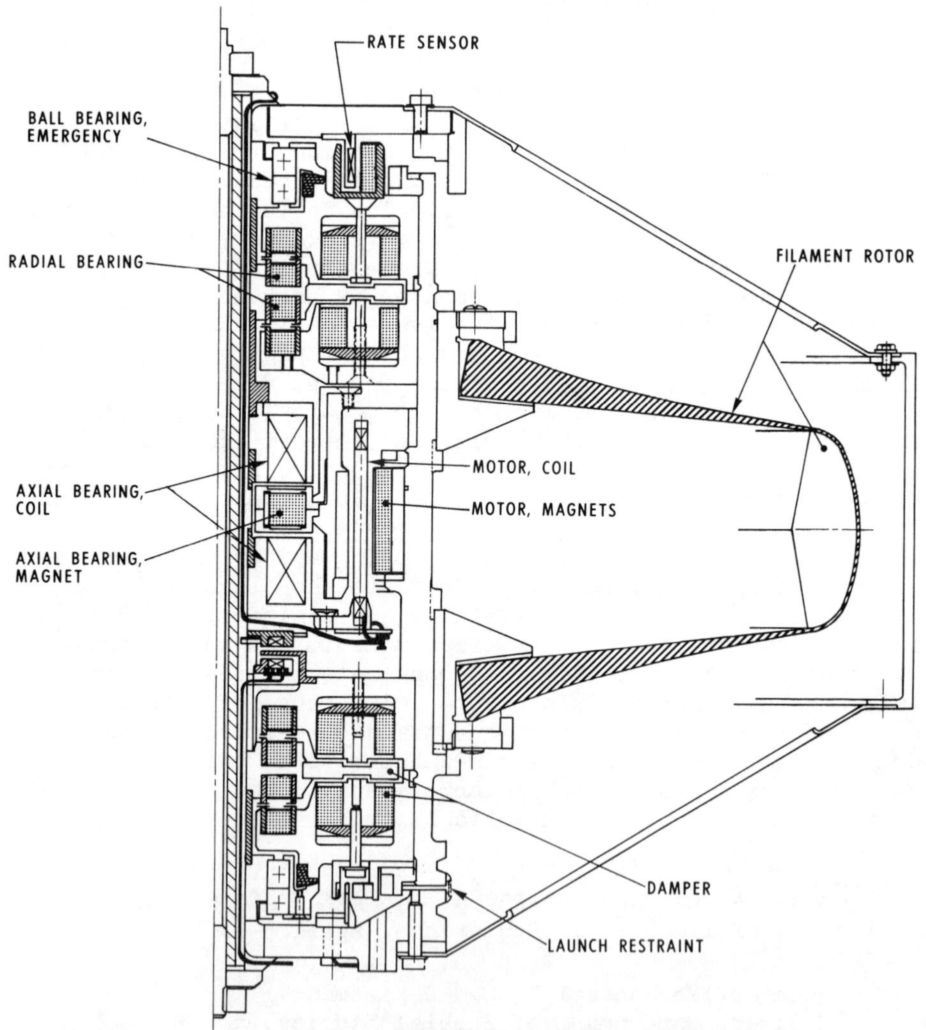

Fig. 1 High-speed momentum wheel configuration.

and rotating elements, since the rotor is supported completely by magnetic fields. Additionally, the rotor speed selected for nominal operation no longer is limited by the wear life of ball bearings, nor do magnetic bearings suffer the marked drag torque increase experienced by ball bearings at high speeds.

Magnetic bearings have very favorable characteristics for high-speed applications. Therefore, the rotor weight and/or size may be reduced without decreasing the angular momentum capacity. The desirability of magnetic rotor suspension instead of ball bearings is clear. High-speed rotor operation using conventional materials is limited by high mechanical stresses. Fiber composite rotors naturally follow if the benefits of high-speed operation are to be maximized. Providing for magnetic bearings and a fiber composite rotor in a single device becomes a significant technical challenge. Development of such a momentum wheel has been undertaken, and the results are reported herein.

Design Specifications

The primary characteristics used for the design of the magnetic bearing wheel were the following: angular momentum, 100 N-m-sec with $\pm 10\%$ operating range; nominal operating speed, 24,000 rpm; control torque, 5×10^{-2} N-m; and design life, 10 yr. The development effort was intended to design and test a momentum wheel using magnetic bearings and a composite rotor that could be adapted to a flight model with little additional effort. Major emphasis was placed on obtaining an optimum relationship among reliability, weight, power, and size to obtain significant advantages in these areas relative to low- and moderate-speed wheel systems utilizing conventional bearings and rotors.

General Concept

Overall Design

Figure 1 shows the overall concept of the wheel and indicates the different subsystems.

Radial Suspension

The radial centering of the rotor is provided by four pairs of magnetic rings, each pair consisting of one ring on the stator and one ring on the rotor. The restoring force is produced between the iron sleeves. The magnetic field is provided by many adjacent radially polarized segments of samarium cobalt, forming a ring. This configuration, shown in Fig. 2,

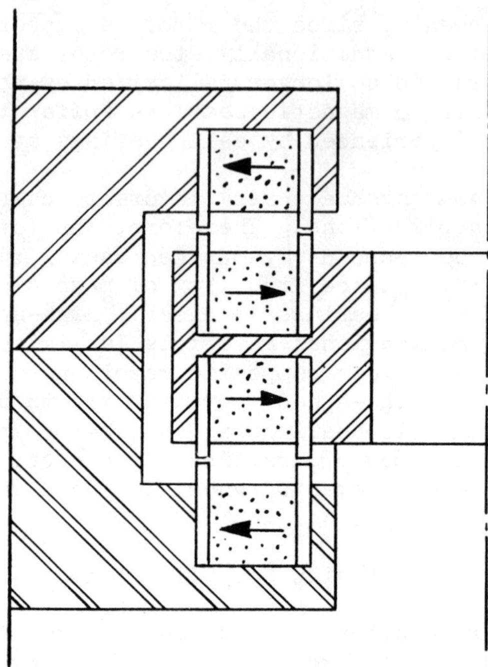

Fig. 2 Radial bearing showing arrangement of permanent magnets.

has been chosen after comparing the characteristics of repulsion systems with those of attraction configurations. The magnetic circuit is as short as possible and produces a high ratio of radial stiffness to mass because of the efficient utilization of the permanent magnet material. The iron sleeves equalize the magnetic field in the air gap, hence minimizing rotational losses. The dimensions of the magnets and iron sleeves are selected to obtain an axial-stiffness-to-radial-stiffness ratio of under 3 for a gap of 0.7 mm, although the radial stiffness remains high even for wide gaps. This bearing configuration permits the smallest ring diameter allowed by the diameter of the central shaft, thus minimizing weight.

The gaps between rotor and stator rings have been chosen as 0.7 mm for reliability. If the rotor contacts the emergency bearings, a gap of 0.5 mm remains between the rotating parts. This prevents any possibility of contact, even under the worst conditions. For the nominal gap of 0.7 mm, the radial stiffness achieved at the centering ring is 160 N/mm, and the axial stiffness is 500 N/mm. The weight of the magnetic material in the centering rings is 350 g.

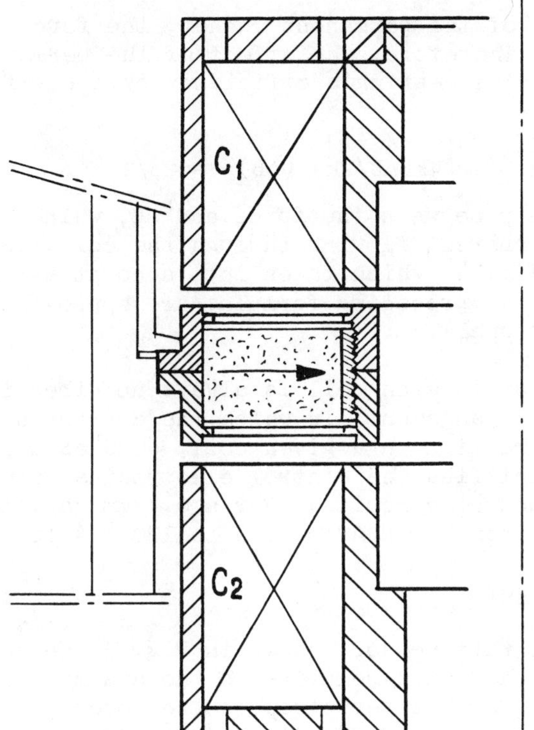

Fig. 3 Axial bearing showing relationship of electromagnets C_1 and C_2 to permanent magnet ring.

Axial Bearing

The axial position of the rotor is maintained through an active servoloop using the axial speed signal provided by an electrodynamic rate sensor. The axial bearing, shown in Fig. 3, consists of two electromagnets on the stator and a samarium cobalt permanent magnet on the rotor. In the absence of current in the coils, the permanent magnet creates fields B_1 and B_2 in gaps e_1 and e_2, respectively. When $e_1 = e_2$, $B_1 = B_2$ and the resultant attractive forces, F_1 and F_2, are balanced.

When a current flows in one coil, the field induced has the same direction as that produced by the magnet in one of the gaps, and the opposite direction of that produced in the other gap. If ΔB is the field created by the coil, then $\Delta F_1 = \Delta F_2 = 2kB\Delta B$. The resulting force is

$$(F_1 + \Delta F_1) + (-F_1 + \Delta F_2) = 4kB\Delta B$$

In the absence of the permanent magnet, the force would be only $2k(\Delta B)^2$. Therefore, the effect of the permanent magnet is to increase the bearing's efficiency by a coefficient whose value is

$$(4kB\Delta B/2k)\ (\Delta B)^2 = 2B/\Delta B$$

The relationship between the force and ΔB, which is proportional to the current flowing through the coil, is the form indicated in Fig. 4, which takes into account saturation and the fact that the preceding formulas are approximations when ΔB is not very small.

The current in each coil is always unidirectional. Although this arrangement, in which the current always flows in the same direction in a given coil, doubles the required copper, it simplifies the control electronics and the provision of redundant circuits. For a 0.7-mm gap, the efficiency varies from 150 N/A for 1 A to 120 N/A for 5 A.

Axial Rate Sensor

The axial rate sensor, shown in Fig. 5, detects the axial rate of the rotor relative to the stator. The output signal is used for the servoloop of the rotor axial position. A ring made of radially magnetized samarium cobalt segments produces a field that moves axially with the rotor relative to a coil made up of 360 turns of enameled copper wire. The

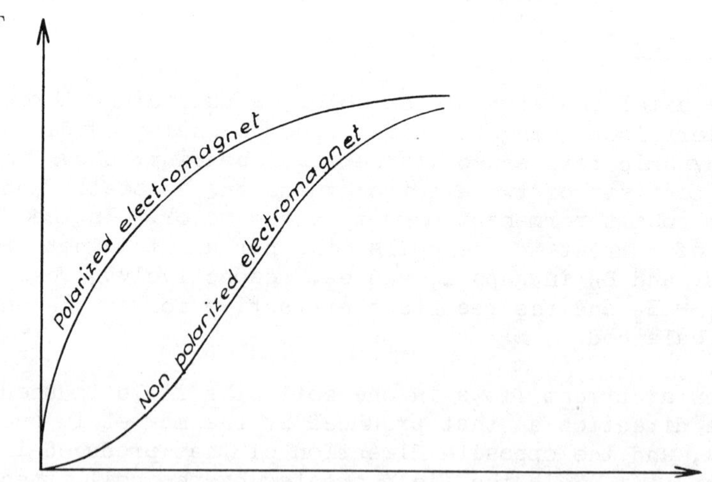

Fig. 4 Force vs field strength for electromagnetic coil with and without the field supplemented by a permanent magnet.

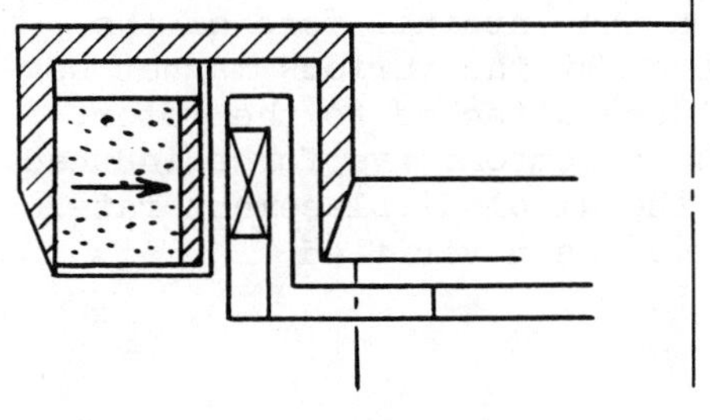

Fig. 5 Axial rate sensor.

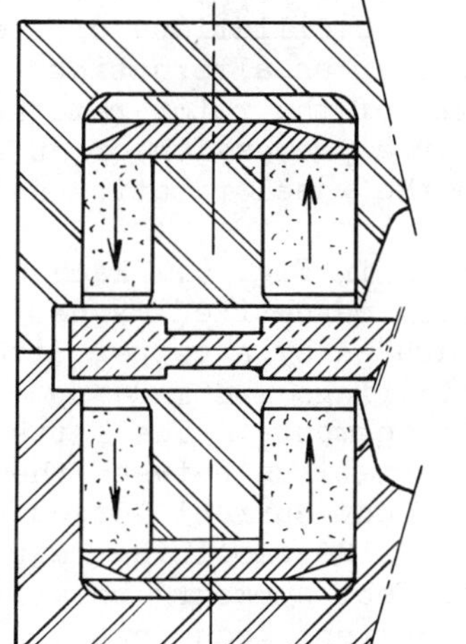

Fig. 6 Eddy current rotor damper configuration.

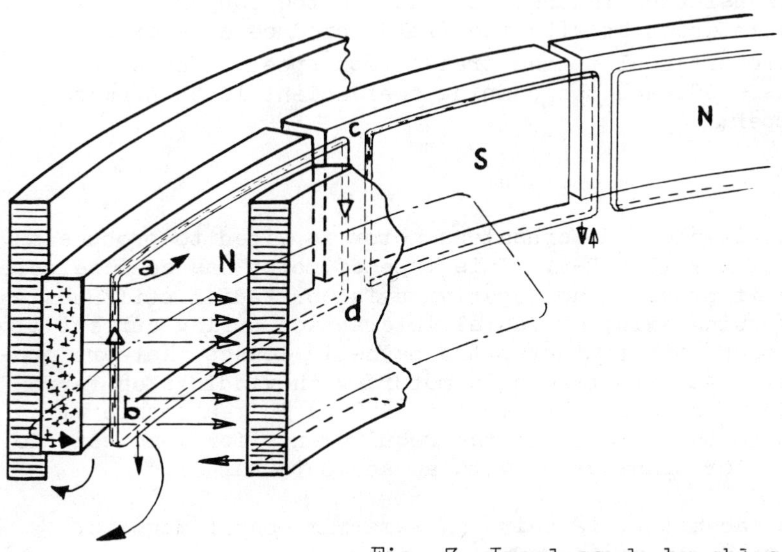

Fig. 7 Ironless dc brushless motor.

magnetic field is confined to the coil by a soft iron magnetic envelope. The soft iron ring on the internal face of the magnets smooths the field produced by the various magnet segments. The envelope and magnet are attached to the rotor, and the coil is stationary. The electromotive force induced in the coil is proportional to the axial displacement rate. The sensitivity, 20 V/(m/sec), has been verified experimentally.

Damping System

The damping system consists of two dampers, one at each end of the shaft, designed to damp the oscillations of the rotor. These oscillations may be either an alternating radial translation, in which the axis of the rotor remains parallel to its nominal direction, or a precession or nutation about an axis perpendicular to the wheel rotation axis.

The damper configuration is shown in Fig. 6. Each of the dampers is made up of four axially magnetized magnetic rings attached to the rotor and separated by a copper disk attached to the stator. The magnetic rings are made up of soft iron rings over samarium cobalt segments. The latter rings are used to smooth the field variations between the adjacent magnet segments so that the eddy current losses due to field variations are minimized. If the smoothing results in a magnetic field in the gap of perfect symmetry relative to the rotor rotation axis, no eddy current loss occurs in the copper, for normal rotor rotation. However, a radial rotor translation induces currents in the copper disk which, through interaction with the field, produce a retarding force proportional to the translation speed. The damping ratio is 0.05, and the damping coefficient is 58 N/(m/sec) per damper.

Motor

The ironless dc brushless motor is sized to produce a torque of 5×10^{-2} N-m. This type of motor has been selected because it provides no negative axial or radial stiffness and no disturbing axial or radial interaction during current flow. In addition, it eliminates the magnetic losses that are produced even at zero torque in motors with windings on ferromagnetic materials. Its weight (450 g) is reasonable with respect to the torque and the requirements for 1-mm gaps to permit rotor translation with an adequate margin.

In the motor, 12 pairs of samarium cobalt magnets, placed with alternate radial polarization, produce a magnetic

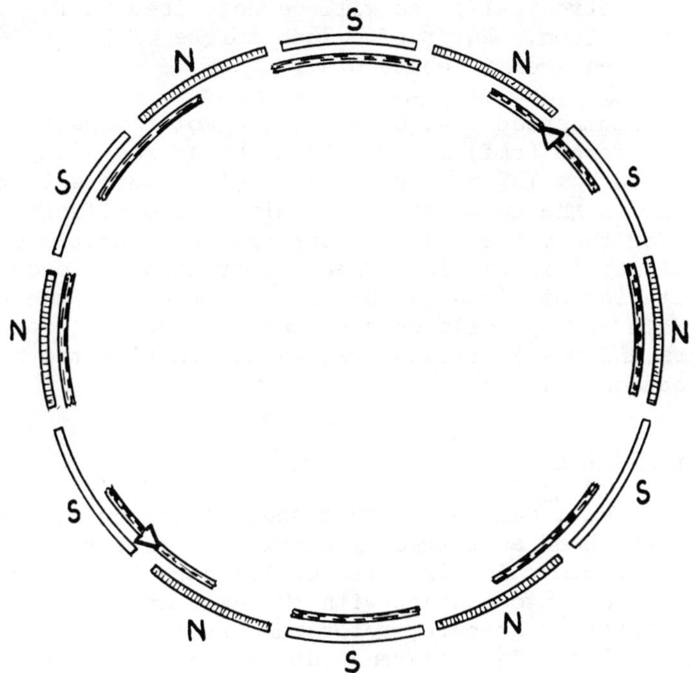

Fig. 8 Arrangement of motor coils and permanent magnets with one coil energized.

field in the interval between them and an inner laminated steel ring. The magnets are bonded to rings of magnetic alloy (AFK502) which close the flux between two adjacent poles north and south. These rings are laminated to minimize the losses produced by eddy currents due to the current variations in the winding. The winding is composed of rectangular coils, impregnated with a rigid resin that is a good heat-conductor, placed in the gap. It is made of insulated wire, consisting of 90 strands of 0.07-mm diam, to limit the eddy current losses in the copper during the induction variations from the rotating magnet segments. When a current passes through the winding, the forces due to the field/current interaction on the axial strands of the coils are added and are in a direction so that they create a rotation torque, whereas the axial forces created by the return strands are cancelled.

Figures 7 and 8 show the arrangement of the four pairs of coils relative to the magnets at a given instant of rotation. Only one pair is active at a given instant, and the pairs are supplied sequentially. The computation is performed electronically, as will be described in the electronics subsection. For the nominal torque of 5×10^{-2} N-m, the current required is about 6 A.

At maximum speed (24,000 rpm), the motor negative electromagnetic force (emf) is 23 V, which requires a supply voltage of 28 V + 10% because of the voltage drops in the winding and in the commutation circuits. The efficiency is over 90% for the motor itself under nominal conditions, and the overall efficiency, including electronics, is over 80%. The commutation utilizes the set of 12 magnets on the rotor and two Hall effect cells on the stator to control four amplifiers and power transistors, which comprise the motor drive electronics.

Emergency Bearings

Bartemp ball bearings with a special treatment for vacuum operation are used as emergency bearings. If for any reason the axial servoloop is disrupted or its capacity is exceeded, the rotor comes into contact with the emergency bearing system, permitting graceful rotor rundown or recovery of magnetic suspension. The clearance at the level of the emergency bearings is adjusted axially within 0.2 mm and radially within 0.4 mm.

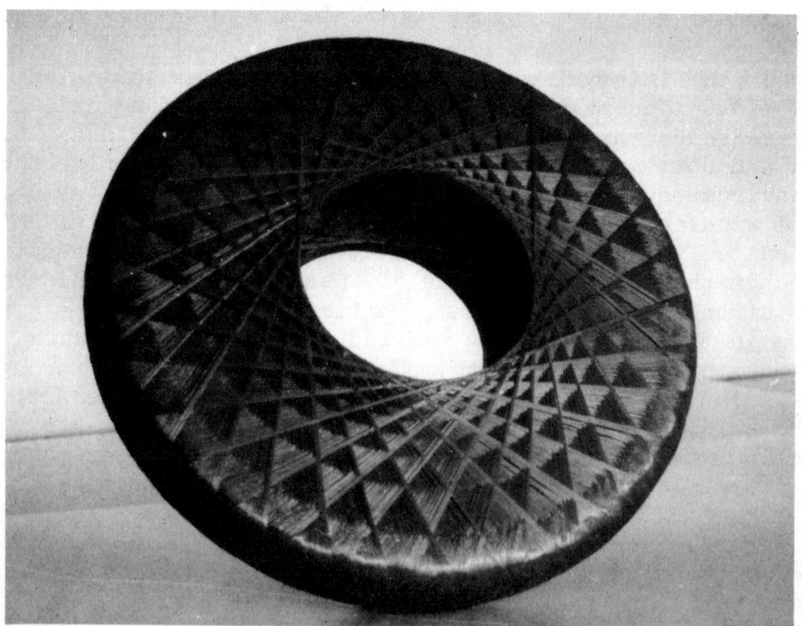

Fig. 9 Cycloprofile filament-mound rotor.

<u>Caging Mechanism</u>

Several methods of caging the rotor assembly during launch have been considered. The main problem is the reliability of the device during launch as well as after device activation, when the wheel is to go into normal operation. The device must retract with none of its parts in a position to damage the wheel by untimely displacement.

Spacers are placed between the rotor and stator. Their height is such that the emergency bearings are not in their extreme position, and the axial magnetic forces tend to bring the rotor away from the spacers when the clamps are not attached. The clamps that connect the rotor with the stator while enclosing two cones are held by a band. This band, which is tightened by means of a cable, is based on the same principle as the marman clamp successfully used many times for spacecraft separation. Unlocking is performed through a simple and reliable pyrotechnic device already used in many space programs. When the band is unlocked, the clamps and the spacers move apart under spring action, and the rotor is released. For the engineering model, the system utilizing the explosive cable cutter has been replaced by a band, which is tightened for vibration tests and released for functional tests.

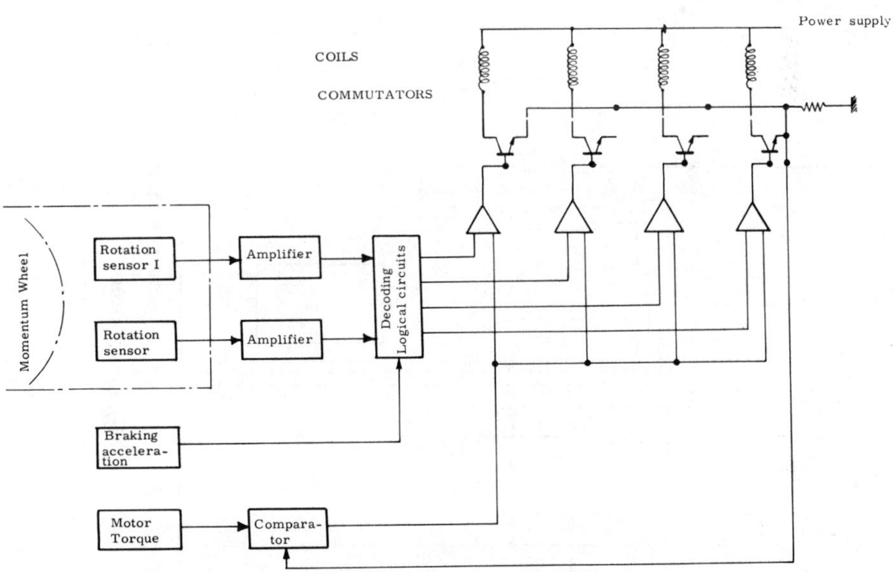

Fig. 10 Motor drive electronics.

Rotor Structure

The rotor consists of three parts: an aluminum alloy hub that serves as the interface between the filament rotor assembly and the rotor shaft, a circumferentially wound filament rim, and the cycloprofile envelope windings that attach the rim to the hub. The rim is wound about a mandrel. After cure, the mandrel and rim are finish-machined. The envelope winding consists of two layers of filament tape. After the tape is cured, the mandrel is washed away. Three complete rotors have been fabricated, one each from glass fiber, graphite fiber, and Kevlar 49 (Dupont de Nemours trademark). The Kevlar rim has a thin layer of graphite upon which all finish-machining is performed.

A preliminary balance is performed on mechanical bearings, whereas the final balance is accomplished on the magnetic bearings. Maximum rotor excursions are within 2 x 10^{-3} mm. The first critical frequency of the free rotor is 450 Hz for glass, 500 Hz for Kevlar 49, and 600 Hz for graphite. The predicted rotor stress for nominal operating conditions is of the order of one-quarter of the ultimate strength. Figure 9 shows the rotor and its wound construction.

Electronics

The electronics consists of motor drive electronics and bearing suspension electronics.

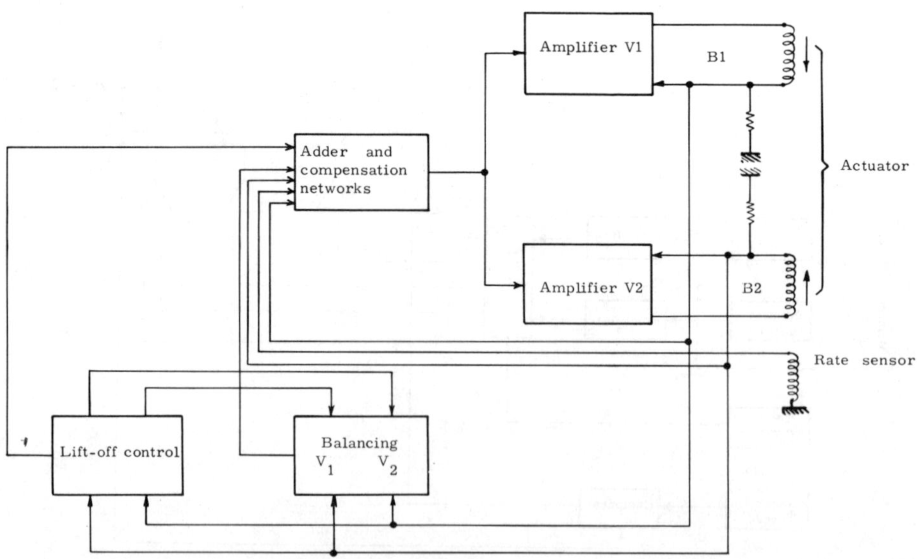

Fig. 11 Magnetic suspension electronics.

Motor Drive Electronics. As shown in Fig. 10, the motor drive electronics consists of rotation sensors, decoding logic circuits, the motor torque control system, and amplifier and coil commutators. Rotation sensors consist of two Hall effect cells delivering an output modulated by the field of the magnets rotating with the rotor. From the two signals delivered by the Hall effect cells and the signal established by the acceleration-braking selector, the logic circuits provide the commutation of the four cells.

Motor torque control comprises one integrated circuit operational amplifier, which compares an input voltage that is proportional to the needed reaction torque with a voltage that is proportional to the motor current. The amplifier output voltage is used as a bias for commutation. Amplifiers and coil commutators consist of four channels, each driving one of the four commutation commands and the bias current delivered by the motor torque control system.

Suspension Electronics. The suspension electronics, shown schematically in Fig. 11, consists of a rate sensor, compensation networks, and a current amplifier for driving the coils in the axial bearings. The axial rate sensor is a coil mounted on the rotor moving in a permanent magnetic field. It has a sensitivity of 20 V/(m/sec). The signal of the sensor is added to the current feedback after being corrected in the compensation network. The output signal of this network is provided to two amplifiers, each energizing one actuator coil.

Current balancing logic for the two amplifiers eliminates the effect of drift. Instead of a switching amplifier using pulse-width modulation, a linear power amplifier has been chosen because it is simpler and requires only a small amount of power to maintain suspension. A current of only 50 mA per coil is needed, with an operating voltage of 7 V.

Table 1 Modulus of elasticity of carbon and Kevlar 49 fibers

Fiber	In fiber direction, kg/mm^2	Cross fiber, kg/mm^2
Carbon	15,000	1000
Kevlar 49	10,000	1000

In this type of suspension, almost no power is needed to suspend the rotor against gravitational forces. Electrical power is used only to stabilize the rotor under dynamic loads and to bias the power transistors. In addition to the suspension electronics, a "liftoff" circuit permits rotor levitation off of the emergency bearings, as during initial turn-on.

Rotor Stress and Modal Analysis

Stress and modal analysis of the high-speed momentum wheel has been performed using an axisymmetric finite-element model. The most important part of the analysis is the cycloprofile, a composite material made of fibers and resin. A special effort is necessary to model an equivalent shell that takes into account its geometrical and physical properties. The parameters of the study are the moduli of the fibers of the cycloprofile. For static analysis, the results are stresses under a constant rotation speed. For modal analysis without centrifugal stresses, the model is clamped or free, and results are given for different Fourier coefficients.

To account for the large statistical variation in the elastic properties of the composite material, the analysis has been performed for different values of the elasticity modulus in the direction of the fiber and in the direction orthogonal to the fiber. For rotors of carbon fiber and Kevlar 49 fiber, the values obtained from the experiments are given in Table 1. It appears from the stress analysis that for both fibers the maximum stress is less than one-quarter of the ultimate tensile strength, and the free frequencies are over 500 Hz. For the clamped modes, the calculated frequencies range from 300 to 500 Hz and are high enough so that they do not interact with the principal structural modes during launch.

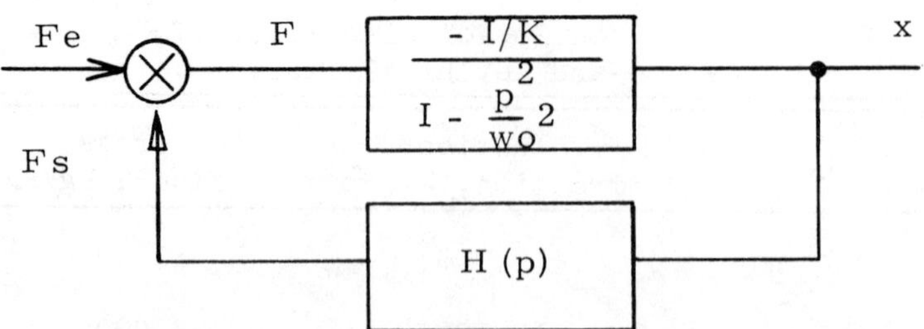

Fig. 12 Axial control system block diagram.

With the free-free model, the first frequencies are 500 and 600 Hz for Kevlar and carbon, respectively. If the rotation speed is 24,000 rpm, coupling between the two frequencies will be negligible. If the speed is 30,000 rpm, coupling should be negligible also because damping of composite material at these high frequencies is substantial, and the introduction of centrifugal stresses that had been neglected in the analysis will stiffen the rotor.

Axial Servoloop

The radial restoring stiffness of the passive magnetic centering rings is accompanied by instability in the axial direction. The imbalance force in the axial direction is a linear function of axial displacement near the equilibrium position. The wheel then is axially subject to a divergent axial force as a consequence of the radial stiffness. It also is subject to various external forces, caused mainly by satellite maneuvers or gravitational forces during ground tests.

The axial equation of motion of the magnetic suspension is

$$M\ddot{x} = Kx + F_e + F_s$$

where
F_e = external forces
F_s = stabilization force
X = axial displacement from the equilibrium position
M = suspended mass
K = imbalance stiffness

The purpose of the servomechanism is to cancel the external forces by bringing the suspended mass into a position so that F_e and Kx balance each other. Axial stability can be obtained by controlling the current to the axial coils to generate forces in the proper direction. Under the effect of force F_e

$$M\ddot{x} - Kx = F_e$$

$$x\left[1 - (M/K)\,p^2\right] = -(F_e/K)$$

The transfer function representing the system is then

$$G(p) = \frac{1/K}{1 - \left(p^2/\omega_0^2\right)}$$

where $\omega_0^2 = K/M$.

Figure 12 is a block diagram of the axial control system where

$$H(p) = K'p \frac{1}{1 - \left(p^2/\omega_0^2\right)} \frac{1}{1 - Z_{1p}} \frac{1 + aZ_{2p}}{1 + Z_{2p}}$$

The transfer function Fc/F_e, representing a stable system, must have an even number of positive poles, whence the introduction of a positive pole in $H(p) \to 1/(1 - Z_{1p})$. So that it does not oscillate continuously under the effect of a permanent force F_e, the loop also must have a static gain that is null; hence $H(p) \to p$. In addition, the current feedback loop includes a phase advance network and a compensation network for the time constant of the coil.

Rotor Dynamics

The primary characteristics of the rotor dynamics are the following: 1) radial stiffness (both bearings and actuator): $K_R = 180 \times 10^3$ N/m; 2) axial stiffness (both bearings and actuator): $K_A = 1100 \times 10^3$ N/m; and 3) transverse stiffness: $K_T = 415$ N-m/rad.

Normal Modes. The normal rotor modes are expressed as follows:

Translation perpendicular to the spin axis:

$$\omega_T = \sqrt{K_R/M}$$

Precession:

$$\omega_P = \frac{1}{2}\left[-\frac{C}{A}\Omega + \sqrt{\left(\frac{C}{A}\Omega\right)^2 + 4\frac{K_T}{A}}\right]$$

Nutation:

$$\omega_A = \frac{1}{2}\left[\frac{C}{A}\Omega + \sqrt{\left(\frac{C}{A}\Omega\right)^2 + 4\frac{K_T}{A}}\right]$$

where

C = principal inertia = 40×10^{-3} kg-m^2
A = transverse inertia = 25×10^{-3} kg-m^2
Ω = spin rate

and $C/A > 1$ for both carbon and Kelvar rotors.

Critical Frequencies. When $\Omega = \omega_p$

$$\Omega = \sqrt{K_T/(A + C)} \cdot (60/2\pi)$$

or 710 and 671 rpm for Kevlar and carbon rotors, respectively. When $\Omega = \omega_T$

$$\Omega = \sqrt{K_R/M} \cdot (60/2\pi)$$

or 1520 and 1480 rpm for Kevlar and carbon rotors, respectively. The nutation frequency never is reached in the present design because $C/A > 1$.

Reliability. For 10 years of operation, the calculated reliabilities for conventional electronic components or with hybrid circuits are indicated in Table 2.

Test Results and Characteristics

The total stiffness measured on the fully assembled wheel are

radial stiffness = 180×10^3 N/m

axial stiffness = 1110×10^3 N/m

Table 2 Calculated reliability of components

Component	Reliability	
	Nonredundant	Redundant
Without hybrid circuits		
Axial control	0.822	0.994
Speed control	0.877	0.985
Overall	0.720	0.980
With hybrid circuits		
Axial control	0.945	0.997
Speed control	0.958	0.998
Overall	0.905	0.995

The damping ratio of the radial translation mode measured on the wheel is

$$C/C_c = 0.05$$

The axial force of the actuator is 100 N for 1 A.

The drag torque during tests is due mainly to windage and eddy currents in bearings, actuator, and dampers. The measured values at different speeds, for a relatively high pressure of 4×10^{-3} Torr, are

$$8{,}000 \text{ rpm} = 4 \times 10^{-3} \text{ N-M}$$

$$16{,}000 \text{ rpm} = 6 \times 10^{-3} \text{ N-m}$$

$$24{,}000 \text{ rpm} = 9 \times 10^{-3} \text{ N-m}$$

The critical transitional frequencies in running up the wheel are as follows:

1) Precission frequency at 700 rpm. This theoretical frequency has not been observed during the tests.

2) Translation frequency at 1500 rpm. After static balancing of the rotor, and amplitude of the rotor translation when passing through this speed remains acceptable (less than $\pm$0.1 mm).

At higher speeds, a conical movement of the rotor appears. After dynamic balancing, the amplitude of this movement at nominal speed is less than $\pm$0.02 mm at the bearings.

Table 3 Power breakdown

Rotor speed, rpm	Function	Reaction torque, W	
		0 N-m	0.05 N-m
8,000	Rotation	4	60
	Suspension	5	5
	Total	9	65
16,000	Rotation	12	110
	Suspension	6	6
	Total	18	116
24,000	Rotation	25	150
	Suspension	8	8
	Total	33	158

Table 4 Estimated characteristics of future high-speed momentum wheels

Angular Momentum, N-m-sec	Speed, rpm	Mass,[a] kg	Idling power	Diameter, mm	Height mm
		Reaction Wheels[b]			
1	1,500	2.5	1	200	120
5	2,500	4	1	350	150
		Momentum Wheels			
20	8,000	7	5	350	170
50	8,000	8.5	5	350	170
100	8,000	11	5	350	200
100	24,000	8.5	12	350	200
		Energy Storage[c]			
2500	30,000	30	15	500	350

[a]These dimensions and mass include the hybrid redundant electronics.
[b]There is no crossing of the bearing critical frequencies.
[c]Two wheels of this type in counter-rotating operation could supply 500 W of energy to a satellite during eclipse period (synchronous orbit) and provide a constant angular momentum (100 N-m-sec, for example) for attitude control.

The maximum permissible slewing rate at 24,000 rpm is 0.01 rad/sec. The total mass of the wheel assembly is 11 kg (7.5 kg for the rotor and 3.5 for the stator). The moment of inertia is 40×10^{-3} kg-m^2 for an angular momentum of 100 N-m-sec at 24,000 rpm. The mass of the electronics package of the engineering model is 2.5 kg, but a considerable weight reduction is possible for a flight model configuration in hybrid circuit technology.

The power breakdown, including aerodynamic drag in a 4×10^{-3}-Torr vacuum, is given in Table 3. This performance is for the initial test only. Improved rotor balance will reduce power consumption. A test without the graphite rotor, which reduces windage and imbalance effects, yielded a power consumption of 13 W at 24,000 rpm.

Future Development Possibilities

The technologies of the magnetic suspension and motor and the overall concept can be adapted to momentum or reaction wheels in general. Composite rotors offer the possibilities of increasing the operating speed and saving weight. With a 30-cm-diam rim, a mass of 1.77 kg at 24,000 rpm and a mass of 9.45 kg at 4500 rpm would be required to provide an angular momentum of 100 N-m-sec. The fiber composite rotor becomes the only weight-efficient solution for larger angular momentum wheels utilized for stabilization or for providing storage of electrical power during eclipse periods, as in the case of two counterrotating wheels. Table 4 considers the feasibility of applying the technology of the wheel developed herein to other types of spacecraft momentum devices.

Conclusion

A substantial development has been accomplished. The first marriage of filament composite rotors and high-speed magnetic bearings has been achieved. Future spacecraft designers now have available the technology to reduce the weight and increase the reliability of a major element in the attitude control system. Broader application of this technology appears inevitable as greater experience is gained.

Acknowledgment

This paper is based upon work performed under the sponsorship of the International Telecommunications Satellite Organization (INTELSAT) under Contract CSC-IS-555. Views expressed are not necessarily those of INTELSAT.

SPACECRAFT APPLICATION OF HEAT PIPES

E. A. Stipandic*
General Electric Space Division,
Valley Forge, Pa.

Abstract

Heat pipes have been used in many spacecraft applications to maintain structure and component thermal control. These applications and the relevant internal heat-pipe design details are reviewed, along with the present and future applications. Other topics discussed are subsystem and system design parameters, which must be considered prior to the implementation of heat pipes on spacecraft. Many flight-proven heat-pipe designs are now available with distinct advantages over conventional thermal control techniques. However, the limitations of these devices must be understood thoroughly and accommodations made in the spacecraft system design, integration, and test.

Introduction

The heat pipe is a simple, high-performance heat-transfer device that transfers high quantities of heat over long distances, without appreciable temperature drops. The basic heat pipe (Fig. 1) consists of a sealed containment vessel, a capillary wick structure, and a saturated working fluid. Heat absorbed at one end (evaporator section) vaporizes the fluid; vapor flows to a cold region of the pipe, where it condenses. The capillary forces in the evaporator section produce a pressure gradient that returns the fluid to the evaporator via the wick structure. These devices are ideal for spacecraft applications: they are highly efficient, contain no moving parts, require no power, are lightweight, and have been flight-proven. Heat pipes may be categorized as follows:

Presented as Paper 76-241 at the AIAA/CASI 6th Communications Satellite Systems Conference, April 5-8, 1976, Montreal, Canada.

*Engineer, Environmental Control.

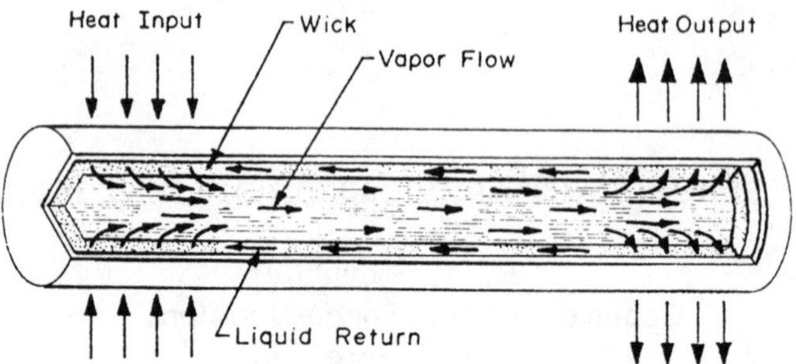

Fig. 1 Heat-pipe schematic.

1) Isothermalizing/high-heat transporters. The isothermalizing feature is used to maintain near temperature equalization within a structural element, equipment mounting platform, or within a component. The high-heat transport feature is used where high electronic thermal dissipations must be transported to and distributed over a large radiator area.

2) Variable conductance. This feature is attained by the addition of a noncondensable gas to the basic heat pipe (Fig. 2). The vapor flow within the pipe forces the noncondensable gas to the cold condenser region. The length of blockage produced by the noncondensable gas is a direct function of the pipe temperature and vapor pressure. As the heat input to the pipe increases, the pipe temperature and pressure increase, thus reducing the length of blockage. The turn-on and turn-off temperatures may be controlled by the condenser thermal conditions, the gas volume, and working condenser volume. The variable-conductance feature is an alternative to controllable heater designs, louvers, and thermal switches. Control may be attained to within $\pm$ 2°C, with power variations greater than 15:1. An active system using a feedback-controlled, low-power heater on the gas reservoir may be employed to attain the most constant control.[1,2]

3) Thermal diode. This feature permits heat to be transported in only one direction (Fig. 3). It is particularly attractive for applications in which a heat source is on/off and the radiator environment experiences a large variation in heat flux.

Previous Spacecraft Heat-Pipe Applications

The following is a summary of the past uses of heat pipes on various spacecraft and the flight performance results:

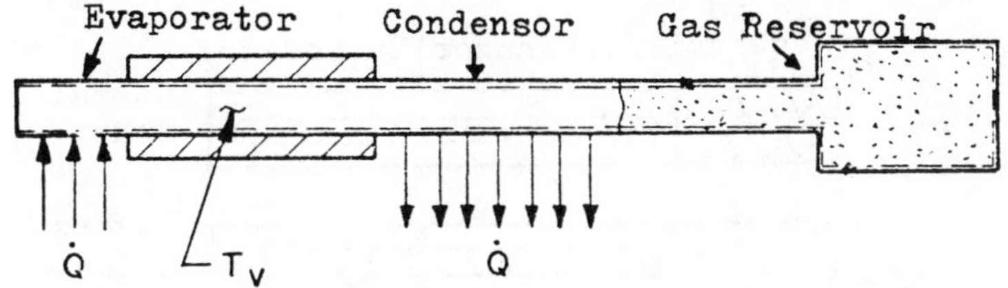

Fig. 2a Variable-conductance heat pipe.

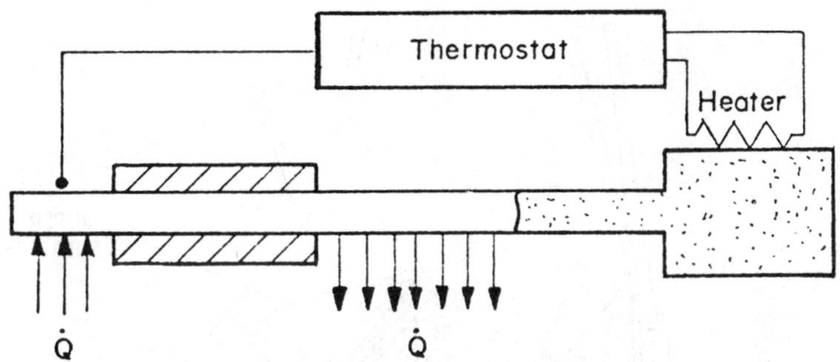

Fig. 2b Feedback variable-conductance heat pipe.

1) Agena/ATS-1 (1967). A water heat pipe attached to the launch vehicle operated successfully.[3] The results of this experiment indicated no change in performance with the absence of a gravitational force.

2) GEOS-II (1968). Two Freon-11, 1-in.-diam, aluminum heat pipes were used to improve the temperature excursions of the primary transponder. Flight performance indicated a significant improvement in transponder maximum and minimum temperatures over those observed for a similar condition on GEOS-I.[4]

3) PAC (1969). Three aluminum, ammonia, arterial heat pipes (identical to those on ATS-5) were used to isothermalize the main equipment platform of the three-axis-stabilized spacecraft. The spacecraft design also incorporated an acetone, flexible stainless-steel heat pipe. Performance of the ammonia pipes was judged successful. However, the heat flux inputs were too low to verify successful operation of the arteries. Insufficient data on the flexible pipe prevented an accurate determination of performance.

4) ATS-5 (1969). Sixteen heat pipes were used to reduce the temperature gradients within the body-mounted solar arrays. The heat

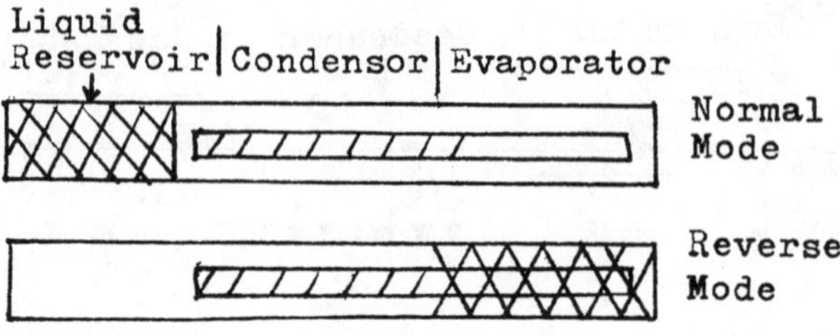

Fig. 3 Diode heat pipe.

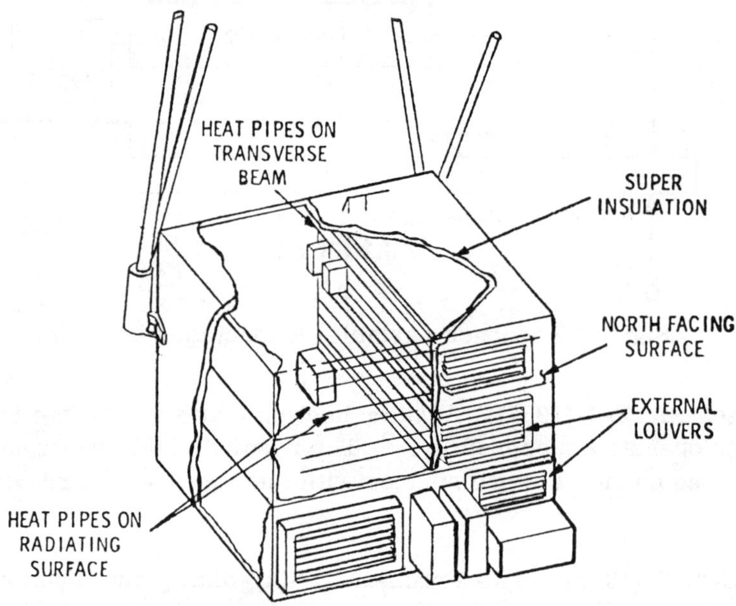

Fig. 4 ATS-6 EVM thermal control subsystem.

pipes were bonded within the cylindrical honeycomb substrates. The aluminum tubing was extruded with two flat sections, which facilitated bonding to the honeycomb facesheets. Unfortunately, thermal performance never was verified in flight because the spacecraft failed to attain three-axis stabilization. The spacecraft dynamics problems were caused, in part, by the damping within the heat pipes.

5) OAO-B (1970). Three Freon-21, aluminum, longitudinal-groove heat pipes were mounted to the inside of the spacecraft cylindrical structure. The purpose of the pipes was to minimize the structure temperature gradients and to validate a general approach to thermal control of large structures. [5] Each heat pipe was attached to the

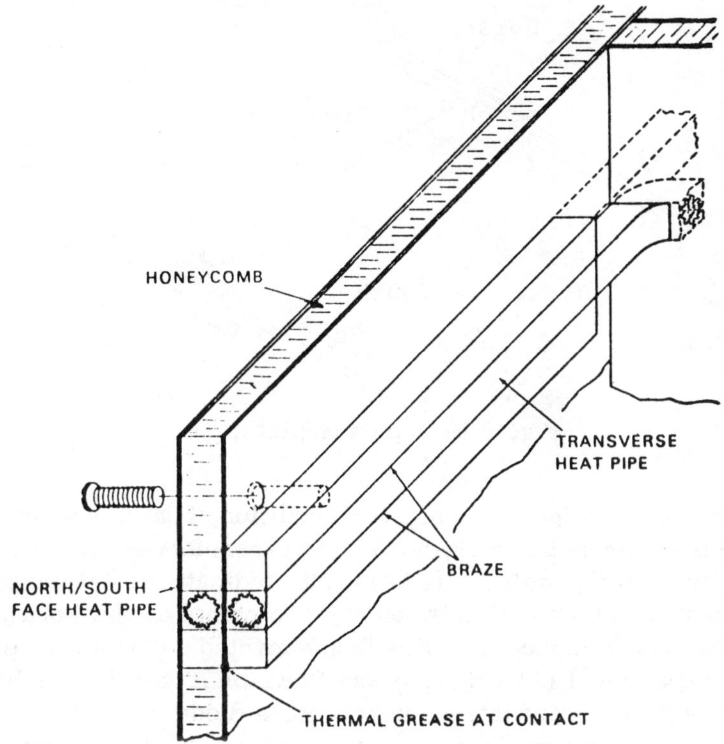

Fig. 5 ATS-6 heat pipe interface.

structure via eight specially contoured saddles. No flight data resulted because of a lunach vehicle failure.

6) OAO-C (1972). Three aluminum ammonia heat pipes, each with a different wick configuration, were mounted to the spacecraft structure in a fashion similar to the OAO-B. [6] The objectives were to isothermalize the structure and to attain a performance comparison among the different wick designs (pedestal artery, spiral artery, longitudinal grooves). Performance for all three designs was considered satisfactory. In addition, a variable-conductance heat-pipe experiment was flown, which controlled the onboard processor computer to 17 ± 3°C. [7] This experiment included a methanol, stainless-steel heat pipe with an integral split fin radiator. This represented the first flight confirmation of the variable-conductance feature. Performance was considered satisfactory.

7) ATS-6 (1974). Fifty-five aluminum, ammonia, longitudinal-groove heat pipes were used to isothermalize the main equipment platforms (Fig. 4). The heat pipes were bonded within the North/

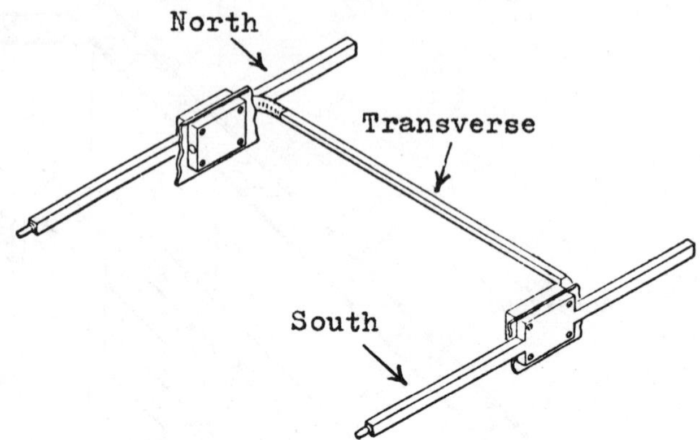

Fig. 6 Heat pipe triplets.

South honeycomb equipment panels. In addition, heat pipes bonded within the transverse beam (Figs. 5 and 6) transferred heat between the North and South panels. Flight results indicate a total temperature gradient less than 4°C between any two points on the North/South panel. This performance is better than expected. The advanced heat-pipe experiment (AHPE) also was flown on ATS-6.[8] It included a feedback variable-conductance heat pipe, a diode heat pipe, a phase-change material package, and a second-surface quartz mirror radiator. The feedback control feature was provided by a 2.8-W heater attached to the gas reservoir. Temperature stability to within $\pm$ 2°C has been obtained with feedback control, as compared to $\pm$ 9.5°C for passive control. Both results were for an input power variation of 0-30 W. Overall performance is considered satisfactory.

8) Skylab (1973). The Orbital Workshop contained 40 heat pipes, each 12 ft. in length, with overlapping joints to form essentially four continuous thermal loops on the spacecraft structure. The primary functions of the heat pipes were to reduce equipment touch temperatures on the sunlit side and to eliminate water condensation areas on the cooler side of the spacecraft. The heat pipes used a pedestal artery wick design with Freon-22 working fluid. Ammonia, unfortunately, could not be used because of toxicity problems. Heat loads in flight were in excess of design conditions, and performance has not been established fully. Some of the Skylab experiments also contained heat pipes that performed satisfactorily. These pipes included two ammonia, aluminum pipes and a copper, water-heat pipe. Slab-type wicks were used in all of the pipes. Water was used for the internal compartment pipe because of toxicity problems.

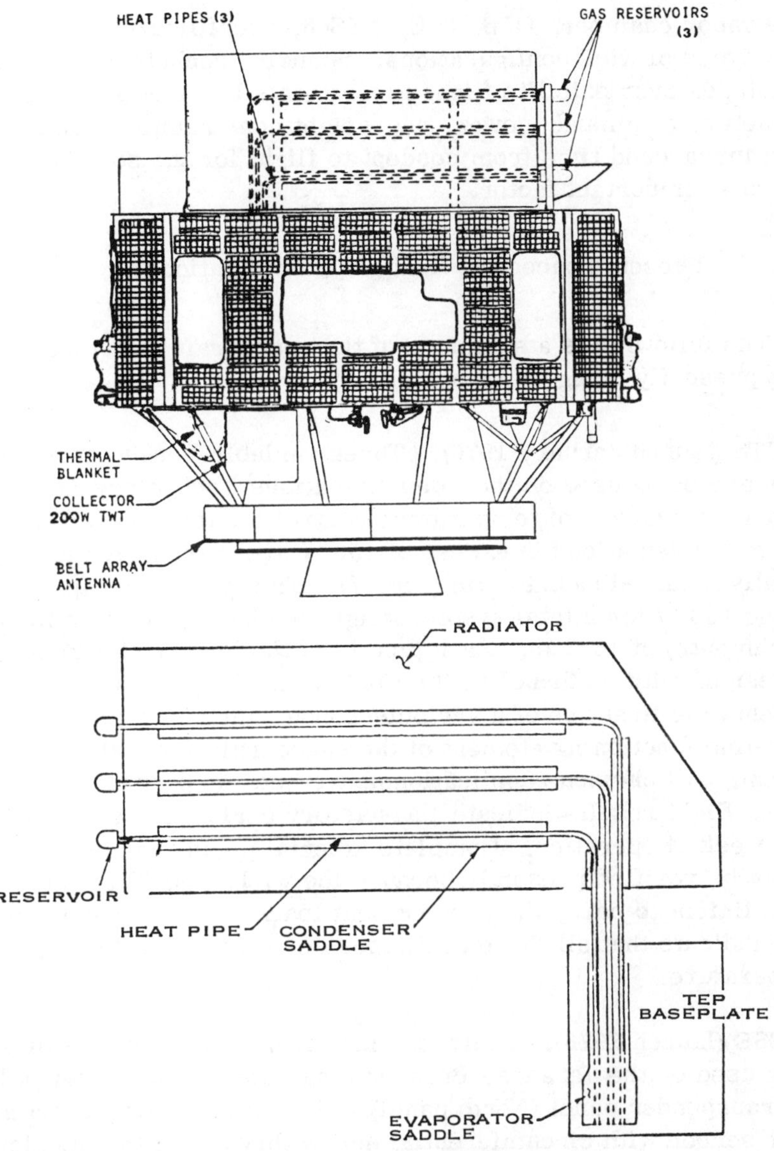

Fig. 7 CTS heat pipes.

9) Sounding-rocket heat-pipe experiment flights (1972-1974). Three sounding-rocket flights, each with approximately 6 min. of zero gravity, have been most timely and valuable in establishing performance capabilities of the latest in American and European heat-pipe technology[9]. Data accumulated on these flights include performance information on the following types of heat pipes: cryogentic, flat-

plate vapor chamber, CTS, IUE, ATS-6, overfilled/underfilled, and many types of wick configurations. Sounding rockets have distince advantages over satellite flight experiments because of less severe telemetry, command, power, and weight constraints. The relatively short turnaround time from concept to flight for the sounding rockets is also an important factor.

Present Spacecraft Heat-Pipe Applications

The following is a summary of the spacecraft heat-pipe applications presently being implemented:

1) CTS (launch January 1976). Three variable-conductance heat pipes are being used on the Communications Technology Satellite as the primary means of temperature control for the 200-W TWT.[10,11] Thermal dissipations in excess of 150 W are transported to an externally mounted radiator fin (Fig. 7). The system is capable of rejecting 196 W for a total system weight (including radiator fin and attachments) of 16.2 lb. Each pipe is of the high-performance arterial design, with methanol as the working fluid. This application represents the first use of a variable-conductance heat-pipe system as an in-line functioning element of the spacecraft thermal control subsystem. All previous applications have been as part of an experiment. Recent flight results indicate satisfactory performance during the first week of operation. Complete arterial performance, however, cannot be verified presently because the wicks and fillets alone may be sufficient to carry the present heat loads. Flight instrumentation does indicate that all three heat pipes do "turn on" at the expected temperature.

2) BSE (launch 1976). Fourteen aluminum, ammonia heat pipes are being used on the Japanese Broadcasting Satellite to isothermalize the transponder panel (North panel).[13] The wick design is a multilayer screen with circumferential screw threads on the interior walls. The major thermal dissipations on the panel are the three 100-W TWT's and their power supplies. The variable-conductance feature is not needed because sufficient radiator area exists on the panel. The heat pipes are bolted onto the solid magnesium panel: 12 internal and 2 external. The internal pipes place serious constraints on the component packaging and panel equipment layout. Successful operation of these heat pipes is required to meet the mission objectives.

3) IUE (launch 1978). Two aluminum, ammonia, longitudinal-groove heat pipes are being used on the International Ultraviolet Explorer to isothermalize the aft equipment deck. The heat pipes form two hoops and are mounted to the exterior of the aft deck via clamps. The groove design is very similar to the ATS-6 pipes, but the tubing has been formed in a hot extrusion process with two flat sections for improved thermal contact. Previous groove tubing had been fabricated in a cold process, with limited control on concentricity and groove dimensions.

Future Spacecraft Heat-Pipe Applications

Since 1967, NASA has funded heat-pipe research and development extensively, with participation from every center. Ames Research Center has managed most of the basic research, whereas Goddard Space Flight Center has been the primary user and has supported the major application efforts. Some of the present NASA efforts that may result in future applications follow:

1) Cryogenic heat pipes. These pipes are operating in the temperature range of 20° to 80°K, with applications to infrared (IR) detector and telescope cooling. Flexible joints are being developed which may be used on gimbals, deployable radiators, and as a means of introducing minimum structural loads into the telescope optics. Diode heat pipes also are being developed for one-way cooling applications.

2) Variable-conductance heat pipes. Variations in control are being investigated which have possible advantages over the standard gas-control technique. The longitudinal groove wick designs are being adapted for variable-conductance pipes in both the cryogenic and room-temperature regimes.

3) Electrohydrodynamics. Heat-pipe basic research is continuing in this area, with no firm spacecraft application in the immediate future.

4) Arterial dynamics. Investigations are continuing in search of a reliable, high-performance artery design that may be used in gas-controlled pipes.

5) SIPS. Efforts are continuing into development hardware for the Small Instrument Pointing System (SIPS) for Shuttle payload applications. The present design requirement is control of the SIPS canister to $20 \pm 1^{\circ}$C over a wide range of environments for a variation

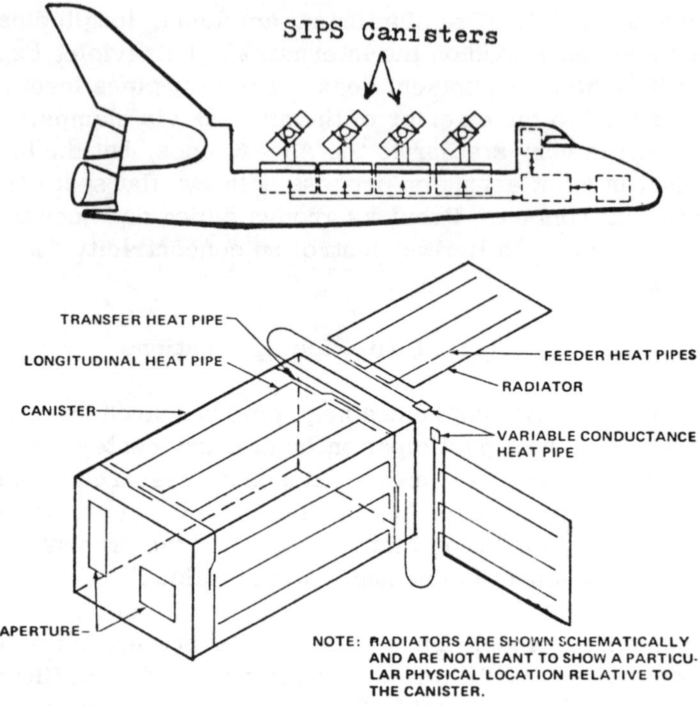

Fig. 8 SIPS heat-pipe concept.

of instruments inside the canister. [12] The instruments vary from a solar experiment with 300-W internal heat load to a large-aperture stellar experiment with 25-W internal load. The general concept of thermal control includes longitudinal and transfer heat pipes on the canister, active feedback variable-conductance heat pipes, and feeder heat pipes on the external radiating fins (Fig. 8).

Past NASA spacecraft heat-pipe concepts investigated which still may have future applications have included leading-edge cooling for the shuttle, solar energy concentrators, RTG cooling, shuttle landing wheel well temperature control, and Brayton cycle radiators. As confidence in heat-pipe performance is attained slowly, more use will be made on future spacecraft. As communications satellites become more powerful, heat pipes will be the only light-weight, effective means of transferring the high-dissipation loads of the transponders. The variable-conductance feature will become used more widely, as it requires very minimal power in the off condition and allows external platform areas, not intended for equipment mounting, to be utilized as heat-radiating areas. Although heat pipes have been very successful in flight operations, many design pitfalls exist. The designer must be fully aware of the limitations of the various heat pipes, as discussed in the following section.

Spacecraft Design Considerations

The following is a summary of the many design parameters that must be examined prior to the incorporation of heat pipes in a spacecraft design. The listing of parameters is arbitrary and not by order of importance:

1) Integration. Heat pipes, if required, should be designed into the system in the early design stages, as was done on ATS-6. This alleviates many of the integration and layout problems, particularly at the heat-pipe/structure interfaces. In order that a heat pipe may carry the maximum heat loads, with the minimum temperature gradients, the heat input/output interfaces must be designed properly and located as near the heat sources as possible. Interface materials (adhesive, grease, RTV, etc.) commonly are used. Easy repair/removal is certainly a desirable feature, provided that it does not compromise the thermal performance. Aluminum heat pipes extruded with flanges offer integration and thermal interface advantages. If heat pipes are to be bonded within honeycomb panels (ATS-5 and 6), then the epoxy core temperature often will dictate the qualification test temperature. This, in turn, dictates the wall thickness and weight. The ideal layout configuration, from a system test standpoint, is with all of the heat pipes in the same plane. This allows all of the heat pipes to be tested under a single spacecraft orientation. Any bends should be avoided, if possible, because they create manufacturing difficulties and make spacecraft testing more difficult. Bends have been shown to have a minor effect on performance. The primary constraint is usually the minimum bend radius of the container tube and wick.

2) Heat transport requirements. These requirements are specified by the thermal designer in terms of a) the axial heat flux requirement, the distance over which an amount of heat must be transported (Watt-centimeters), and b) the radial heat flux requirement, the maximum heat input over a given area (Watts per square centimeter). In addition, the thermal designer must define the maximum allowable condenser blockage due to overfill and gas generation. Once these requirements are defined, the actual heat-pipe design may proceed by

defining the following: wick material, working fluid, containment vessel size and material, fill criteria, and the detailed manufacturing processes.

3) Environmental testing constraints. Heat pipes may constrain the spacecraft thermal vacuum tests severely because of the limited performance capability of the heat pipes when tilted against gravity. The OAO-B thermal vacuum test support hardware was constrained by the minimum tilt capability of the Freon heat pipes. The CTS heat-pipe system required a water-cooled tube to simulate the heat-pipe operation for certain spacecraft orientations during solar simulation testing. A major ATS-6 concern was the heat-pipe tilt performance while on the launch pad.

4) Weight. Heat pipes are the lightest available means of transporting heat over long distances. For short distances (2-5 cm), normal metals usually are more efficient.

5) Safety. Because of the very high pressures encountered in some heat pipes, an adequate pressure qualification program usually will be required. Toxicity is a concern with some working fluids, particularly ammonia. Precautions must be taken in all areas of manufacture and integration once the pipe has been charged with the toxic fluid.

6) Analysis techniques. NASA has made available computer codes that may be used for heat-pipe parametric design studies. These include both conventional and variable-conductance heat pipes. Techniques exist for incorporating heat pipes in the overall spacecraft thermal mathematic models.

7) Qualification testing. This testing includes thermal performance, vibration, lifetime verification, and pressure qualification. Thermal performance usually is verified as a function of fluid fill, temperature, and tilt. Vibration testing often is required on each artery wick heat pipe but may be waived for certain tolerant designs, such as the longitudinal groove. In some cases, proof pressure tests are more severe than vibration because of the pressures reached during high-temperature adhesive cure operations. To insure materials compatibility, accelerated life tests are conducted at elevated temperatures simulating expected lifetime conditions.

Conclusions

In the heat pipe, the spacecraft thermal designer now has a very useful tool for many applications: 1) structure and component platform isothermalizing; 2) one-way heat-transfer device (thermal diode); 3) constant temperature control of components with large dissipation variations (variable-conductance pipes); and 4) heat exchangers.

References

[1]Kirkpatrick, J., "Variable Conductance Heat Pipes," International Concerence on Heat Pipes, Oct. 1973, Stuttgart, West Germany.

[2]Bienert, W., Brennan, P., and Kirkpatrick, S., "Feedback Controlled Variable Conductance Heat Pipes," AIAA Paper 71-421, 1971; also Progress in Astronautics and Aeronautics: Fundamentals of Spacecraft Thermal Design, Vol. 29, edited by J. W. Lucas, The MIT Press, Cambridge, Mass., 1972, pp. 463-485.

[3]Deverall, J. E., Salmi, E. W., and Knapp, R. J., "Heat-Pipe Performance in a Zero Gravity Field," Journal of Spacecraft and Rockets, Vol. 4, Nov. 1967, pp. 1556-1557.

[4]Harkness, R. E., "The GEOS-II Heat Pipe System and Its Performance in Test and Flight," T. G. 1049, 1969, Johns Hopkins Univ., Applied Physics Lab.

[5]Bilenas, J. A. and Harwell, W., "Orbiting Astronomical Observatory Heat Pipes Design, Analysis, and Testing," Paper 70-HT/Spt-9, 1970, American Society of Mechanical Engineers.

[6]Harwell, W., Edelstein, F., McIntosh, R., and Ollendorf, S., "Orbiting Astronomical Observatory Heat Pipe Flight Performance Data," AIAA Paper 73-758, 1973; also Progress in Astronautics and Aeronautics: Thermophysics and Spacecraft Thermal Control, Vol. 35, edited by R. G. Hering, The MIT Press, Cambridge, Mass., 1974, pp. 445-465.

[7]Kirkpatrick, J. P. and Marcus, B. D., "A Variable Conductance Heat-Pipe Flight Experiment," AIAA Paper 71-411, 1971; also Progress in Astronautics and Aeronautics: Fundamentals of Spacecraft Thermal Design, Vol. 29, edited by J. W. Lucas, The MIT Press, Cambridge, Mass., 1972, pp. 505-527.

[8]Kirkpartick, J. P. and Brennan, P. J., "Advanced Thermal Control Flight Experiment," AIAA Paper 73-757, 1973; also Progress in Astronautics and Aeronautics: Thermophysics and Spacecraft Thermal Control," Vol. 35, edited by R. G. Hering, The MIT Press, Cambridge, Mass., 1974, pp. 409-430.

[9]McIntosh, R., Ollendorf, S., and Harwell, W., "The International Heat Pipe Experiment," AIAA Paper 75-726, May 1975, Denver, Colo.

[10]Stipandic, E., Gray, A., and Gedeon, L., "Thermal Design and Test of a High Power Spacecraft Transponder Platform," AIAA Paper 75-680, May 1975, Denver, Colo.

[11]Mock, P. R. and Marcus, B. D., "Communications Technology Satellite: A Variable Conductance Heat Pipe Application," AIAA Paper 74-749, July 1974; also Journal of Spacecraft and Rockets, Vol. 12, Dec. 1975, pp. 750-753.

[12]Harwell, W. and Canaras, T., "Feasibility Study to Use Heat Pipes to Control Temperatures of SIPS Canister," Contract NAS 5-22334, Aug. 1975, NASA Goddard Space Flight Center.

[13]Scollon, T., private discussions, Jan. 1976.

Chapter II – Spacecraft Attitude Control Subsystems

DISCRETE-TIME ATTITUDE CONTROL OF FLEXIBLE SPACECRAFT

Kent R. Folgate *

General Electric Space Division,
Valley Forge, Pennsylvania

Abstract

Two technology trends make this topic timely. First, the evolution of spacecraft to those with large lightweight deployed structures. This is particularly noticeable in the area of communication satellites. Whereas the current commercial and military operational systems are dominated by spin stabilized spacecraft with relatively rigid solar array drums, the next generation of Intelsat, Telesat, Westar and DSCS spacecraft will be three-axis stabilized with deployed sun tracking arrays. In addition, the FLTSATCOM and NASA Tracking Data Relay Satellites will have multiple large deployed antennas. The other trend is towards the replacement of older all analog control systems by those with on-board digital processors. The ATS-6 satellite has already demonstrated in orbit the versatility and performance of such a system.

This paper presents key results of an intensive effort to develop an efficient basic approach and supporting design tools to permit synthesis of attitude control systems that must accommodate spacecraftwith low frequency lightly damped structural modes.

I. Background

Several years ago, the author led a project that was initiated to develop a lightweight, reliable, adaptable controller capable of

Presented as Paper 76-262 at the AIAA/CASI 6th Communications Satellite Systems Conference, April 5-8, 1976, Montreal, Canada.

*Chief Engineer, Satellite Control Systems.

yielding high-performance attitude control in the presence of low-frequency, lightly damped spacecraft structure. The approach selected to meet this ambitious set of requirements was to implement a low-bandwidth, discrete-time controller that employs a dual-bandwidth state estimator. Several factors favored the low-bandwidth approach. It involved minimum risk, since proven "rigid-body" designs just employ controllers with bandwidths that are low relative to the existing resonant frequencies of the structure. In addition, the cost of structural testing is reduced significantly if the controller design requires knowledge only of the lower bound of resonant frequencies rather than their precise location.

A very low-bandwidth controller creates an interesting design challenge. Stability compensation necessarily must occur at a low frequency. This can be difficult to implement using conventional series compensation, because of the large time constants involved. If low gain is used to lower the bandwidth, attitude standoff errors become large. Combating these errors with integral compensation or lag networks results in a conditionally stable system prone to acquisition difficulties. The dual bandwidth state estimator was chosen because it circumvents these difficulties while simultaneously dealing with the diverse requirements of acquisition and precision

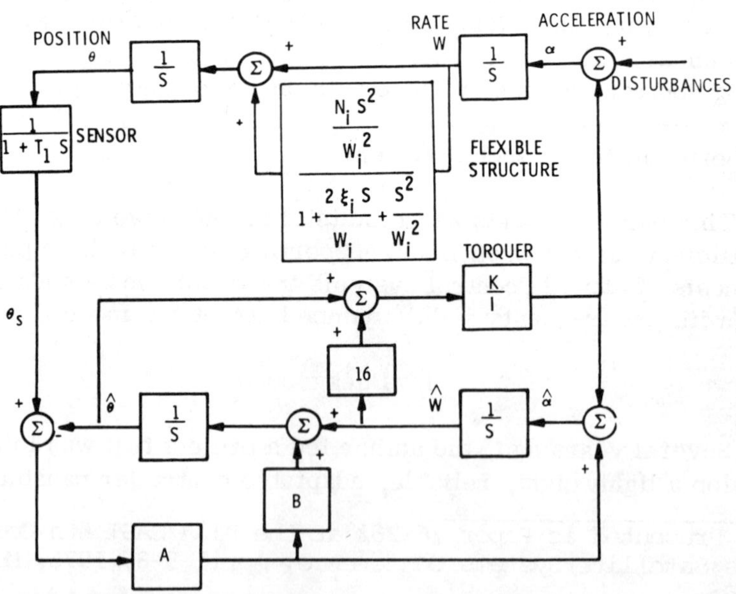

Fig. 1 Analog state estimator stabilization.

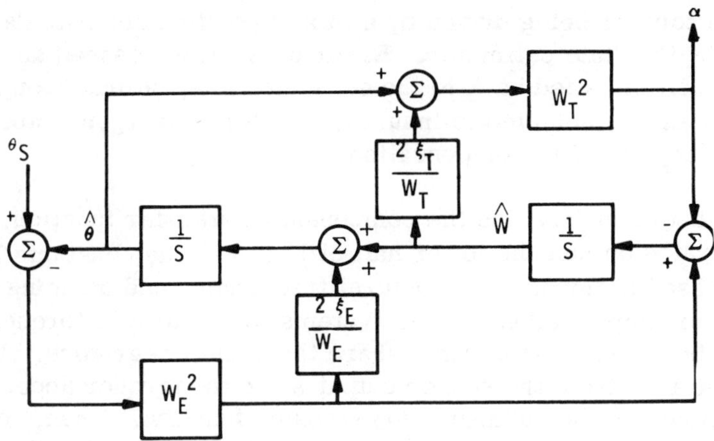

Fig. 2 Stability compenstation element.

pointing. Weight, power, and volume efficiencies were realized by choosing a discrete-time controller.

The preceding choice results in a sampled data control system; i.e., one where the plant is a continuous system, whereas the controller is a discrete-time system. Topics of general interest to be reported include 1) the method of selecting controller parameter values to achieve stability compensation; 2) the results of linear analysis performed on the sampled data system; and 3) the novel approach used to build an all-digital, nonlinear simulation that runs extremely fast even in the presence of discontinuous inputs and lightly damped flexible structure modes.

II. State Estimator Provides Versatile Stability Compensation

The characterizing properties of state estimator stabilization are independent of whether the implementation is analog or discrete time. Since more readers are familiar with all continuous systems, the transfer function resulting from employing an analog state estimator will be developed and examined. Discrete-time implementation then will be discussed.

Figure 1 shows a single-axis example of a control system using analog state estimator stabilization. The upper part of the diagram represents the plant and has flexible structure modes introduced between the position and rate variables. The lower half of the diagram shows a torquer being driven by a mix of position and rate data generated by the state estimator. Error between the sensed and estimated position is fed back to the estimator integrators through constants A and B, selected to insure estimator convergence and overall control loop stability compensation.

To better understand the compensation transfer function, Fig. 2 concentrates on just the lower half of Fig. 1. The constants have been reexpressed in terms of the natural frequencies and damping factors of the two simple second-order systems (estimator or torquer as indicated by the subscripts) that share the same integrators. Let the transmittance from the sensor output θ_S to the torquer acceleration α be defined as the compensation transfer function. Then, for the illustrated state estimator stabilization, the transfer function is

$$\frac{\alpha}{\theta_S} = \frac{W_T^2\, W_E^2 \left\{1 + \left[(2\,\xi_T/W_T) + (2\,\xi_E/W_E)\right] S\right\}}{W_C^2 \left\{1 + (2\,\xi_C\, S/W_C) + (S^2/W_C^2)\right\}} \tag{1}$$

$$W_C^{\,2} = W_T^{\,2} + W_E^{\,2} + (2\,\xi_T)\,(2\,\xi_E)\,W_T\,W_E \tag{2}$$

$$\xi_C = (\xi_T\,W_T + \xi_E\,W_E)\,/\,W_C \tag{3}$$

The subscript E pertains to the simple second-order system parameters when considering only the estimator loops, the subscript T when considering only the torquer loops, and the subscript C when combining both.

Note that, when there is any appreciable separation between the torquer and estimator loop natural frequencies, the compensation transfer function rapidly approaches the following:

$$\frac{\alpha}{\theta_S} \cong \frac{W_L^{\,2} \left\{1 + \left[(2\,\xi_L/W_L) + (2\,\xi_H/W_H)\right]\; S\right\}}{\left\{1 + (2\,\xi_H/W_H)\,S + (S^2/W_H^{\,2})\right\}} \tag{4}$$

where the subscripts L and H refer to lower and higher component natural frequencies, respectively.

Examination of the approximate transfer function (4) reveals that, without the lead break, the total open-loop transfer function has a magnitude of unity and a phase shift of 180° at a frequency near W_L, as long as W_L is chosen small compared to the other compensation parameter W_H, the sensor frequency $(1/T_1)$, and the structure resonances. Stability compensation with the lead term then is obtained easily by choosing ξ_L sufficiently positive. The quadratic rolloff at the higher frequency W_H is also an important characteristic of this stabilization scheme. It performs the important function of rolling off the open loop gain intentionally after crossover so that the low-frequency, lightly damped structural resonant peaks do not protrude up high enough to cause instability. Inherent signal processing lags, i.e., sensor time constants and buffer amplifier rolloffs, no longer automatically provide this protection, as they do in systems with only high-frequency resonances.

Proper parameter selection enables the lower stabilization frequency role to be accomplished within <u>either</u> the torquer or estimator loops. During acquisition, sensor range is more important than accuracy. It is also essential that a system employing derived rate have this information as soon as the sensor comes out of saturation. Thus, for acquisition, the low-frequency stabilization is accomplished by using a low-gain K in the torquer loop, whereas the quadratic rolloff is generated by the higher-bandwidth estimator loop. For precision control, sensor range is given up for accuracy, and the torquer/estimator loop frequency roles are interchanged. In this case, low-frequency stabilization is achieved with the estimator, and a higher torque gain K reduces standoff error while simultaneously generating the quadratic rolloff.

III. Discrete-Time Approach Simplifies Implementation

The previous section used an analog controller example to present basic state estimator stabilization concepts. Each nonredundant control axis would require several operational amplifiers (opamps) to generate the estimated state variables. Using digital electronics permits the opamps to be replaced by memory locations and all algebraic functions to be accomplished by simple summing. A single set of algorithms may be time-shared among all three control axes. Additional benefits accrue by using a true discrete-time controller as opposed to merely emulating an analog design with high-repetition-rate, short-sample interval digital electronics.

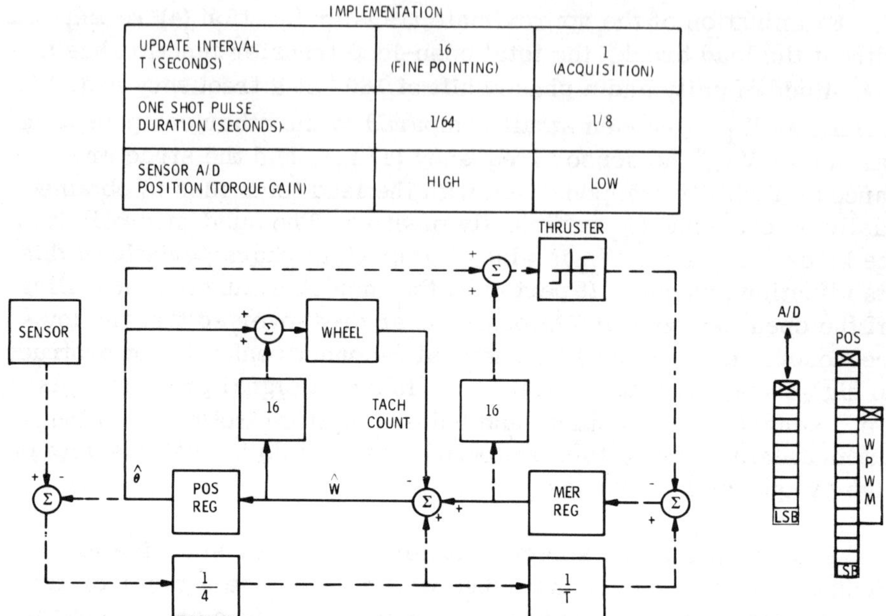

IMPLEMENTATION

UPDATE INTERVAL T (SECONDS)	16 (FINE POINTING)	1 (ACQUISITION)
ONE SHOT PULSE DURATION (SECONDS)	1/64	1/8
SENSOR A/D POSITION (TORQUE GAIN)	HIGH	LOW

Fig. 3 Discrete-time implementation of state estimator stabilization.

Figure 3 illustrates a discrete-time implementation. Both a momentum exchange actuator (the wheel) and an external torquing device (the thruster) are included in the design. The output of the position register (pos reg) is the estimated attitude $\hat{\theta}$. The other register (mer) contains an estimate of the momentum equivalent rate of the system, i.e., the rate of vehicle would have if the wheel were not operating. Note that the wheel is driven with a combination of position and rate information (estimated rate $\hat{W}$ is the difference between the mer value and the wheel tachometer, whereas the thruster control law uses position and mer data. When the wheel is not operating, the thruster control law is effectively a conventional position plus rate deadband test (since mer = $\hat{W}$ when the tachometer output is zero), but, when the wheel is operating, the thruster control law performs the momentum unloading function even when vehicle attitude and rates remain near zero.

The solid lines in Fig. 3 indicate operations performed each 1-sec minor cycle period. The dashed line operations are performed each update interval. For acquisition, the update period coincides

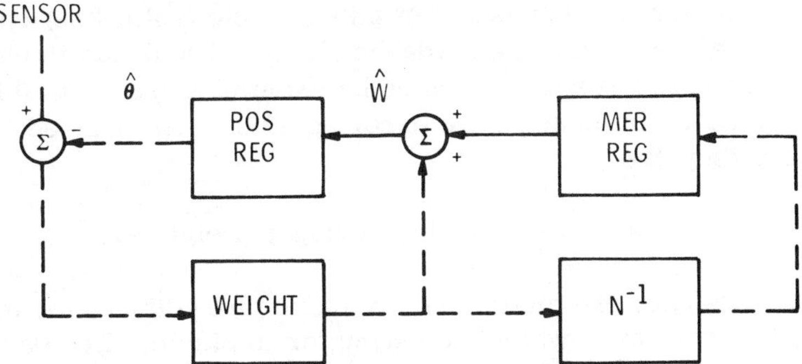

Fig. 4 Discrete-time state estimator.

with the minor cycle period, and the torque loop gain (determined by the error necessary to produce full duty cycle to the wheel pulse width modular [WPWM] drive circuits) is set low. For fine control, a single command reduces the update frequency to once each 16 sec while simultaneously increasing the torque loop gain by changing the scale of the sensor data relative to the WPWM. In order to reduce processor requirements to only shifting and adding, the control law, estimator, and gain parameters were restricted to binary multiples.

Clarifying Fig. 3 further, the sequence of events is as follows. The number of tachometer pulses accumulated since the execution of the previous minor cycle is subtracted for the current mer value to obtain the estimated rate $\hat{W}$, which is added into the position register (rectangular integration). If it is an updating cycle, the updating operations are executed. Finally, the sum of the estimated position and 16 times the estimated rate is loaded into the WPWM register for execution over the next second. An updating sequence includes the following. A weighted error signal is formed by subtracting the position register data from the sampled sensor output and dividing the result by four. The weighted error signal first is added into the position register and then weighted further by the inverse of the update interval prior to summing into the mer register. The estimated position and 16 times the mer value are summed to form the thruster control variable subjected to the deadband test. Note that, although the wheel control signal is recomputed each minor cycle, the thruster control signal is recomputed and executed only during update cycles.

Figure 4 shows a more general form of the discrete-time state estimator by itself. An arbitrary weight factor is applied to the error signal, and any number N of minor-cycle (solid line) operations are permitted between updates (dashed line). The dynamic characteristics of the estimator are revealed by examining the roots (eigenvalues) of the defining sequence error equation. As derived in the Appendix, they are

$$\text{Roots} = 1 - \text{weight} \pm \sqrt{\text{weight (weight} - 1)} \qquad (5)$$

These results are plotted on a Z-plane plot (Fig. 5) for the range of weight factors over which the estimator is stable. Besides showing the degree of damping of the estimator (as indicated by the magnitude of the roots), the polar angle reveals the damped natural frequency per update. For example, with a weight factor of 1/4, the estimator response to an initial unit error is as shown in Fig. 6.

The 30 deg/update damped natural frequency characteristic is clearly evident. Changing the update period only modifies the time scale of the transient response. This interesting property simplifies the checkout of an estimator configured to have a single weight factor with the dual bandwidth realized entirely by update period change.

Control system design usually employs linear analysis techniques to select controller parameters and simulation methods initially to evaluate the influence of system nonlinearities. Trying to employ S-plane analysis to a sampled data system is cumbersome in all but the simplest situations. This is because it is difficult to replace the discrete-time controller with a representative continuous model. The following approach was selected as a much improved alternative.

First, a single-axis linear model is formed for the sampled data system. The differential equations characterizing the continuous variables of the system are converted into a system of sequence equations which propagates their state forward over the minor cycle interval of the controller. The wheel drive signal is considered a constant input. Next, the system state vector is augmented to include the two controller variables, estimated position, and wheel drive signal, which are recomputed each minor cycle. The mer register is introduced as a constant input. The augmented system is propagated forward N times, where N is the number of minor cycles per update interval. Finally, the propagated system is augmented again to include the mer register as a state variable, and the updating operations are modeled. This fully augmented set of sequence equations defines the total system's transition from one update to the next. The

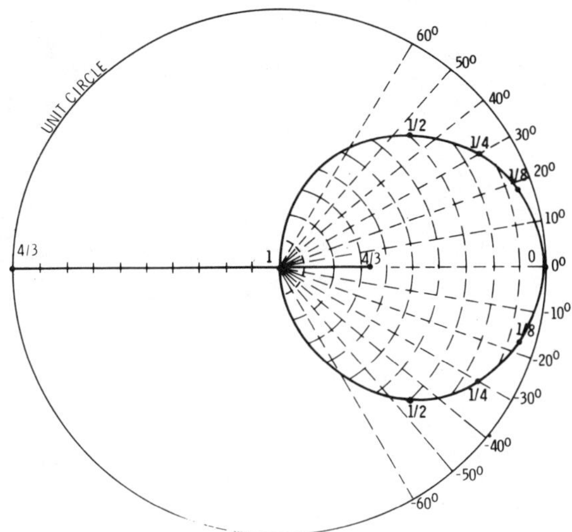

Fig. 5 Estimator root locus as a function of weight.

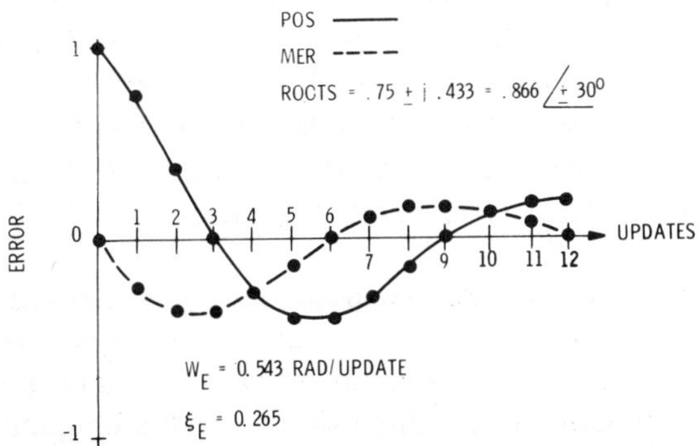

Fig. 6 Estimator response to unit error.

eigenvalues of the transition matrix characterize the stability and transient response of the sampled data system. These are solved for as a function of input parameters of interest such as vehicle inertia, sensor gain and time constants, minor cycle time, update period, torquer gain, wheel time constant, and estimator weight factor. Of particular interest is examining these roots as a function of flexible structure mode parameters.

With a flywheel loop configured as previously described, Fig. 7 plots the magnitude of a complex pair of lightly damped ($\xi = 0.0005$)

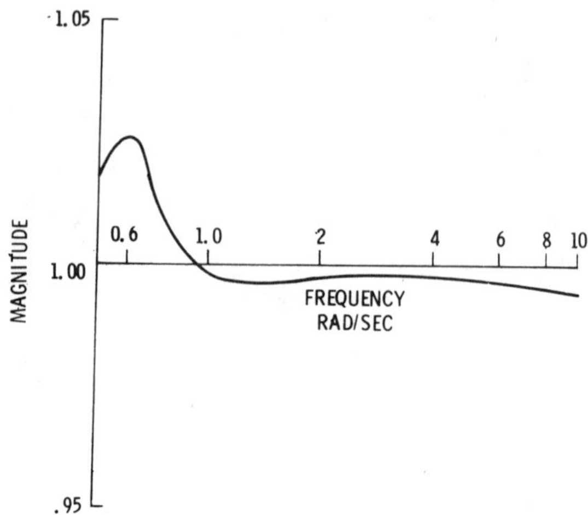

Fig. 7 Z-plane magnitude of structural roots vs frequency, 1-sec update (low gain).

structural roots as a function of their natural frequency, with the controller operating in the 1-sec update acquisition mode. The modal gain used (N_i in Fig. 1) is 2.0, resulting in 63 dB of peaking at resonance. The other roots of the system are not plotted, since they remain stable (magnitude less than unity) and do not migrate much from their rigid-body locations. Total system stability is indicated as long as the lowest structural frequency is greater than 0.95 rad/sec.

Figure 8 reveals an interesting property of the discrete-time controller. The gain in the torquer loop has been increased by a factor of 16 and all structural damping removed ($\xi = 0$). The response is not unstable at all frequencies. When the resonant frequency is low enough to interact with the dominant rigid-body roots, instability results. Next, there is a transition region where highly damped frequency bands may appear. Finally, the root magnitudes converge to their open-loop value of $\exp(-\xi W)$. Several open-loop nonzero damping lines are shown dashed in Fig. 8. Increasing either the resonant structure modal gain or the torquer loop gain accentuates the peaks and valleys in the transition zone, as is seen by comparing Figs. 7 and 8.

With the damping reset to the assumed nominal ($\xi = 0.0005$) and the torque loop gain left high, Fig. 9 shows the magnitude of the resonant structure root frequency sweep after increasing the update per-

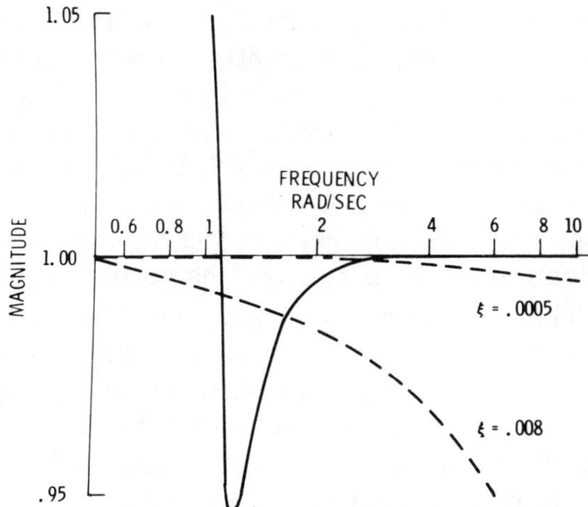

Fig. 8 Z-plane magnitude of structural roots vs frequency, 1-sec update (high gain).

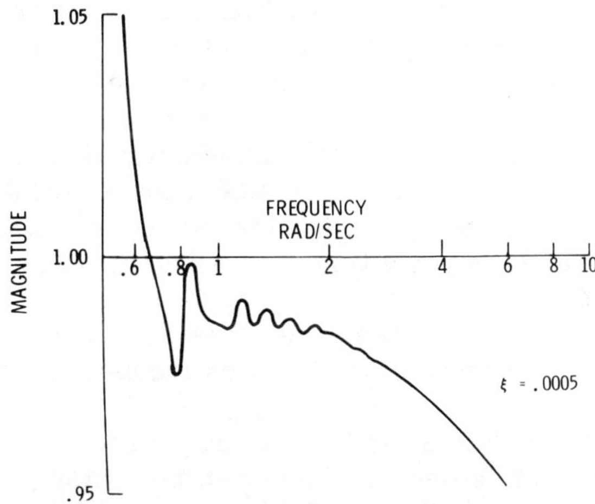

Fig. 9 Z-plane magnitude of structural roots vs 16-sec update (high gain)

iod to 16 sec. This controller configuration corresponds to the fine pointing mode commanded after the completion of acquisition. By comparing the results of Figs. 8 and 9, it is clear that decreasing the updating frequency has the same effect as increasing the resonant structure damping.

Figure 10 shows a Z-plane locus of the dominant rigid-body roots of the 16-sec update mode as a function of normalized gain exterior to the controller. This includes items like sensor gain change or inertia reduction due to fuel depletion, etc. Within the controller, digitally executed constants are invariant, and the system has low sensitivity-to-torquer characteristics, since it is inside a high-gain feedback loop. The roots characterizing the system transient response are dominated completely by the estimator for nominal gain and below. This is observed by comparing the implemented (weight = 1/4) estimator roots from Fig. 5 with the dominant system roots in Fig. 10. Physically, this occurs because the high-gain torquer loop rapidly removes indicated errors so that the response of the system is limited by the estimator. For higher than nominal gain, the dominant system roots no longer follow the path taken by the estimator roots in Fig. 5 (with weight = gain/4 used as the parameter) because the torquer loop begins to influence the roots as the estimator bandwidth increases.

The analyses of this section concentrate on the easily linearized flywheel loop; however, stability analysis results using the classical contactor describing function for the thruster also correlate well with simulation and hardware test results. Although this control loop will not be discussed, the following clarification of Fig. 3 is included. The implementation of the integrated thruster acceleration feedback into the mer register is achieved by adding (or subtracting) an appropriate stored number any time a one-shot thruster pulse is fired.

V. Discrete-Time Simulation Significantly Reduces Running Time

Digital simulation has its own unique set of problems. Since a digital computer is itself a discrete-time system, it handles discrete-time models easily. To simulate continuous systems, the approach traditionally taken has been to integrate numerically the differential equations describing the dynamics of the system. This followed naturally, since the earliest control systems were all continuous and because of the extensive use of analog computers for simulation prior to the development of high-speed, cost-competitive digital machines.

Numerical integration is not very attractive when either of two situations is present, discontinuous inputs or high-frequency (short-time constant) dynamics. Either of these requires drastic reduction of the integration step size with the corresponding increase in running.

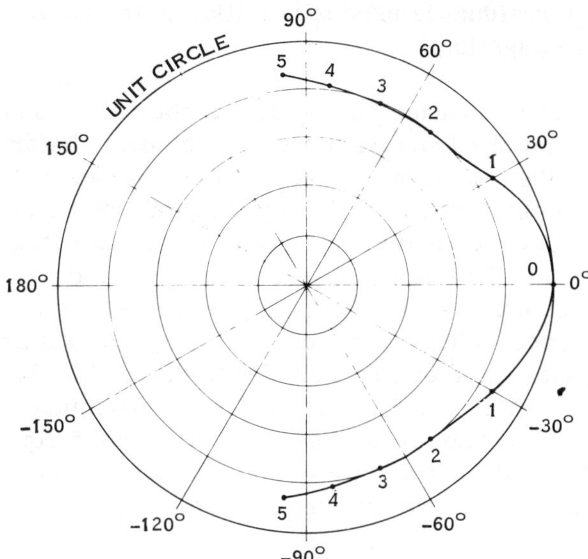

Fig. 10 Dominant rigid body roots as a function of normalized gain, 16-sec update.

time and cost. Both of these occur with the system described in this paper: discontinuous inputs because of the discrete-time controller, and relatively high-frequency dynamics when simulating flexible structure modes. Running times in the latter case become truly prohibitive.

To deal with these problems, all numerical integration was replaced by discrete-time propagation. The key to the approach is appropriate introduction of system nonlinearities. As in the previous section, constant transition and input propagation matrices (sequence equations) are derived from the linear differential equations of the system's continuous state variables. These are used to propagate the continuous states forward for the time interval corresponding to the minor cycle time of the controller. It is assumed that the forcing inputs remain constant for this interval. Propagated state variables include the system angular momentum, wheel momentum, wheel rotor angle, integral of vehicle rate, sensor outputs (due to time constants), and two states per flexible structure mode included. Since the propagation has been linear, system nonlinearities must be introduced before the information is used further. Sensor nonlinearities, such as saturation, are incorporated. The scaled integer portion of the wheel rotor angles are used as the tachometer counts, and any noninteger residual is used to reinitialize the rotor angle states for the next propagation.

Complete attitude-time history is maintained by accumulating the incremental rotations from each propagation cycle before resetting the integral of the individual control axis angular rates to zero. The transformation characterizing the incremental rotation also is used to rotate the angular momentum vector before reinitializing it for the next propagation. This is the discrete-time method of including the major dynamics nonlinearity resulting from choosing a vehicle-fixed rotating reference. The integer controller operations are executed and finite register size effects included to determine the control actuator simuli for the next minor cycle. Environmental disturbance torques and actuator nonlinearities (such as wheel torque-speed characteristics and friction) are included when forming the forcing function inputs for the next propagation interval.

This all-attitude, nonlinear discrete-time simulation approach was used first almost two years ago for performance evaluation studies. Results obtained with it showed excellent correlation with the conventional numerical integration simulation that it replaced. When flexible structure modes were introduced, it ran thousands of times faster than the conventional program, thereby permitting studies that otherwise would have been economically unfeasible. The dynamic portion also was adapted later to control a three-axis-driven table used for real-time spacecraft hardware dynamic testing.

VI. Summary

To date, the primary application of low-bandwidth state estimators has been to improve system performance by extracting information from noisy sensor data. This paper suggests a new role, i.e., a versatile stabilization element for control systems. It is particularly well suited to deal with systems that must accommodate low-frequency, lightly damped structural modes. In deference to the conventional servo analyst, the estimator is introduced first as a stability compensation element in a continuous control system. The flexibility of interchanging low-frequency stabilization and higher-frequency rolloff responsibilities between the estimator and torquer loops is demonstrated. The unique characteristics of discrete-time implementation are described and stability analysis results discussed. Finally, an all-discrete-time digital simulation approach is presented.

It effectively avoids the increased running time and cost penalties incurred by numerical integration simulations when either high-frequency local dynamics or regularly spaced discontinuous inputs are present.

Appendix: Discrete-Time Estimator Dynamics

The linear single-axis propagation of spacecraft angular position (pos) and rate (mer) forward over N 1-sec minor cycle intervals, with an assumed constant disturbance acceleration D applied, is

$$\begin{bmatrix} \text{pos}\ (K+1) \\ \text{mer}\ (K+1) \end{bmatrix} = \begin{bmatrix} 1 & N \\ 0 & 1 \end{bmatrix} \begin{bmatrix} \text{pos}\ (K) \\ \text{mer}\ (K) \end{bmatrix} + DN \begin{bmatrix} N/2 \\ 1 \end{bmatrix} \tag{A1}$$

The estimator, as configured in Fig. 4 and ignoring sensor dynamics, propagates the estimated states forward as follows:

$$\begin{bmatrix} \hat{\text{pos}}\ (K+1) \\ \hat{\text{mer}}\ (K+1) \end{bmatrix} = \begin{bmatrix} 1 & N \\ 0 & 1 \end{bmatrix} \begin{bmatrix} \hat{\text{pos}}\ (K) \\ \hat{\text{mer}}\ (K) \end{bmatrix} + \text{weight}\ \ \text{pos}\ (K+1) - \left[\hat{\text{pos}}\ (K) + N \cdot \hat{\text{mer}}\ (K)\right]\Big\} \begin{bmatrix} 1 \\ 1/N \end{bmatrix} \tag{A2}$$

Let the error vector between the actual and estimated variables be defined as

$$\begin{bmatrix} \text{epos} \\ \text{emer} \end{bmatrix} = \begin{bmatrix} \text{pos} \\ \text{mer} \end{bmatrix} - \begin{bmatrix} \hat{\text{pos}} \\ \hat{\text{mer}} \end{bmatrix} \tag{A3}$$

The estimator dynamics are characterized by the propagation of the error vector. Combining the last three equations yields

$$\begin{bmatrix} \text{epos}\ (K+1) \\ \text{emos}\ (K+1) \end{bmatrix} = \begin{bmatrix} (1 - \text{weight}) & N\ (1 - \text{weight}) \\ -\text{weight}/N & (1 - \text{weight}) \end{bmatrix} \begin{bmatrix} \text{epos}(K) \\ \text{emer}(K) \end{bmatrix} + \frac{DN}{2} \begin{bmatrix} N\ (1 - \text{weight}) \\ (2 - \text{weight}) \end{bmatrix} \tag{A4}$$

The roots of the transition matrix are

$$\text{roots} = 1 - \text{weight} \pm \sqrt{\text{weight}\ (\text{weight} - 1)}$$

Examining the roots reveals that the estimator is stable (root magnitudes less than unity) as long as

$$0 < \text{weight} < 4/3 \tag{A6}$$

A stable estimator converges the error vector to

$$\begin{bmatrix} \text{epos } (K \to \infty) \\ \text{emer } (K \to \infty) \end{bmatrix} = \text{DN} \begin{bmatrix} \text{N } (1 - \text{weight})/\text{weight} \\ 1/2 \end{bmatrix} \tag{A7}$$

SPACECRAFT ATTITUDE CONTROL
USING A MAGNETICALLY SUSPENDED MOMENTUM WHEEL

R. C. Quartararo*
Rockwell International, Downey, Calif.

Abstract

An analysis is presented of the attitude dynamics of a spacecraft containing a momentum wheel supported by the relatively soft, springlike forces provided by magnetic bearings. System performance is established in terms of settling time and the response amplitudes due to a jet impulse and rotor unbalance. Nondimensional plots of performance parameters are presented as functions of suspension stiffness and damping for configurations typifying a momentum-bias-controlled spacecraft and a dual-spin spacecraft. The analytical results are applied to an example spacecraft to show that realistic magnetic bearing properties result in good performance. The analysis also suggests that magnetic bearings may be used to enhance system performance for some spacecraft configurations by providing nutation damping, isolation of wheel unbalance effects, and avoidance of certain critical speeds.

Nomenclature

$\mathcal{A}$ = I_{V11}/I_{W11}; vehicle nondimensional transverse inertia
$\underline{b}_1,\underline{b}_2,\underline{b}_3$ = set of orthogonal unit vectors fixed in B
B = label for spacecraft body
$\mathcal{B}$ = complex amplitude of β
C_W = location of mass center of W
$\mathcal{C}$ = I_{W33}/I_{W11}; nondimensional spin inertia of W
$\mathcal{D}$ = $L\, m_W\, (1 + m_W/m_V)\, \delta_1/I_{W11}$; nondimensional static unbalance of W
e = $e_1 + ie_2$
e_1, e_2 = ε_1/L, ε_2/L; nondimensional translational displacement of W in B

Presented as Paper 76-265 at the AIAA/CASI 6th Communications Satellite Systems Conference, April 5-8, 1976, Montreal, Canada.

*Member of the Technical staff, Space Division.

F_1, F_2, F_3 = components of external force vector

$\mathcal{F}_i$ = $m_W\ L\ F_i/(m_V\ I_{W11}\ \bar{\dot{\phi}}^2)$, where $i = 1,2,3$; nondimensional force

h = $I_{W33}\ \bar{\dot{\phi}}$; nominal angular momentum of W

i = $\sqrt{-1}$

I_{V11}, I_{V33} = transverse and spin moments of inertia for the vehicle in its nominal condition; measured relative to 0 and referred to $\underline{b}_i$

I_{W11}, I_{W33} = transverse and spin moments of inertia for W; measured relative to O_W and referred to $\underline{w}_i$

I_{W13} = product of inertia for W; measured relative to O_W and referred to $\underline{w}_i$

$\mathcal{I}$ = I_{W13}/I_{W11}; nondimensional dynamic unbalance

K_{A1}, K_{A2} = magnetic suspension damping and spring constants, respectively, for axial ($\underline{b}_3$) motion

K_{R1}, K_{R2} = magnetic suspension damping and spring constants, respectively, for radial ($\underline{b}_1$ and $\underline{b}_2$) motion

$\mathcal{K}_{A1}$ = $2\ K_{A1}\ L^2/(I_{W11}\ \bar{\dot{\phi}})$; nondimensional axial damping constant

$\mathcal{K}_{A2}$ = $2\ K_{A2}\ L^2/(I_{W11}\ \bar{\dot{\phi}}^2)$; nondimensional axial spring constant

$\mathcal{K}_{R1}$ = $2\ K_{R1}\ L^2/(I_{W11}\ \bar{\dot{\phi}})$; nondimensional radial damping constant

$\mathcal{K}_{R2}$ = $2\ K_{R2}\ L^2/(I_{W11}\ \bar{\dot{\phi}}^2)$; nondimensional radial spring constant

L = distance from O_W to shaft end

m_W, m_V = mass of W and mass of vehicle

M_1, M_2 = components of torque applied to vehicle

$\mathcal{M}_i$ = $M_i/(I_{W11}\ \bar{\dot{\phi}}^2)$, where $i = 1,2$; nondimensional torque

O = nominal location of vehicle and wheel mass centers in B

O_W = nominal location of wheel mass center in W

s = Laplace operator

s(), c() = sin( ) and cos( )

t, T = time and settling time

$\mathcal{T}$ = complex amplitude of Θ

W = label for momentum wheel

$\underline{w}_1, \underline{w}_2, \underline{w}_3$ = set of orthogonal unit vectors fixed in W

β = $\beta_1 + i\ \beta_2$

β_1, β_2 = small-angle rotations about $\underline{b}_1$ and $\underline{b}_2$ locating $\underline{w}_3$ relative to B

$\underline{\delta}$ = $\delta_1\ \underline{w}_1 + \delta_3\ \underline{w}_3$; small displacement vector from O_W to C_W

$\underline{\varepsilon}$ = $\varepsilon_1\ \underline{b}_1 + \varepsilon_2\ \underline{b}_2 + \varepsilon_3\ \underline{b}_3$; small displacement vector from O to O_W

Θ = $\Theta_r + i\ \Theta_y$

$\theta_r, \theta_p, \theta_y$ = roll, pitch, and yaw displacements of B, respectively

μ = $m_W (1 + m_W/m_V) L^2/I_{W11}$; nondimensional wheel mass
τ = $\dot{\phi}t$; nondimensional time
$\dot{\phi},\dot{\phi}^*$ = constant and variational portions of spin speed of W

Introduction

The development of magnetic suspension bearings for application to spacecraft attitude control wheels is reaching an advanced stage. In a review of hardware development efforts over the past several years, Henrikson et al.[1] note that several types of magnetic bearings have been developed to support momentum wheels. Both Henrikson and Meeks[2] suggest that these devices will be applied to a flight spacecraft in the near future.

Interest in magnetic bearings stems from several potential advantages that they have over conventional mechanical bearings, which result from the elimination of mechanical contact and no need for lubrication[1-3]: life is increased, drag torques are reduced, environmental operating ranges are increased, and design flexibility is increased. In addition to these advantages, analytical results reported in this paper suggest other unique benefits achievable with magnetic bearings.

The potential advantages of magnetically suspended control wheels are offset partially by a new requirement for electrical power to drive the bearing assembly and the increased weight of a single-bearing assembly. These disadvantages result from the need for actively controlled magnetic suspension. Practical magnetic materials cannot achieve passively stable suspension in three axes; therefore, magnetic bearings require electromagnets, displacement sensors, and electronics to control motion in at least one axis. Improvements in the active control method have reduced significantly the power required to operate advanced magnetic bearings,[1] and the weight disadvantage is eliminated if the design requires no single point failure.[2]

The development of low-power, lightweight magnetic bearings has led to the use of radial-passive, axial-active bearings with relatively low suspension stiffness.[2,3] By way of comparison, designs of this type have developed passive radial suspension stiffnesses in the 1000- to 5000-lb/in. range, whereas actively controlled axes produce suspension forces 10 times as stiff,[1] and conventional mechanical bearings are approximately 100 to 1000 times as stiff.

The low support stiffness and small shaft-to-bearing gap (approximately 0.005 to 0.015 in.) inherent with present magnetically suspended wheels raise questions about the behavior of a spacecraft employing such a device for attitude control. Will the system be stable? Will attitude jet firings cause the shaft to contact the bearing? Will wheel unbalance produce excessive motion?

This paper addresses these questions for a spacecraft stabilized by a magnetically suspended momentum wheel. First, settling time and amplitudes of response due to wheel unbalance and attitude jet impulses are determined. These performance parameters are expressed as nondimensional functions of suspension stiffness and damping. Plots are presented for two vehicle geometries, one representing a standard momentum bias configuration and the other representing a dual-spin spacecraft. Then, the nondimensional results are used to determine suspension parameters needed to insure good performance of a typical configuration. Finally, the required suspension parameters are compared with those achieved by presently available magnetic bearings.

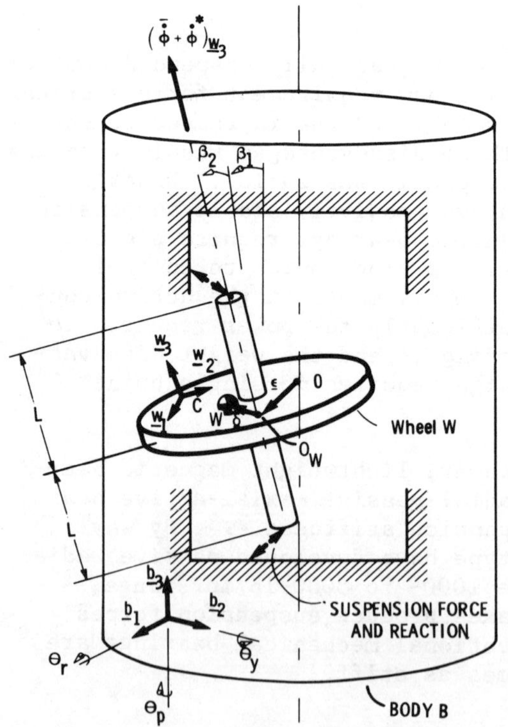

Fig. 1 Space dynamic model.

The open literature contains little on the attitude behavior of a spacecraft containing a softly suspended momentum wheel. A paper by Cretcher and Mingori,[4] dealing with the closely related problem of a flexibly connected dual-spin spacecraft, provides some insight to the questions addressed here. In fact, their paper was very valuable in suggesting the type of results to be pursued in this study.

Model

The analysis is based on the simple, two-body model of a momentum bias spacecraft illustrated in Fig. 1. Main body B is axisymmetric, and wheel W is very nearly axisymmetric. Magnetic suspension interaction forces resist all motions of W relative to B, except rotation about the spin axis of W.

Two sets of orthogonal unit vectors are used. The $\underline{b}_i$ ($i = 1, 2, 3$) unit vectors are fixed in B, with $\underline{b}_3$ parallel to the centerline of B and oriented in the same direction as the nominal bias vector. The $\underline{w}_i$ unit vectors are fixed in W. $\underline{w}_3$ is parallel to the wheel's intended symmetry axis and is directed such that $\underline{w}_3 = \underline{b}_3$, when W is undisplaced relative to B.

Equations of motion are linearized about an undisplaced, unperturbed configuration referred to as the "nominal condition." It is defined as follows: B is stationary in inertial space; W is axisymmetric with its mass center C_W located at point O_W; W is centered in its bearings, so that $\underline{w}_3 = \underline{b}_3$ and the mass centers of both bodies are coincident at point O; W provides momentum bias vector $h\underline{w}_3$ by spinning at constant rate $\bar{\phi}$ ($h\underline{w}_3 = I_{W33}\,\bar{\phi}\,\underline{w}_3$), and all magnetic suspension forces equal zero.

Small displacements and perturbations are measured with respect to the nominal condition. Inertial angular displacements of B are expressed as roll (Θ_r), pitch (Θ_p), and yaw (Θ_y) rotations about axes parallel to $\underline{b}_1$, $-\underline{b}_3$, and $\underline{b}_2$, respectively. Translational motion of W relative to B is given by the small vector $\underline{\varepsilon} = \varepsilon_1\underline{b}_1 + \varepsilon_2\underline{b}_2 + \varepsilon_3\underline{b}_3$, and the relative angular displacements of $\underline{w}_3$ relative to $\underline{b}_3$ are measured by the small rotations β_1 about $\underline{b}_1$ and β_2 about $\underline{b}_2$. Small changes in the spin speed of $\underline{W}$ are given by $\dot{\phi}^*$, so that the total spin velocity is $\dot{\phi}\,\underline{w}_3 = (\bar{\phi} + \dot{\phi}^*)\,\underline{w}_3$. A small static unbalance of W is given by $\underline{\delta} = \delta_1\underline{w}_1 + \delta_3\underline{w}_3$, and a small dynamic unbalance is admitted by allowing one nonzero product of inertia, I_{W13}, measured in the W-fixed coordinate frame.

Relative displacements ε_1, ε_2, ε_3, β_1, and β_2 are resisted by magnetic bearing suspension forces. It is assumed that suspension forces are applied to points on the intended spin axis of W at the ends of the shaft. Radial forces (those oriented perpendicular to $\underline{b}_3$) are assumed to be proportional to minus the relative velocity of the shaft end times K_{R1} and minus the relative displacement of the shaft end times K_{R2}. Similar terms in the axial ($\underline{b}_3$) direction are assumed to have coefficients K_{A1} and K_{A2}.

The assumption that magnetic bearings provide suspension forces linearly proportional to displacement is substantiated by test data reported by Sabnis et al.[3] However, the assumption of linear viscous damping is suspect. The same test data indicate that passive velocity-dependent forces exist, but that they increase with amplitude. Even so, the assumption of viscous damping is maintained to ease the analysis. However, the reader is reminded that this assumption is deficient in some respects.

Equations

Equations of motion are derived from four vector equations. Newton's second law is written for W and again for the total vehicle, and the rotational equation of motion, $\underline{M} = \dot{\underline{H}}$, also is written for W and the vehicle. Expansion of the vector equations, dropping all terms involving products of variational and/or perturbation terms and assuming that no external pitch torque is applied, results in the following equations:

Roll-yaw

Vehicle:

$$I_{V11}\,\ddot{\theta}_r + I_{W11}\,\ddot{\beta}_1 + h\,\dot{\theta}_y + h\,\dot{\beta}_2 = \dot{\bar{\phi}}^2\, I_{W13}\, s(\dot{\bar{\phi}}t) + M_1 \tag{1a}$$

$$I_{VII}\,\ddot{\theta}_y + I_{W11}\,\ddot{\beta}_2 - h\,\dot{\theta}_r - h\,\dot{\beta}_1 = -\dot{\bar{\phi}}^2\, I_{W13}\, c(\dot{\bar{\phi}}t) + M_2 \tag{1b}$$

Wheel:

$$I_{W11}\,\ddot{\theta}_r + I_{W11}\,\ddot{\beta}_1 + h\,\dot{\theta}_y + h\,\dot{\beta}_2 + 2\,K_{R1}\,L^2\,\dot{\beta}_1 + 2\,K_{R2}\,L^2\,\beta_1 = \dot{\bar{\phi}}^2\, I_{W13}\, s(\dot{\bar{\phi}}t) \tag{1c}$$

$$I_{W11}\ \ddot{\theta}_y + I_{W11}\ \ddot{\beta}_2 - h\ \dot{\theta}_r - h\ \dot{\beta}_1 + 2\ K_{R1}$$

$$L^2\ \dot{\beta}_2 + 2\ K_{R2}\ L^2\ \beta_2 = \bar{\dot{\phi}}^2\ I_{W1}\ \ c(\bar{\dot{\phi}}t) \tag{1d}$$

Wheel translation

Radial:

$$\left(1 + \frac{m_W}{m_V}\right) \ddot{\varepsilon}_1 + \frac{2\ K_{R1}}{m_W}\ \dot{\varepsilon}_1 + \frac{2\ K_{R2}}{m_W}\ \varepsilon_1$$

$$= \bar{\dot{\phi}}^2 \left(1 + \frac{m_W}{m_W}\right)\ \delta_1\ c(\bar{\dot{\phi}}t) - \frac{1}{m_V}\ F_1 \tag{2a}$$

$$\left(1 + \frac{m_W}{m_V}\right) \ddot{\varepsilon}_2 + \frac{2\ K_{R1}}{m_W}\ \dot{\varepsilon}_2 + \frac{2\ K_{R2}}{m_W}\ \varepsilon_2$$

$$= \bar{\dot{\phi}}^2 \left(1 + \frac{m_W}{m_W}\right)\ \delta_1\ s(\bar{\dot{\phi}}t) - \frac{1}{m_V}\ F_2 \tag{2b}$$

Axial:

$$\left(1 + \frac{m_W}{m_V}\right) \ddot{\varepsilon}_3 + \frac{2\ K_{A1}}{m_W}\ \dot{\varepsilon}_3 + \frac{2\ K_{A2}}{m_W}\ \varepsilon_3 = \frac{1}{m_V}\ F_3 \tag{2c}$$

The pitch equations are not presented, since they reduce to $\dot{\phi} = \bar{\dot{\phi}}$ and $\theta_p = 0$, for zero initial conditions.

Equations (1) and (2) are put in a form better suited for parametric studies by making them nondimensional. This is accomplished by substituting the nondimensional time $\tau = \bar{\dot{\phi}}\ t$ (i.e., τ is a measure of time equal to the wheel's angular displacement measured in radians) into Eqs. (1) and (2) and rearranging terms to obtain the following:

Roll-yaw

Vehicle:

$$A\ \mathring{\mathring{\theta}}_r + \mathring{\mathring{\beta}}_1 + C\ \mathring{\theta}_y + C\ \mathring{\beta}_2 = \mathcal{I}\ s\tau + M_1 \tag{3a}$$

$$A\ \mathring{\mathring{\theta}}_y + \mathring{\mathring{\beta}}_2 - C\ \mathring{\theta}_r - C\ \mathring{\beta}_1 = -\mathcal{I}\ c\tau + M_2 \tag{3b}$$

Rotor:

$$\mathring{\mathring{\theta}}_r + \mathring{\mathring{\beta}}_1 + C\,\mathring{\theta}_y + C\,\mathring{\beta}_2 + K_{R1}\,\mathring{\beta}_1 + K_{R2}\,\mathring{\beta}_1$$
$$= I\ s\tau \tag{3c}$$

$$\mathring{\mathring{\theta}}_y + \mathring{\mathring{\beta}}_2 - C\,\mathring{\mathring{\theta}}_r - C\,\mathring{\beta}_1 + K_{R1}\,\mathring{\beta}_2 + K_{R2}\,\beta_2$$
$$= -I\ c\tau \tag{3d}$$

Rotor translation

Radial:

$$\mu\,\mathring{\mathring{e}}_1 + K_{R1}\,\mathring{e}_1 + K_{R2}\,e_1 = \mathcal{D}\ c\tau - F_1 \tag{4a}$$

$$\mu\,\mathring{\mathring{e}}_2 + K_{R1}\,\mathring{e}_2 + K_{R2}\,e_2 = \mathcal{D}\ s\tau - F_2 \tag{4b}$$

Axial:

$$\mu\,\mathring{\mathring{e}}_3 + K_{A1}\,\mathring{e}_3 + K_{A2}\,e_3 = -F_3 \tag{4c}$$

In these equations, a circle over a variable is used to indicate a derivative with respect to nondimensional time, $s\tau = \sin(\tau)$ and $c\tau = \cos(\tau)$.

System Performance

Equations (3) and (4) now are used to determine spacecraft system performance as a function of the nondimensional magnetic bearing suspension parameters K_{A1}, K_{A2}, and K_{R1} and K_{R2}. Pitch motion is not analyzed, because it is not affected by suspension parameters in the linearized equations. Roll and yaw motions are assumed to be controlled by an attitude control system that fires a jet to correct momentum vector misalignment when a deadband threshold is exceeded. Motion inside the deadband is examined by analyzing the stability of motion, settling time, and response due to wheel unbalance. The effect of attitude control jet firings is established by analyzing the response to a single jet firing impulse.

Roll-Yaw Motion

Evaluation of roll-yaw motion is performed at discrete, physically significant values for the two nondimensional configuration parameters appearing in Eqs. (3). The wheel's nondimensional spin inertia, $C = I_{W33}/I_{W11}$, ranges from zero for a rod-shaped wheel to two for a disk. Most wheels used in

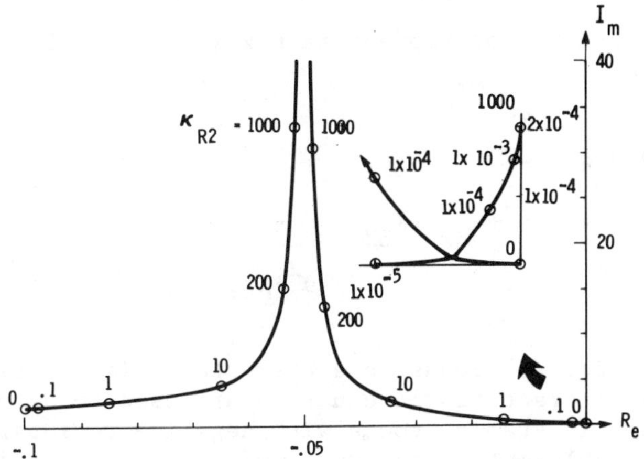

Fig. 2 Roll-yaw roots for $A = 1x10^4$, $C = 2$, and $K_{R1} = 0.1$ (complex conjugate roots not shown).

momentum-bias spacecraft approximate a disk; therefore, only $C = 2$ is considered. The vehicle's normalized transverse inertia, $A = I_{V11}/I_{W11}$, ranges between one and infinity. Typically, momentum-bias spacecraft lie in the range from $A = 1 \times 10^2$, for relatively large rotors, to $A = 1 \times 10^6$, for relatively small rotors. Results are presented for $A = 1 \times 10^4$, which is thought to represent a standard momentum-bias spacecraft. (It also is possible to make some comments about system behavior over the $A = 1 \times 10^2$ to $A = 1 \times 10^6$ range based on performance parameters computed at the two extremes, but not plotted in this paper.)

Performance also is determined for a configuration with a very large wheel. $A = 2.5$ is selected, because it is representative of a dual-spin spacecraft. The Fig. 1 model is a somewhat inadequate representation of dual-spin spacecraft, because present configurations do not have coincident mass centers of the two bodies. However, performance based on $A = 2.5$ is adequate to provide some understanding as to the applicability of magnetic bearings to a dual spinner. (Although plotted large-wheel results mostly deal with $A = 2.5$, calculations made at $A = 2$ and $A = 1.67$ with $C = 2$, and not presented in this paper, are used to make observations about behavior on both sides of the ratio $C/A = I_{W33}/I_{V11} = 1$, which divides stable from unstable configurations in some vehicles.)

<u>Stability</u>. Stability of roll-yaw motion is determined from <u>Eqs. (3)</u>. First, Laplace transforms of the equations are obtained assuming zero initial conditions. Then, the

determinant of the coefficient matrix is set equal to zero, resulting in

$$\begin{vmatrix} A\,s & C & s & C \\ -C & A\,s & -C & s \\ s^2 & C\,s & s^2+K_{R1}s+K_{R2} & C\,s \\ -C\,s & s^2 & -C\,s & s^2+K_{R1}s+K_{R2} \end{vmatrix} = 0 \qquad (5)$$

Expansion of Eq. (5) leads to a characteristic equation. Roots of the characteristic equation are used to determine if the system is stable, roots with negative real parts indicating stable motion and roots with positive real parts indicating instability.

Roots of the characteristic equation are presented as a function of K_{R2} for $A = 1 \times 10^4$ in Fig. 2, and for $A = 2.5$ in Fig. 3. Figure 2 typifies results computed over the ranges $A = 1 \times 10^2$ to $A = 1 \times 10^6$ and $K_{R1} = 0.01$ to $K_{R1} = 100$, and Fig. 3 typifies plots obtained by examining a few representative values of K_{R1} with $A = 1.67$, 2.0, and 2.5.

Results of the root calculations suggest that roll-yaw motions are stable for all physically realizable parameter values. This conclusion is in agreement with Cretcher and Mingori's results[4] and may be interpreted in terms of dual-spin stability criteria. Energy sink analyses have shown that energy dissipators mounted on the despun platform of a dual-spin spacecraft (body B, here) always tend to stabilize roll-yaw motions, whereas the effect of wheel-mounted dissipators

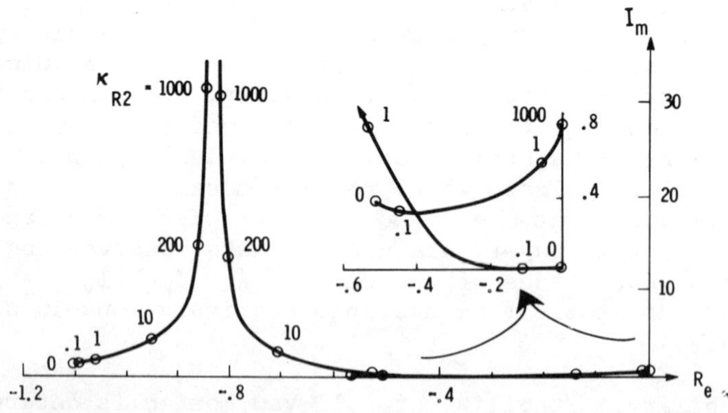

Fig. 3 Roll-yaw roots for $A = 2.5$, $C = 2$, and $K_{R1} = 1$ (Complex conjugate roots not shown).

depends upon the inertia ratio $C/A = I_{W33}/I_{V11}$. If the ratio is greater than one, the energy dissipator on the wheel is stabilizing, and, if the ratio is less than one, the dissipator is destabilizing. Thus, the root calculations imply that, for the Fig. 1 model, energy dissipation in the bearings acts as if it is a platform-fixed effect, and therefore always stabilizes roll-yaw motions.

The stability analysis also indicates that, as K_{R2} approaches ∞, one root approaches $0 + i\lambda$, where $\lambda = C/A$ is the nondimensional precession frequency of a momentum-bias spacecraft with rigid bearings. Furthermore, eigenvectors computed for the roll-yaw equations indicate that free rotations with large values of K_{R2} consist of the superposition of three characteristic motions. One is a low-frequency precessional motion of the vehicle, which is virtually undamped. It appears as an inertial coning motion of the vehicle's pitch axis at frequency λ, with the wheel's spin axis tilted in its bearings; the spin axis has almost no velocity relative to B. The other two characteristic motions involve higher-frequency, damped oscillations of W relative to B. Thus, motion following a transient is characterized by high-rate oscillations of the wheel relative to B, which damps out quickly, plus a slow precessional motion of the spacecraft, which persists for a very long time with the wheel tilted in its bearings. A nutation damper is needed to eliminate the persisting motion.

Settling Time. A good measure of relative stability is provided by the settling time. The settling time for root i is

$$T_i = -1/\mathrm{Re}(s_i)$$

where $\mathrm{Re}(s_i)$ is the real part of the ith rooth. Thus, T_i is the time required for the characteristic motion associated with the ith root to decay to 1/e times its initial value. The system settling time is defined to be the maximum value of all of the individual settling times.

It is anticipated that standard momentum-bias configurations will require the normalized stiffness to be relatively large, in order to prevent shaft-bearing contact following a jet firing. Therefore, the system time constant approaches infinity (as the nutation root approaches the imaginary axis). In this case, K_{R1} and K_{R2} are selected on the basis of their ability to (among other things) damp wheel relative motion. Damping of precessional motion is left to a nutation damper. Thus, the measure of relative stability for $1\times10^2 \leq A \leq 1\times10^6$ is selected to be the relative motion settling time, which is

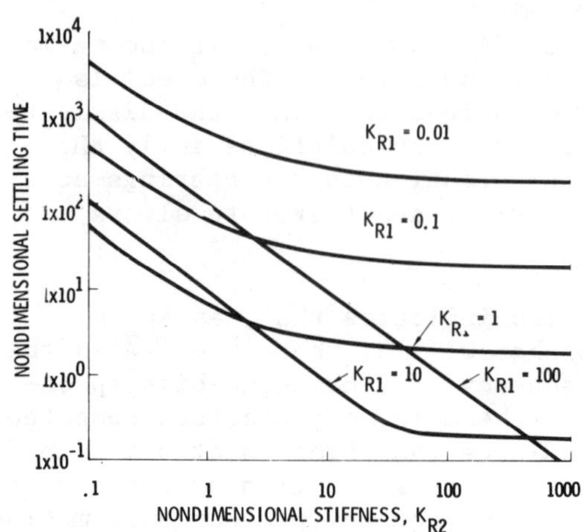

Fig. 4 Rotor relative roll-yaw setting time for $A = 1x10^4$, $C = 2$.

the maximum of the settling times associated with the two roots approaching each other as K_{R2} becomes very large. Plots of wheel relative roll-yaw motion are presented in Fig. 4 for $A = 1x10^4$.

Settling times for the large-wheel case are presented in a different manner. For this case, the ability of magnetic bearings to damp all motion seems promising; therefore, settling times for the two roots nearest to the imaginary axis are plotted together in Fig. 5. System settling time is equal to the maximum value of T at a specified value of K_{R2}.

Impulse Response. Attention now is turned to determining roll-yaw response to an impulsive torque applied to B, such as might result from firing an attitude control jet. Equations (3) are used to describe the vehicle's response with $I = 0$ (no dynamic unbalance). The equations are forced by a unit, nondimensional torque impulse applied about the $\underline{b}_2$ axis at $\tau = 0$; i.e., $M_1 = 0$, $M_2 = \delta(\tau)$.

Two new variables (normalized beta angle and normalized nutation angle) are introduced to aid in presentation of the

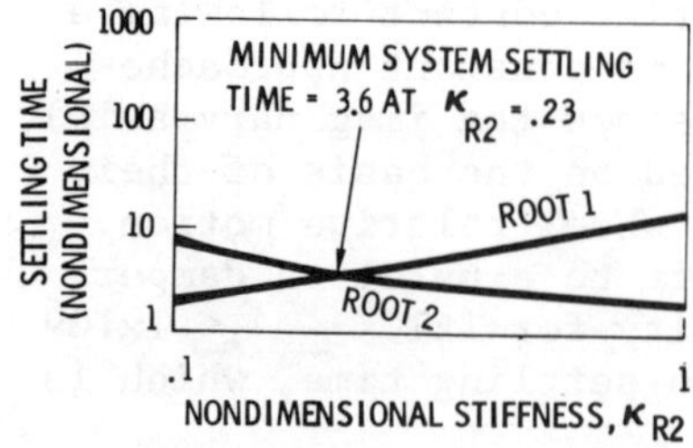

Fig. 5 System roll-yaw settling times for large-wheel vehicles with $A = 2.5$, $C = 2$, $K_{R1} = 1$.

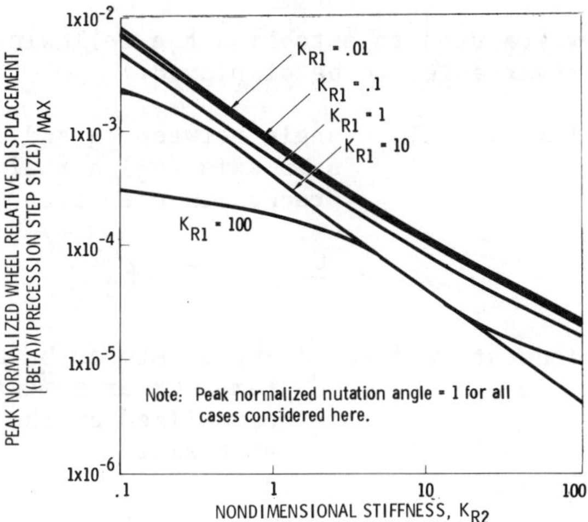

Fig. 6 Peak roll-yaw response to a nondimensional torque impulse for $A = 1\text{x}10^4$, $C = 2$.

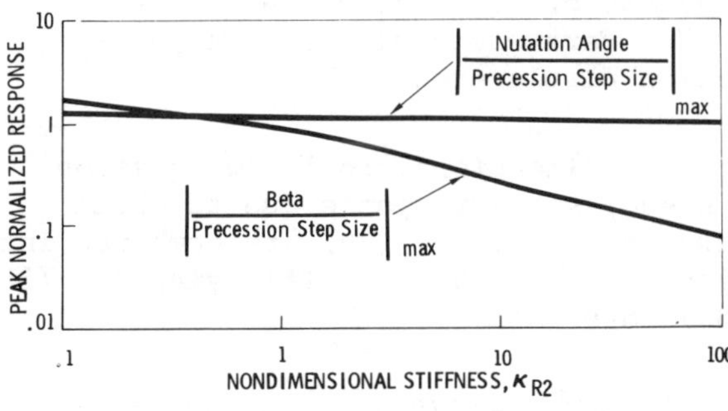

Fig. 7 Peak roll-yaw response to nondimensional torque impulse for $A = 2.5$, $C = 2$, $K_{R1} = 1$.

impulse response data. These variables are defined in terms of two other new variables:

precession step size = inertial angular displacement of the system's momentum vector

= A/C for a unit nondimensional impulse

nutation angle = angle between $\underline{b}_3$ and the system angular momentum vector

= $\left[\theta_r^2 + (\theta_y - A/C)^2\right]^{1/2}$, following $M_2 = \delta(\tau)$ impulse

These terms now are used to establish the following equations for the two new variables to be displayed:

normalized beta angle = angle between $\underline{b}_3$ and the wheel's spin axis ($\underline{w}_3$) normalized by the precession step size

$$= \frac{C}{A}\left(\beta_1^{\,2}+\beta_2^{\,2}\right)^{1/2} \tag{6}$$

normalized nutation angle = angle between $\underline{b}_3$ and the system angular momentum vector normalized by the precession step size

$$= \frac{C}{A}\left[\theta_r^{\,2}+\left(\theta_y-\frac{A}{C}\right)^2\right]^{1/2} \tag{7}$$

Peak roll-yaw responses to a unit M_2 impulse, as established by computer simulations of Eqs. (3, 6, and 7), are presented in Figs. 6 and 7. Parameters plotted are peak-normalized beta and nutation displacements.

Unbalance. Next, the nondimensional roll-yaw equations are evaluated to obtain steady-state response due to wheel dynamic unbalance. First, Eqs. (3a) and (3b) are combined into one complex equation (with $M_1 = M_2 = 0$) by multiplying Eq. (3b) by i and adding it to (3a) to obtain

$$A(\mathring{\mathring{\theta}}_r+i\mathring{\mathring{\theta}}_y) + (\mathring{\mathring{\beta}}_1+i\mathring{\mathring{\beta}}_2) + C(\mathring{\theta}_y-i\mathring{\theta}_r) + C(\mathring{\beta}_2-i\mathring{\beta}_1) = I(c\tau+is\tau)$$

Substitution of the complex variables $\theta=\theta_r+i\theta_y$ and $\beta=\beta_1+i\beta_2$ into the complex equation of motion produces

$$A\,\mathring{\mathring{\theta}} + \mathring{\mathring{\beta}} - i\,C\,\mathring{\theta} - i\,C\,\mathring{\beta} = I\,e^{i\tau} \tag{8a}$$

Similar treatment of Eqs. (3c) and (3d) produces

$$\mathring{\mathring{\theta}} + \mathring{\mathring{\beta}} - i\,C\,\mathring{\theta} - i\,C\,\mathring{\beta} + K_{R1}\,\mathring{\beta} + K_{R2}\,\beta = I\,e^{i\tau} \tag{8b}$$

Equations (8) are the complex roll-yaw equations of motion. Their steady-state solution is of the forms $\theta = T\,e^{i\tau}$ and $\beta = B\,e^{i\tau}$, where T and B are complex constants. Substitution into Eqs. (8), division by $e^{i\tau}$, and simultaneous solution for T

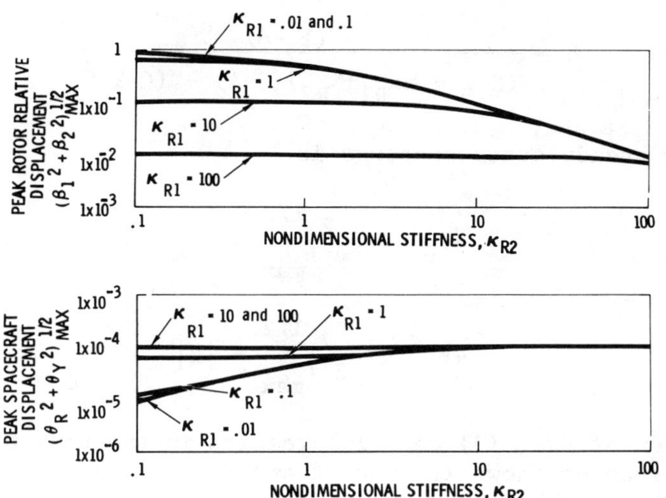

Fig. 8 Peak steady-state response due to wheel dynamic unbalance for $A = 1 \times 10^4$, $C = 2$, $I = 1$.

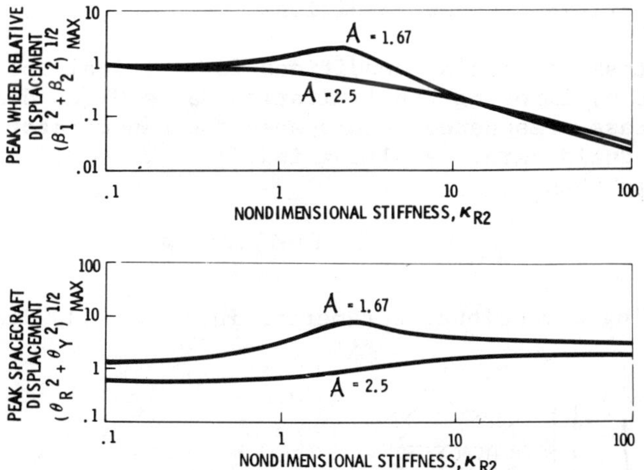

Fig. 9 Peak steady-state roll-yaw response due to wheel dynamic unbalance for $C = 2$, $I = 1$, $K_{R1} = 1$.

and B produce

$$T = \frac{I\ (i\,K_{R1} + K_{R2})}{(C-1 + iK_{R1} + K_{R2})(C-A) - (C-1)^2} \tag{9a}$$

$$\mathcal{B} = \frac{\mathcal{I}\ (1 - \mathcal{A})}{(\mathcal{C}-1 + i\mathcal{K}_{R1}+\mathcal{K}_{R2})(\mathcal{C}-\mathcal{A}) - (\mathcal{C}-1)^2} \tag{9b}$$

The peak steady-state response is

$$\left(\theta_r^{\ 2} + \theta_y^{\ 2}\right)^{1/2}_{max} = |\mathcal{T}| \tag{10a}$$

$$\left(\beta_1^{\ 2} + \beta_2^{\ 2}\right)^{1/2}_{max} = |\mathcal{B}| \tag{10b}$$

Evaluation of Eqs. (9) and (10) results in the peak steady-state responses shown in Figs. 8 and 9.

Review of these figures raises an interesting question. Why are there no characteristic resonance peaks (critical speeds) for the standard momentum-bias configuration? This question is answered by evaluating the equations for $\mathcal{T}$ and $\mathcal{B}$ to establish critical speed conditions.

The task of finding conditions for critical speeds is simplified by imposing the limitation $\mathcal{K}_{R1} = 0$ (no damping). For this case, resonance occurs when the denominator in Eqs. (9) equals zero, resulting in $|\mathcal{T}| = |\mathcal{B}| = \infty$. This condition occurs when

$$\mathcal{K}_{R2} = [(\mathcal{C}-1)\ (\mathcal{A}-1)]/(\mathcal{C}-\mathcal{A}) \tag{11}$$

Substituting dimensional parameters into Eq. (11) leads to

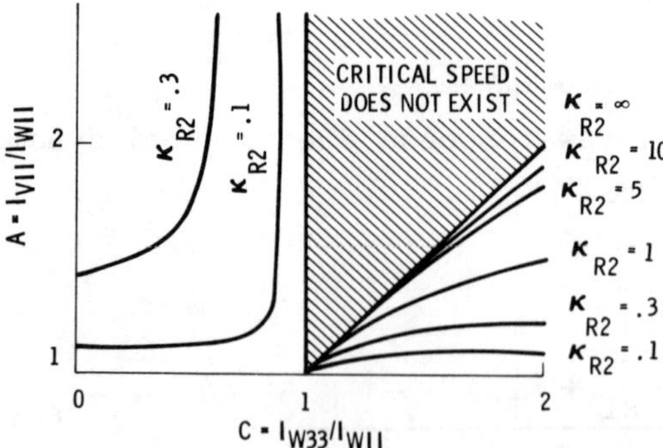

Fig. 10 Roll-yaw critical speed conditions with $\mathcal{K}_{R1} = 0$.

Table 1 System performance with $K_{R1} = 0.1$, $K_{R2} = 1.0$

Performance measure	Predicted performance: Nondimensional	Predicted performance: Dimensional
Roll yaw:		
Relative motion settling time	$\tau \approx 100$	T = 0.4 sec
Impulse response amplitude	$\lvert\beta\rvert$/(precession step size) ≈ 0.001	$\lvert\beta\rvert = 1.1 \times 10^{-4}$ deg
Unbalance response amplitude	$\lvert\beta\rvert_{max} \approx 5 \times 10^{-5}$	$\lvert\beta\rvert_{max} = 2.9 \times 10^{-3}$ deg
Translation:		
Relative motion settling time	$\tau \approx 60$	T = 2.4 sec
Impulse response	$\lvert e_3\rvert_{max} \approx 4 \times 10^{-6}$	$\lvert\varepsilon_3\rvert_{max} = 4 \times 10^{-6}$ ft
Unbalance response	$\lvert e\rvert_{max} \approx 2 \times 10^{-6}$	$\lvert\varepsilon\rvert_{max} = 2 \times 10^{-6}$ ft

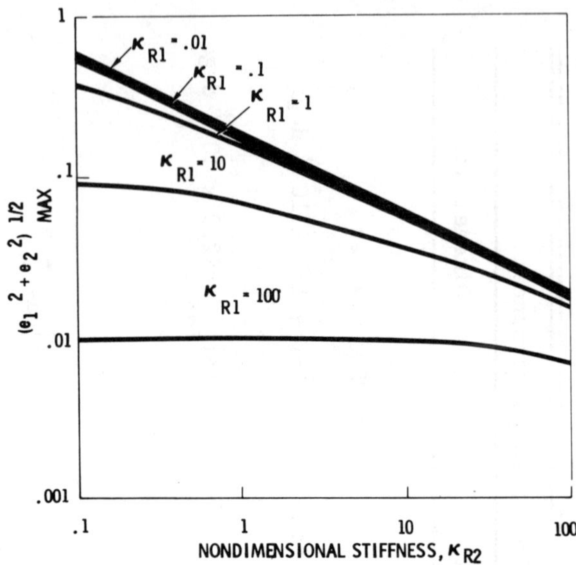

Fig. 11 Peak wheel translational response due to a unit nondimensional force impulse for μ = 30 (standard configuration), $F_1 = \delta(\tau)$.

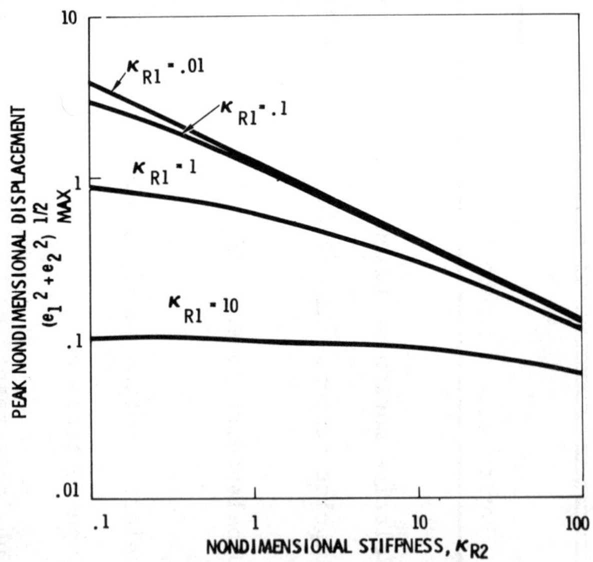

Fig. 12 Peak wheel translational response due to a unit nondimensional force impulse for μ = 0.6 (large-wheel configuration), $F_1 = \delta(\tau)$.

$$\bar{\phi}^2_{crit} = \frac{2\, K_{R2}\, L^2 (I_{W33} - I_{V11})}{I_{W11}(I_{W33}-I_{W11})(I_{V11}-I_{W11})} \tag{12}$$

Equation (12) indicates the wheel speed at which roll-yaw resonance occurs in the presence of dynamic unbalance, with no bearing damping.

The nondimensional critical speed condition, Eq. (11), is evaluated in Fig. 10. Parameters that satisfy Eq. (11) and result in resonance are plotted for the physically realizable parameter ranges $\mathcal{A} \geq 1$ and $0 \leq \mathcal{C} \leq 2$. The figure indicates that, in the region $\mathcal{C} > 1$, $\mathcal{A} > \mathcal{C}$, Eq. (11) is not satisfied, and therefore critical speeds do not exist. This is why resonance peaks appear only in Fig. 8 and 9 for the one case $\mathcal{A} = 1.67$, $\mathcal{C} = 2$. Thus, large-wheel (dual-spin) vehicles with $I_{W33}/I_{V11} > 1$ are the only practical configurations to display the dynamic unbalance resonance.

Rotor Translational Motion

Now, the nondimensional translational equations, Eqs. (4), are subjected to the same analytical procedure applied to the roll-yaw equations. This time the analysis is much easier, because three uncoupled second-order differential equations are involved. To make matters simpler yet, the three translational equations are enough alike that only one has to be solved, and solutions to the remaining two are obtained by similarity.

Evaluation of Eq. (4a) is carried out in a number of texts, such as Thomson.[5] Standard solutions produce the plots presented in Figs. 11 - 16. In these plots, results are presented for nondimensional wheel masses of $\mu = 30$ and $\mu = 0.6$, representing typical values for standard and large-wheel configurations, respectively.

Unlike the roll-yaw case, a translational resonant condition can occur with standard momentum-bias configurations. The static unbalance term $\mathcal{D}$ forces the radial equations of motion in a resonant condition (critical speed) when

$$\mathcal{K}_{R2} = \mu \tag{13}$$

if there is no damping. Conversion of Eq. (13) to dimensional parameters results in the following critical speed equation for radial translation:

$$\bar{\dot{\phi}}^2_{crit} = \frac{2\,K_{R2}}{m_W\left[1 + (m_W/m_V)\right]} \tag{14}$$

That is, resonance occurs when Eq. (14) is satisfied if $K_{R1} = 0$.

Typical Suspension Requirements

The nondimensional results of the foregoing section now are applied to a typical momentum-bias spacecraft in order to establish magnetic bearing stiffness and damping requirements. Parameters for the vehicle to be studied are selected to correspond to the simplified model and to typify a 2000-lb-class momentum-bias spacecraft.

Vehicle parameters are

$I_{V11} = 200$ slugs-ft^2, $I_{V22} = 300$ slugs-ft^2

$m_V = 62$ slugs

Wheel parameters are

$I_{W11} = 0.02$ slugs-ft^2, $I_{W33} = 0.04$ slugs-ft^2

$I_{W13} = 1 \times 10^{-6}$ slugs-ft^2

$m_R = 0.6$ slugs, $L = 1.0$ ft, $\bar{\dot{\phi}} = 250$ rad/sec (2387 rpm)

$\delta_1 = 2 \times 10^{-6}$ ft

shaft-to-bearing gap = 0.001 ft

These parameters are realistic. They produce an angular momentum of 10 ft-lb-sec and a precession period of approximately 125.7 sec (based on rigid bearings); the nonzero unbalance terms correspond to the maximum allowable values for a typical spacecraft wheel; and the shaft-to-bearing gap size is approximately equal to the values cited in Refs. 1-3.

A single attitude control jet pulse is assumed to provide $\int M_2\,dt = 0.02$ ft-lb-sec and $\int F_2\,dt = 0.01$ lb-sec. These values correspond to an impulse resulting from firing a 1-lb thruster aligned with, and offset 2 ft from, the vehicle's symmetry axis for 0.010 sec. One pulse precesses the momentum vector 0.11°.

The nondimensional parameters corresponding to the preceding values are $A = 1 \times 10^4$, $C = 2$, $\mu = 30.3$, $\mathcal{D} = 60.6 \times 10^{-6}$, $\mathcal{I} = 50 \times 10^{-6}$, $\int M_2 \, d\tau = \int M_2 \dot{\phi} \, dt = 4 \times 10^{-3}$, and $\int F_2 \, d\tau = \int F_3 \dot{\phi} \, dt = 1.94 \times 10^{-5}$. For these values, the nondimensional performance curves show that adequate system behavior is achieved with suspension parameters of $\mathcal{K}_{A1} = \mathcal{K}_{R1} = 0.1$ and $\mathcal{K}_{A2} = \mathcal{K}_{R2} = 1.0$ or greater. The dimensional bearing parameters for these values are $K_{A1} = K_{R1} = 0.021$ lb/in/sec and $K_{A2} = K_{R2} = 52.0$ lb/in. These values are well within the range of present magnetic bearing technology. In fact, the simplest bearing type (radial-passive, axial-active) has reported stiffnesses of up to 5000 lb/in., and one configuration achieved damping coefficients in the 0.16-to 0.6-lb/in./sec range.[3]

Performance of the example configuration is obtained from the nondimensional curves and summarized in Table 1. The data indicate that peak radial displacement of a shaft end relative to its bearings is equal to 5% of the gap, and the peak axial displacement is less. Table 1 results also indicate that relative roll-yaw motion decays reasonably fast for this system, which has a 125.7-sec precession period with rigid bearings.

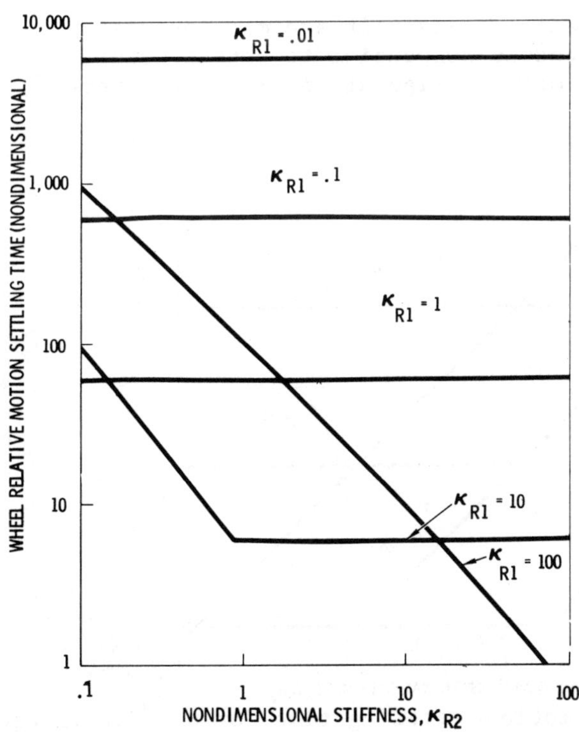

Fig. 13 Wheel relative motion settling time for $\mu = 30$ (standard configuration).

Special Characteristics

Magnetic bearings have special characteristics which, if exploited, may improve overall system performance. The non-dimensional performance curves suggest that magnetic suspension can provide nutation damping and isolation of spacecraft motion from wheel unbalance effects. In addition, it may be possible to avoid critical speeds during wheel speed changes by actively adjusting the bearings' stiffness. Some observations on the ability of magnetic bearings to perform these functions are extracted from the foregoing analysis.

Nutation Damping

The potential effect that soft radial suspension may have on nutation is examined first by considering the typical momentum-bias vehicle discussed in the previous section. For that case, optimum nutation damping results if K_{R2} is reduced to a little less than 1×10^{-4} (see Fig. 2). The lowered spring rate provides a good system settling time of about 800 sec. However, such a low stiffness results in unacceptable overall performance, one problem being that a single jet pulse results in the shaft contacting its bearings.

Large-wheel (dual-spin) spacecraft are more promising candidates for using magnetic suspension to obtain nutation damping. A brief study of a configuration with parameters

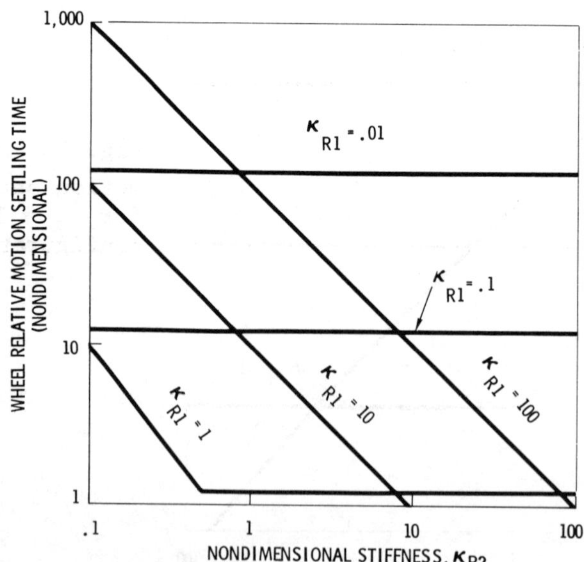

Fig. 14 Wheel relative motion settling time for $\mu = 0.6$ (large-wheel configuration).

I_{V11} = 300 slugs-ft^2, I_{W11} = 120 slugs-ft^2

I_{W33} = 240 slugs-ft^2

$\bar{\dot{\phi}}$ = 6.28 rad/sec (60 rpm), L = 1 ft, shaft-to-bearing gap = 0.001 ft

indicates that, if K_{R1} = 31 lb/in./sec and K_{R2} = 45.4 lb/in. (these values produce the minimum settling time condition indicated in Fig. 5), a system settling time of 0.57 sec is achieved, and reasonably sized jet impulses do not cause shaft-bearing contact. For this configuration, the system performance parameters are excellent, and the bearing stiffness is achieved easily by passive bearings. On the other hand, the bearing damping of K_{R1} = 31 lb/in./sec is much greater than the maximum value measured by Sabnis[3] (0.6 lb/in. /sec) and may not be achievable. However, it undoubtedly is possible to find a combination of achievable stiffness and damping values which results in good system performance.

Unbalance Isolation

Momentum-bias-stabilized spacecraft having extremely tight pointing requirements might experience unacceptably large attitude jitter due to wheel unbalance. In such cases, it may be possible to use magnetic bearings with stiffness values specifically selected to limit unbalance-induced jitter.

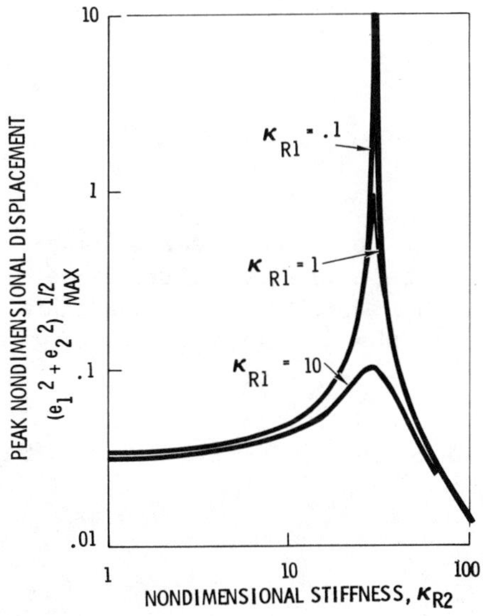

Fig. 15 Wheel peak steady-state translation response due to wheel static unbalance for μ = 30 (standard configuration), $\mathcal{D}$ = 1.

Figures 8 and 9 illustrate the effects of dynamic unbalance. They show that reducing the bearings' damping and stiffness tends to lower the amplitude of vehicle jitter and increase the amplitude of wheel relative motion. Thus, selection of bearing parameters so as to limit attitude jitter must be based on an analysis in which reduced vehicle attitude amplitude is traded for increased wheel motion. Although such a study is not conducted here, one conclusion is clear: the attitude jitter caused by dynamic unbalance with magnetic bearings is less than, or equal to, that with conventional bearings, for standard momentum-bias spacecraft. This is demonstrated in Fig. 8, where attitude jitter with conventional bearings is approximated by assuming $K_{R2} \to \infty$.

The amplitude of attitude jitter due to rotor static unbalance cannot be established from the analysis presented in this paper. The assumption of coincident mass centers uncoupled rotational and translational equations, so that the response of more general, coupled configurations cannot be determined. However, one obvious observation can be made: parameters should be selected to avoid the resonant condition at $K_{R2} = \mu$.

Critical Speed Avoidance

Under some conditions, a wheel speed change may cause the system to pass through a critical speed. Two possibilities

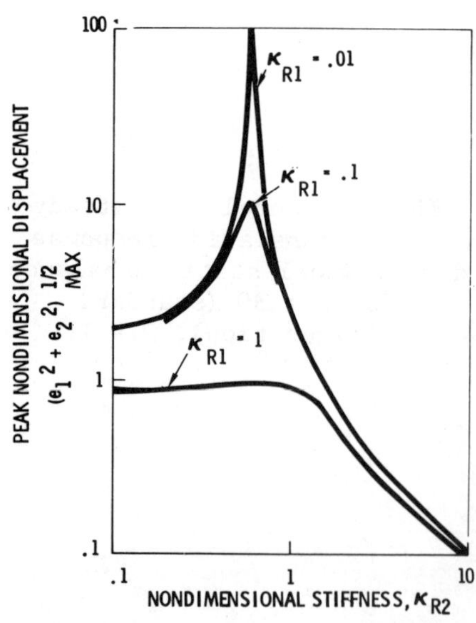

Fig. 16 Wheel peak steady-state translation response due to wheel static unbalance for $\mu = 0.6$ (large-wheel), $\mathcal{D} = 1$.

are suggested by the foregoing results: 1) if $K_{R2} = \mu$, translational resonance of the wheel results from static unbalance; and 2) wheel dynamic unbalance results in resonant roll-yaw motions if Eq. (11) is satisfied. In both cases, it should be possible to change K_{R2} actively (say, by activating an electromagnet when the wheel speed nears a critical speed), so that the resonance peak is shifted and peak response is limited. Once the wheel speed passes beyond the potential critical value, K_{R2} then can be returned to its original value.

Conclusions

A detailed analysis of a relatively simple dynamic model has provided some insight into the behavior of a momentum-bias spacecraft with a magnetically suspended wheel. Nondimensional performance curves, which may be used to predict performance of many spacecraft, have been developed. Application of the curves to an example configuration has shown that good performance can be obtained with easily achievable magnetic suspension parameters. Computations performed over a wide range of parameters indicate that this conclusion is much more general than is indicated by the limited results presented.

The analytical results suggest that magnetic bearings may be used to enhance spacecraft performance by providing nutation damping, isolation from wheel unbalance effects, and avoidance of certain critical speeds. These advantages, together with considerations of long life and low resistance torques, should lead to the selection of magnetic bearings over mechanical bearings for some future missions.

References

[1]Henrikson, C. H., Lyman, J., and Studer, P. A., "Magnetically Suspended Momentum Wheels for Spacecraft Stabilization," AIAA Paper 74-128, Jan. 30 - Feb. 1, 1974, Washington, D. C.

[2]Meeks, C. R., "Magnetic Bearings - Optimum Design and Application," International Workshop on Rare Earth Cobalt Permanent Magnets, Oct. 14-17, 1974, University of Dayton, Dayton, Ohio.

[3]Sabnis, A. V., Dendy, J. B., and Schmitt, F. M., "A Magnetically Suspended Large Momentum Wheel," Journal of Spacecraft and Rockets, Vol. 12, July 1975, pp. 420-427.

[4]Cretcher, C. K. and Mingori, D. L., "Nutation Damping and Vibration Isolation in a Flexibly Coupled Dual-Spin Spacecraft, " Journal of Spacecraft and Rockets, Vol. 8, Aug. 1971, pp. 817-823.

[5]Thomson, W. T., Vibration Theory and Applications, Prentice-Hall, Englewood Cliffs, N. J., 1965, pp. 36-40, 51-52, 99-100.

SEASONAL EFFECTS ON THE PERFORMANCE OF EARTH HORIZON SENSORS

J. V. Flannery* and J. L. Alward+

TRW Defense & Space Systems, Redondo Beach, Calif.

Abstract

Earth horizon sensors, operating in the infrared spectral band, commonly have been the primary attitude reference for Earth-orbiting spacecraft. These devices do not always achieve the on-orbit accuracy that is predicted by analysis or extrapolation of laboratory test data. One major reason for the discrepancy is that inadequate attention has been paid to the modeling or simulation of radiance gradients, internal to the Earth's disk. These gradients change significantly with season, and so Earth sensors on many operational spacecraft operating at synchronous altitudes have exhibited attitude bias errors that are a strong function of time of year. In order to achieve the higher accuracies required for future applications, Earth sensor design parameters must be optimized to bound this error source. This paper demonstrates that the realized gradients can be modeled, and, when combined with the transfer function for the sensor under evaluation, the predicted performance is within the range of performance observed on orbit.

Introduction

Data from several satellites, operating at synchronous altitudes, has indicated Earth sensor errors that are higher than expected. The magnitude of these errors is highly correlated with the time of year, suggesting a strong dependence of Earth sensor performance on Earth radiance levels and distributions. Fig. 1 shows representative data from one Intelsat IV satellite, taken in 1971 - 1972 time frame (Ref. 1). The error extremes for each 10-day interval are plotted on this figure. In general, these observations may be noted by inspection of Fig. 1.

Presented as Paper 76-267 at the AIAA/CASI 6th Communications Satellite Systems Conference, April 5-8, 1976, Montreal, Canada

*Section Head.

+Member of Technical Staff.

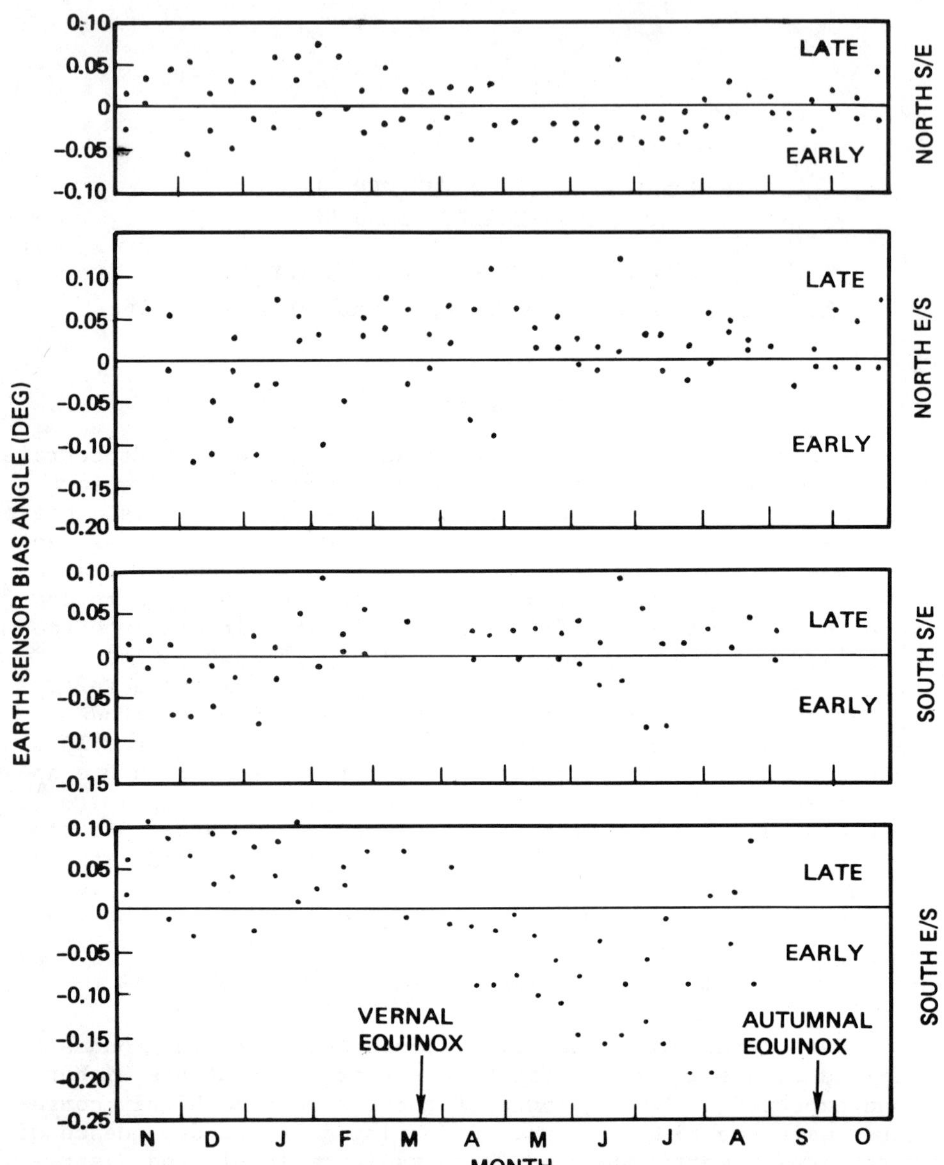

Fig. 1 Bias errors for sensors on INTELSAT IV F2 S/C.

° Sensor errors exhibit a seasonal bias component, roughly sinusoidal with time, and a random component.

° The polarity of the bias component is such as to produce a shorter apparent scanned chord length for the colder hemisphere than that for the warmer hemisphere (North S/E horizon detection late (+) and North E/S detection early (-) during summer months).

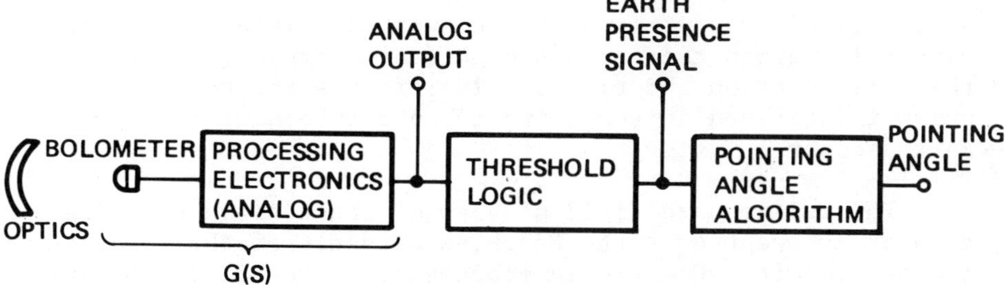

Fig. 2 Earth horizon sensor block diagram.

° Random errors are greater when viewing the colder hemisphere (North sensor October - March, South sensor April - September).

° The horizon crossing indication for scans from space-to-Earth (S/E) is more accurate than those established with an Earth-to-space (E/S) scan.

These same general observations apply for one of the TRW spin stabilized spacecrafts, which uses a slightly different Earth sensor. The sensors on both satellites operate in the 14 - 16μ CO_2 spectral band, and are of the horizon crossing indicator class of Earth sensors. It will be shown in this paper that, through proper modeling of sensor characteristics and Earth radiance profiles, the predicted sensor performance is in good agreement with that observed in orbit. It will also be shown that gradient errors can be reduced significantly through proper optimization of Sensor design parameters.

Earth Sensor Operation

Two representative Earth sensors are considered. One is a horizon crossing indicator (HCI) of the class employed on the Intelsat IV spin stabilized spacecraft. The other sensor for evaluation is a scanning Earth sensor designed for use on a three-axis stabilized spacecraft. Fig. 2 shows the general block diagram of a complete Earth sensor. Not all of these functions are incorporated in Earth sensor hardware. Some functions are implemented in other electronics boxes or in the ground software. The thermistor bolometer detects the radiation emanating from the Earth, producing an electrical signal that is amplified and filtered by the electronics. Next, the analog signal passes through the threshold detector. The output of this detector signals when the field of view of the Earth sensor intercepts the limb of the Earth. On most sensors, this output is high whenever the field of view of the sensor is within the Earth's disk (Earth presence). Finally, the time duration and time phasing of this Earth presence pulse is utilized in computing the attitude angles of the Earth sensor.

The 1.5° square field of view of the horizon crossing indicator is swept over the Earth as a result of the rotation of the spacecraft. Electronic processing, internal to the unit, produces a positive pulse at the space/Earth horizon crossing and a negative pulse at the Earth/space crossing (Fig. 3). This analog waveform is processed further through threshold detection logic to establish the instants of time at which the horizon crossings occurred. These instants occur when the analog voltage is approximately 50% of its peak value. The time duration of the ESA output pulse conveys pointing information as to the displacement of the scan plane from the Earth (cross scan direction). The time phasing of this output pulse conveys pointing information in the scan plane direction.

The 0.6° diam. field of view of the TRW scanning Earth sensor scans back and forth across the Earth as a result of motion of an internal scanning mirror (Fig. 4). In the computation of pointing errors, the timing of the two horizon crossings for both the forward scan and the backward scan is utilized. The threshold level changes automatically, as a function of the slope of the analog signal, such that the indicated horizon crossing instants are invariant to radiance level changes. The spectral range for this sensor is 13.8 to 17.3μ, as compared to the 14.5 to 15.5μ range for the horizon crossing indicator.

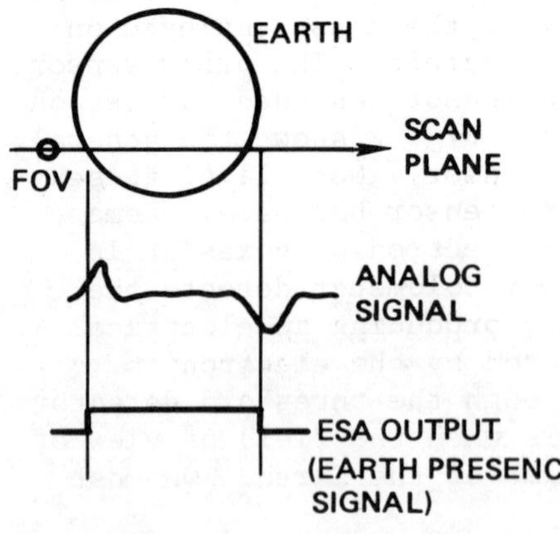

Fig. 3 Signal characteristics for horizon crossing.

Earth Radiance Data

Data are available from the Nimbus and the Tiros programs which show Earth radiance as a function of time and geographical location. Two radiance maps, Figs. 5 and 6 (from Ref. 2), were selected as reflecting representative July and January radiance conditions. These two test cases were utilized in computing radiance profiles intercepted by the Earth sensor's field of view as it sweeps across the Earth.

Notice that the equivalent blackbody temperature is relatively constant for each latitude and note the presence of a significant North-South gradient. This North-South gradient generates the seasonal bias component observed on orbit. Unfortunately, the intersection of the Sensor's field of view with the Earth does not follow a locus of constant latitude. (See dashed lines in Fig. 5.) Instead, the scan intersects lower latitudes at the center of its scan, and higher latitudes near the horizon crossings. This sweep, over different latitudes, generates gradients in the radiance profiles presented to the Earth sensor.

Also, note that the East-West gradients are larger in the colder hemisphere. The magnitude and polarity of the gradients change markedly with longitude. It is believed that the temporal changes in these East-West gradients cause the relatively large random variations in Earth sensor errors.

In order to demonstrate the effect of gradients on Earth sensor performance, these two radiance maps are modeled along with Earth sensor characteristics in the computer simulation described below.

Computer Simulation

The output of the sensor can be computed by appropriately modeling: the convolution of the ESA field of view with the

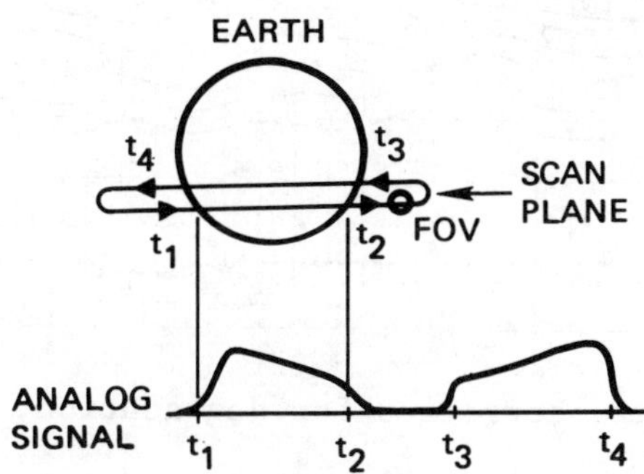

Fig. 4 Signal characteristics for TRW scanning earth sensor.

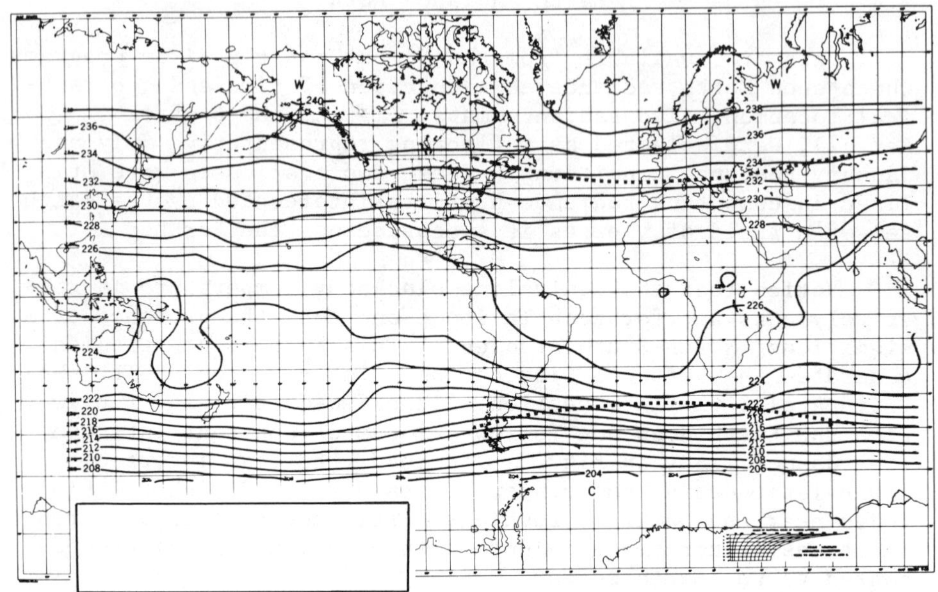

Fig. 5 Tiros VII 15μ isotherms, corrected for degradation, July 13 - 22, 1963.

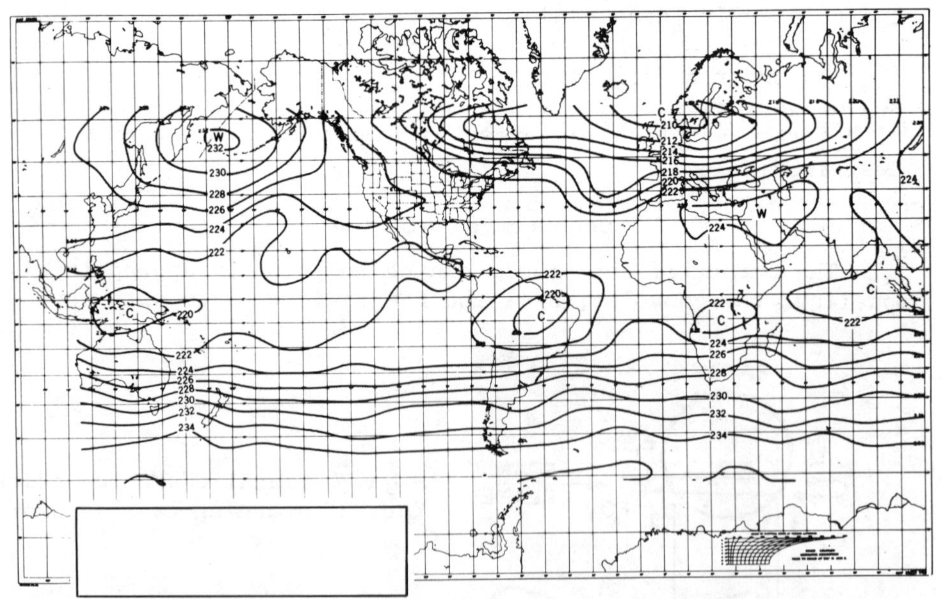

Fig. 6 Tiros VII 15μ isotherms, corrected for degradation, January 6 - 15, 1964.

Earth; the electronic transfer function for the analog processing electronics; the edge locator algorithm; and, finally, the pointing angle algorithm.

Temperature data was read from maps for geographical locations separated by 10° increments in longitude and latitude. The conversion between temperature and radiance then was made by integrating the Planck blackbody curve over the spectral band of the particular Earth sensor. Finally, these radiance values were entered into the computer, along with the appropriate longitude and latitude labeling. Radiances at points intermediate to the grid points were calculated through linear interpolation relationships.

The effective radiance input to the bolometer is calculated by integrating the contributions from each 0.2° x 0.2° portion of the sensors' field of view. For each elemental portion, the longitude and latitude at the point of intersection with the Earth are computed. The radiance associated with that geographical location is used, weighted by the normalized response for that field-of-view element, to determine each contribution.

Transfer functions for the radiance channel, including both the detector and electronics response functions, are given below. The numbers have been normalized such that the time functions are in millisecond units.

$$G(S) = \frac{KS}{(S + .25)\ (S + 1)^3} \qquad \text{(HCI)}$$

$$G(S) = \frac{KS^2}{(S + .001)^2\ (S + .67)\ (S + 1.4)} \qquad \text{(Scanner)}$$

The bolometer signal is processed through the transfer function for the electronics to determine the analog output of the unit. Outputs were derived for various satellite locations (18) for the two periods July 1963 and January 1964. The angles (chord lengths) between the space-to-Earth threshold crossing and the Earth-to-space crossing were computed for each case. These lengths were compared to those that would have been obtained if the Earth were of uniform radiance over the whole disk. Since one-half-chord deviations are associated more nearly 1:1 with attitude deviations in the cross-scan direction, Fig. 7 shows the computed one-half-chord deviations from the reference uniform Earth. The expected performance for the HCI sensor is indicated by the solid line and that for the scanner is shown as a dotted line.

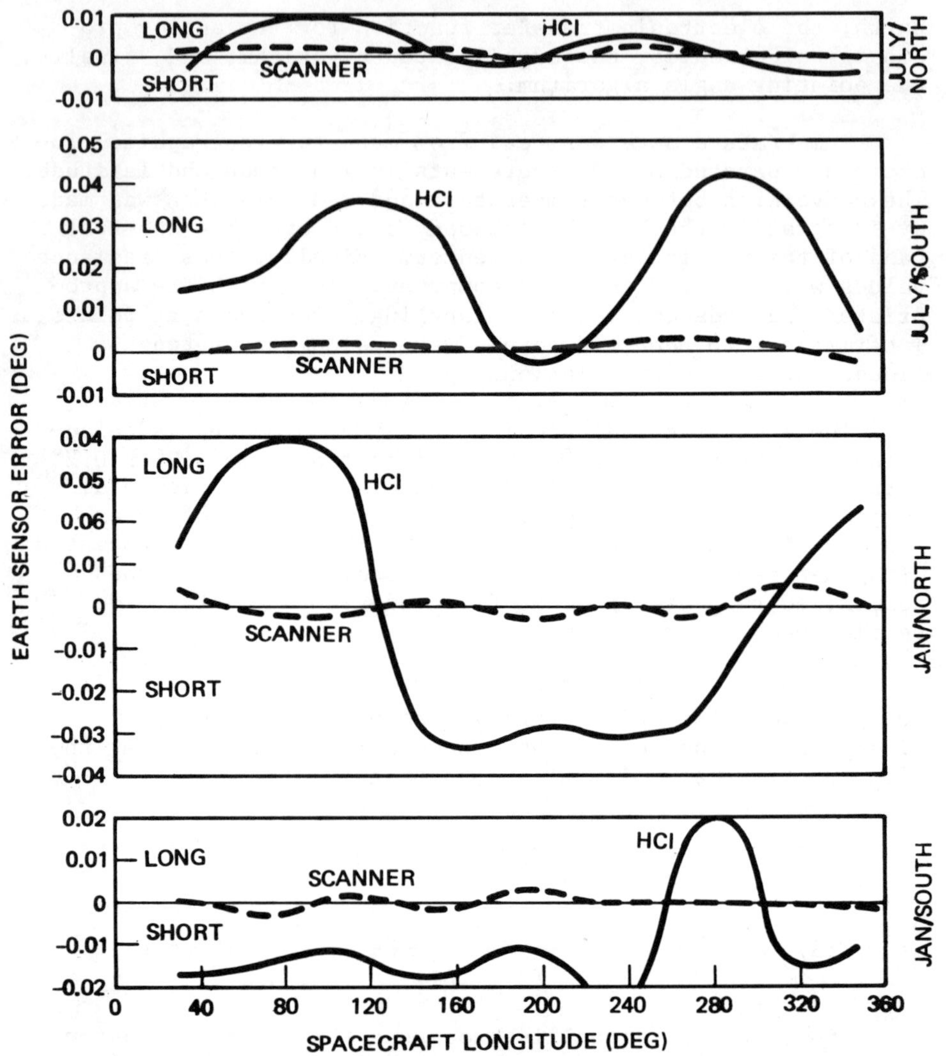

Fig. 7 Computer simulation results.

Performance Comparison

The computer simulation of the horizon crossing indicator shows results comparable to that observed in orbit. The polarity of the predicted error (Fig. 7) can be + or -, depending on the spacecraft longitude. Actual on-orbit performance (Fig. 1) exhibits a relatively large random component such that both + and - errors may be experienced in a short time interval. Errors of up to ± 0.04° are predicted (Fig. 7) for the sensor viewing the colder hemisphere, and less than

Table 1 Comparison of Intelsat IV data[2] with simulation model data

Data category	North			South		
	Jan. mean -July mean	σJan.	σJuly	Jan. mean -July mean	σJan.	σJuly
Intelsat F_2 340.48° long.	-0.044°	0.036°	0.010°	0.039°	0.040°	0.031°
Intelsat F_3 336.1° long.	-0.040°	0.041°	0.007°	0.055°	0.012°	0.032°
Intelsat F_7 330.9° long.	-0.017°	0.043°	0.004°	0.074°	0.010°	0.031°
HCI simulation Atlantic region	0.018°	0.010°	0.003°	-0.001°	0.003°	0.017°
Scanner simulation Atlantic region	0.004°	0.002°	0.001°	0.001°	0.001°	0.002°

± 0.02° for the warmer hemisphere. Flight data (Fig. 1) also indicates much larger errors for the sensor viewing the colder hemisphere.

In contrast, the gradient induced errors predicted for the scanning Earth sensor are about ten times less than that for the HCI. Fig. 7 shows that the maximum error for this sensor is about .004°. This demonstrates that the susceptibility to radiance gradients can be greatly different for different sensor configurations.

Data (Ref. 3) has also been reduced for Intelsat IV spacecrafts F3 and F7, which again shows the same general characteristics of cyclic bias errors and random errors. Table 1 summarizes the error statistics for the various spacecraft, along with the predicted error statistics. Temporal variations in errors were utilized in the computation of rms deviations for the various spacecraft sensors. The rms deviations, using simulation results, were calculated utilizing error variations over a local region of spacecraft longitude values. Both the observed data and the simulation data indicate that the rms deviations are at least three times greater for the colder hemisphere than for the warmer hemisphere.

Performance Optimization

The computer simulation demonstrated that gradient-induced errors can be as large as 0.04°. The exact magnitude and sign of the error is a function of the design characteristics of the sensor and also the radiance profile that is generated as the Earth sensor's field of view crosses the Earth. Since the radiance map changes relatively slowly and is primarily cyclic on a yearly basis, the error will also be slowly time varying for the synchronous altitude applications. For lower altitude missions, or those involving elliptical orbits, the error will be cyclic with orbit period, since different areas of the Earth will be scanned. Larger errors can be expected if the scan path is nominally North/South rather than East/West, since the scan is more nearly orthogonal to the isotherms. Gradient errors should be reduced for operation at lower altitudes, since the input gradient rate (%/degree of rotation) is reduced.

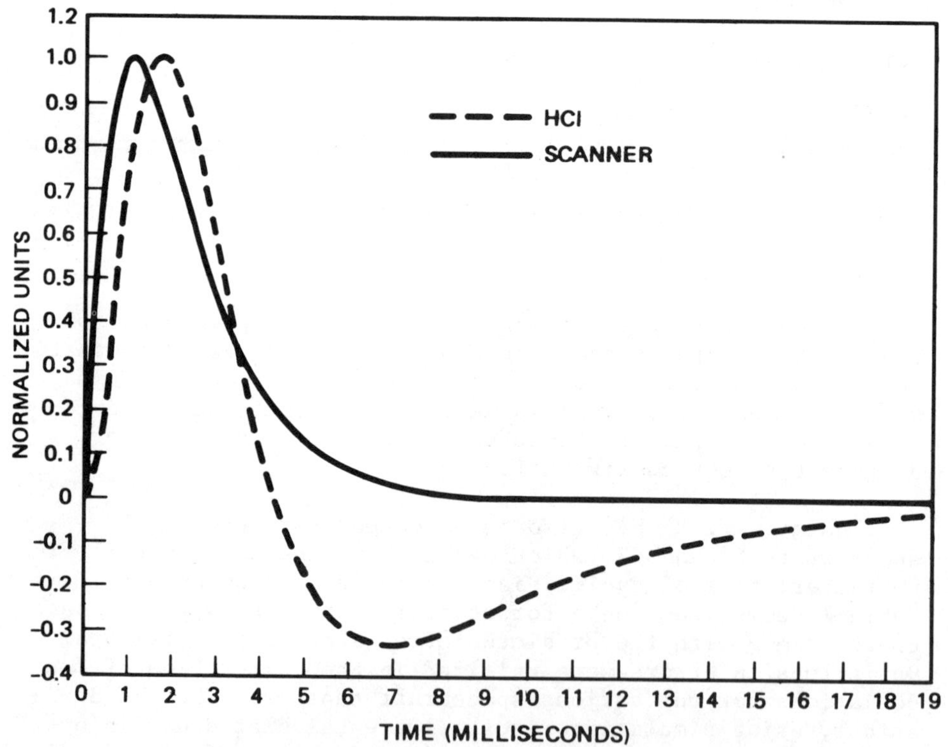

Fig. 8 Normalized impulse reponse.

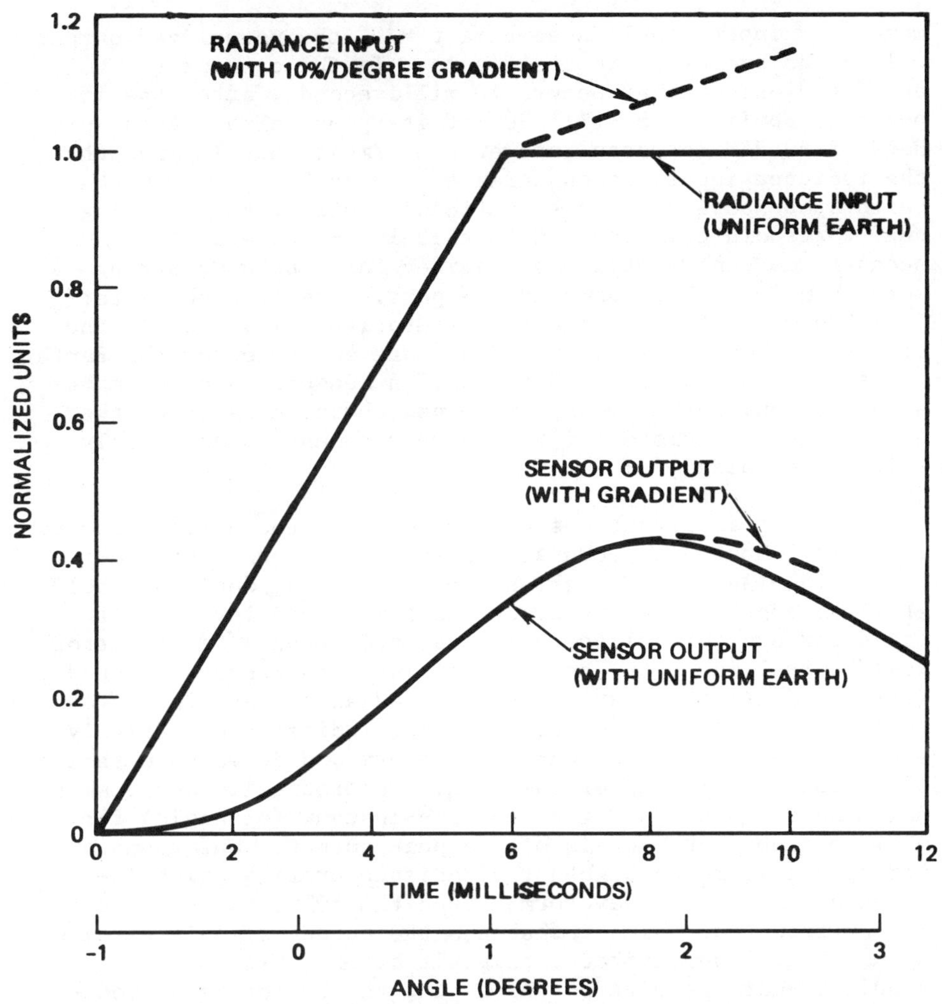

Fig. 9 Horizon crossing indicator output.

Earth sensors can be designed to operate with accuracies in the .01° to .05° range by optimizing sensor performance assuming realistic radiance distributions. Many of the sensor's design parameters affect its sensitivity to edge gradients: field of view, transfer function, horizon locator, and pointing angle algorithm. The influence of some of these parameters is discussed below.

The response of the electronics circuitry to a single impulse input can be determined from its transfer function. Fig. 8 shows the impulse response of both the scanner and the HCI unit. The significance of these responses is that for a momentary input, applied at time t = 0, the normalized output will be as indicated in the figure. In the case of the HCI unit, the output that occurs 10 milliseconds, after the impulse is applied, is still 20% of its peak value. At a spin rate of 60 RPM (or scan rate of 0.36°/ms), this implies that the radiance input, encountered as far as 3.6° in from the Earth/space horizon, can significantly effect the trailing edge threshold crossing. Note that in the case of the scanner, no contribution is realized for inputs occurring more than 10 milliseconds in the past. The long characteristic recovery time of the representative HCI sensor is the primary reason why gradient errors are so large for the Earth/space horizon crossing. Optimization measures should emphasize shortening the impulse response characteristic of the electronics, consistent with the requirement to adequately filter out noise.

Another important design parameter is the horizon locator criteria. Most high accuracy sensors employ a first order correction (AGC, % of peak, slope correction, dual threshold etc.) to account for different input radiance levels. If fixed threshold level operation is used, then a finite error will be realized over the range of input radiances expected. (Threshold crossing early for high radiances, and later for low radiances.) The horizon crossing indicator effectively samples the peak of the output waveform and develops a threshold level that is 50% of the peak. Theoretically there should be no shift in the horizon crossing instant (or angle) for different absolute levels of the peak output. Unfortunately, errors are introduced when the radiance profile which influenced the output waveform around the 50% crossing is different from that which influences the output signal near its peak. Fig. 9 demonstrates this situation. Without any gradient near the edge of the Earth, the output is as indicated with the solid line. In the presence of a +10%/degree edge gradient, the peak will be higher, but the signal around the 50% point is unaffected by the gradient. In this latter case, though, the new 50% point will be displaced in angle from the situation without any gradients. For the conditions shown, the error shift is 0.02°. Optimization of the horizon locator criteria would involve bringing the threshold point and the AGC sample closer in time and angle, so that both points on the output waveform are more nearly influenced by the same radiance profile.

The other design parameters, field of view and pointing angle algorithm are also expected to play an important role in the susceptibility to radiance gradients. It is believed that these parameters and also those previously discussed are best optimized by making end-to-end evaluations of sensor performance both through computer simulations and through laboratory testing.

References

[1]Slabinski, V., Comsat Corporation, Correspondence to J. Flannery, TRW Systems, 4 February 1974.

[2]Kirk, R. J., Watson, F. F., Brooks, E. M., and Carpenter, R. O'B., "Infrared Horizon Definition," April 1967, prepared by Honeywell Inc., Minneapolis, Minn., for NASA Langley Research Center.

[3]Unpublished logs from ground station for Intelsat IV Spacecrafts F2, F3, and F7, period 1972 through 1975, Comsat Corp.; data includes Earth sensor chord length data as a function of time.

Chapter III – Spacecraft Electrical Power, and Propulsion Subsystems

DESIGN AND TEST EXPERIENCE IN LIGHTWEIGHT DEPLOYABLE SOLAR ARRAY SYSTEMS

E. Quittner,* H.F. Borduas,+ and S.S. Sachdev≠

Spar Aerospace Products Ltd., Toronto, Ontario, Canada
and
S. Ahmed§

Department of Communications, Ottawa, Ontario, Canada

Abstract

This paper describes and compares four types of lightweight deployable solar array mechanical systems. These systems have been designed for three-axis stabilized communications spacecraft. The four designs are characterized by the type of substrate and its manner of support as follows: 1) full-length tensioned flexible; 2) peripheral frame-tensioned flexible; 3) peripheral frame-supported semiflexible; and 4) rigid panel. Of these, the first design was tested fully and is flight operational on the Communications Technology Satellite, whereas the remaining three are in various stages of preliminary design and test. Design and test principles, as well as experiences, where applicable, are discussed. The paper concludes with a tabulation of the mechanical characteristics and a comparison and extrapolation of performance parameters: 1) power per unit mass; 2) power per unit stowage volume; and 3) power per unit stacking height, and their respective rates of change for the four types of arrays.

I. Types of Designs

Deployable solar arrays have been and are being designed to satisfy the electrical power needs of advanced three-axis stabilized communications satellite in geostationary orbit. Those needs are, at present, mainly determined by the spacecraft communications and control functions and may include pro-

---

Presented as Paper 76-285 at the AIAA/CASI 6th Communications Satellite Systems Conference, April 5-8, 1976, Montreal, Canada.

*Member of Senior Staff.
+Stress-Dynamics Engineer.
≠Mechanical Systems Engineer.
§Research Scientist.

pulsion in the future. Large deployable solar arrays, therefore, will continue to be an integral part of spacecraft design.

In Canada, design of deployable solar arrays for communications spacecraft was begun during the design of the Communications Technology Satellite (CTS), which was launched in January 1976 on a Delta launch vehicle. The CTS arrays are the first flexible arrays to be flown for a nonmilitary communications spacecraft. Most of the Canadian solar array design and test work has been associated with this program.[1-4] Exploratory work on other designs started almost immediately after the test phase of the CTS array was completed. This paper summarizes the CTS design and test experience and relates it to the other designs. In total, four solar array designs for a spacecraft based on a Delta launch vehicle are discussed and compared. The four designs are characterized by the type and by the manner of support of the substrate. We can distinguish three types of substrates: 1) flexible, 2) semiflexible, and 3) rigid.

Flexible substrates can be tensioned over the entire length of the array and supported at the inboard and outboard ends, a design represented by the CTS array. Flexible substrates also can be tensioned segment by segment within the discrete panels of a hinged, articulated, supporting rigid framework. An array of this kind has been designed and a breadboard model fabricated; a development test program is, at present, underway.

Substrates of the semiflexible kind have been developed by TRW Systems Inc.[5] These, although not tensioned, also require the peripheral support of a hinged, articulated framework. A design of this type, based in part on the TRW substrate and the rigid frame concepts referred to above, was executed in Canada. A development test program is also underway.

Rigid substrates require, as a rule, no supporting framework. A design of this type was carried out for the Department of Communications, for a proposed Canadian multipurpose, three-axis stabilized spacecraft.

The four types of arrays discussed and compared in this paper have been designed for power needs of 0.75 to 1.0 kW at the end of a 7-yr mission life. The spacecraft is of a CTS type subject to the constraints of the Delta launch vehicle. Whereas such similarity of basic constraints makes comparisons interesting, no definite conclusions can be made regarding the final merit of these designs. Different basic constraints of different missions affect each design significantly. In fact, the best design for a given mission can be established only within the context of an integrated spacecraft/solar array design effort. Reliability, cost and weight, stowage volume, and modularity, as well as (at least for the next decade) equal adaptability to both an expendable launch vehicle and space-shuttle-based missions, are likely to be the ranking criteria. Nevertheless, an adequate basis for reasonable extrapolation has been established, and an attempt in that direction has been made in this paper.

II. Design Problem

Three separate and well-defined requirements shape the design of each of the solar arrays discussed: 1) stowing an array of a relatively large deployed size in a limited volume without cell damage during the initial mission phases, 2) deploying the array with sufficient control without damage to the array or spacecraft, and 3) providing a deployed state performance compatible with spacecraft requirements, i.e., power, thermal, reaction control, and attitude control.

Additional requirements that may affect the design are the need to demonstrate array deployment and to demonstrate the stability of sun tracking on ground. Also, provision of power during the transfer orbit spinning phase may be a requirement. This requirement is satisfied readily in the rigid frame and rigid substrate designs by utilizing the outboard panel as a source of power. In the full-length flexible substrate array, the spacecraft body panels and, if necessary, the jettisonable thermal covers on the stowed array sides of the spacecraft can be utilized. The ground test requirements usually pose no additional design constraints, with the probable exception of designs of the articulated frame type, featuring uncoördinated panel deployment.

The controlling factors during the three mission requirements just discussed are as follows:

1) In the stowed configuration, launch-vehicle-induced spacecraft vibration and acoustic environments are the controlling factors. 2) During the deployment phase, the controlling factors are the residual spacecraft rates. 3) During the deployed phase, power, stability, and thermal interface with the spacecraft are the primary requirements.

III. Test Philosophy

The division of testing of solar arrays conforms, in general, to the three-phase division of array operation: stowage, deployment, and the deployed phase. In the testing of the complete system, the total functional substrate/cell assembly, or its simulated representative, will have to be included.

Stowage

Particular emphasis in this phase is placed upon the survival of the array and the potential damage incurred by the cells and interconnects. Typical tests are the process of stowage itself, sustained stowage (where critical for survival or subsequent release), steady-state acceleration, sinusoidal and random vibration, acoustic exposure, and thermal vacuum. For arrays designed to supply power in the stowed condition, additional thermal vacuum tests are carried out.

Deployment

Particular attention in this phase is placed upon the successful operation of the moving and motion-controlling components of the array. Typical tests are release and pyrotechnic shock (where appropriate), deployment, and, depending on the type of array, final tensioning or latch-up. These tests are carried out under nonrepresentative 1-g conditions with the array either deployed vertically and counterbalanced or deployed horizontally and suspended neutrally with respect to the direction of deployment. In either case, interactions with the gravitational field are unavoidable and will modify array behavior. Aerodynamic effects, of some significance during the release and deployment of certain types of arrays, can be eliminated by testing in a vacuum chamber.

Deployed Phase

Particular attention in this phase is placed upon factors affecting 1) cell performance, and 2) array dynamics and stability. Typical tests are measurement of cell temperatures and electrical performance, verification of array normal modes of vibration and their frequencies, and, depending upon the type of array, stability under array stepping. Here the gravitational effects are overriding, and special techniques of analysis [6] are required.

IV. Flexible Full-Length Tensioned Substrate System of the CTS

This system has been reported extensively in earlier publications [1, 2, 4]. As it is a forerunner of the other designs, a brief summary of its main features, including the

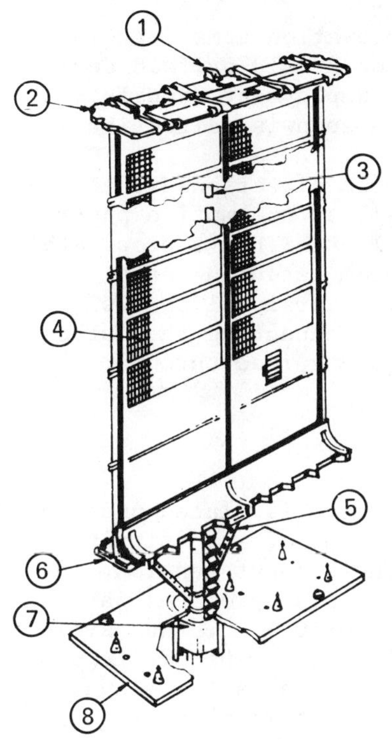

Fig. 1 CTS deployable solar array.

analyses and tests, is presented. An illustration of the deployed array as a panel-mounted modular unit is given in Fig. 1.

Stowage

The arrays are stowed accordion-fashion against the North and South panels of the spacecraft. Each array is sandwiched between two honeycomb panels, the outboard pressure panel and the inboard pallet, and subjected to a nominal 2-psi (12.8-kPa) pressure. Protection of the solar cells and interconnects from the launch and apogee motor firing vibration environment is provided by polyurethane foam inserted between folds. Two polyurethane foam pads bonded to the pressure panel and pallet provide additional cushioning to the array. Thus, the stowage problem is solved most efficiently, the entire package occupying a small volume. The stowage pressure was based on the friction forces required to contain the stowed array.

Deployment

The deployment is accomplished in three stages. The first stage consists of raising the stowed array as a module from the spacecraft panel until the elevation arms lock in position. The second stage consists of a fold-by-fold deployment. The third stage, at the end of deployment, consists of the tensioning of the substrate in the spanwise direction.

Control in the axial and transverse directions relative to the direction of deployment, as well as the motive power for deployment, is supplied by a single BI-STEM (a proprietary storable tubular extendible member) boom extending at a nominal rate of 1 in/sec (2.5 cm/sec). Torsional control is provided by means of two control lines connecting the outboard pressure panel to the inboard pallet and guided through tabs attached to the folds of the array.

Deployed Phase

Once deployed, the flexible substrate is placed under a nominal tension of 8 lb. (35.6 N) by means of a tensioning mechanism at the tip of the array. The blanket tension, in conjunction with the boom bending stiffness, produces natural frequencies calculated to be in excess of 0.1 Hz, thus avoiding potentially uncontrollable flexible array/spacecraft dynamic interactions. The inboard end of the array is elevated to a distance of approximately 30 in. (0.76 m) above the spacecraft panel, thereby providing a view factor to space in excess of 0.82 to the heat-radiating surfaces of the spacecraft.

Supporting Analyses

Analyses of the type performed for the CTS and described in what follows typically are required to substantiate designs presented in this paper.

Tradeoff Studies

Tradeoff studies have been conducted in two stages: 1) flat-pack *vs* roll-up arrays; and 2) flat-pack arrays, having a split substrate with a single boom nominally in the plane of the substrate, *vs* a one-piece substrate with a single offset boom behind the substrate. As a result of these studies, the last-mentioned flat-pack configuration was chosen on grounds of least weight, best stowage volume economy, growth potential, and superior boom thermal stability.

Stowage Analyses

Static pressure distribution and dynamic pressure variation of the stowed array were studied for both spacecraft and subsystem qualification test installations. A finite-element plate model for the structural elements and a spring mass representation of the stowed blanket-interleaf system have been employed for this study. Development sample test data were used to provide input parameters, whereas subsystem tests confirmed the soundness of the analysis and design.

Deployment Analyses

The key analysis of deployment was concerned with the stability of the array and potential for cell damage during deployment from a spacecraft with an initial spin rate of 2 rpm and small residual roll and pitch rates. Sensitivity studies, in order to assess the effects of variations of different factors (e.g., unequal array extension rates, blanket tension, initial spacecraft angular rates, etc), also have been performed.

Deployed Array Analyses

For this phase, two types of analyses were carried out: 1) design-type analyses, and 2) performance-type analyses.

1. Design-Type Analyses. These were carried out in order to gain confidence in the system under normal and failure conditions. These were the analysis of tension distributions in the blanket and their effect on array natural frequencies; the analysis of boom strength and stability in the event of a seizure-type failure of the tip tension mechanism; and the analysis of the stability of the array stepping as a consequence of stepper motor and array dynamic interactions. This last analysis was particularly revealing, in that it indicated the importance of damping, be it mechanical or electrical, in the stability of stepper motor operation for this application.

2. Performance-Type Analysis. As part of the CTS solar array technology and attitude control system experiments[7], analyses of the CTS solar array were carried out at the Communications Research Centre. These investigations were to serve three purposes: 1) to insure the structural and functional integrity of the array in its operating environment; 2) to understand the interaction of the flexible array with the spacecraft attitude control system; and 3) to develop an analytical model of the array, supplemented by ground test[4] and flight data, for the future design of the generic class of spacecraft represented by CTS. Some of the results of this last effort are reported in Ref. 6.

Testing

The tests on the integrated CTS solar array subsystem, organized according to the three phases of array operation and according to the subsystem model designation, are shown in Table 1.

Dynamic tests were also performed on a deployed array suspended vertically in a vacuum chamber[4] in order to determine: 1) natural frequencies and mode shapes, 2) modal damping, 3) response to dynamic inputs that simulated attitude acquisition maneuvers, thruster firings, and other specialized mission events, 4) automatic sun acquisition and tracking performance, and 5) stepping performance.

V. Flexible Rigid-Frame Supported Tensioned Substrate System

This design evolved shortly after the CTS, to which it also is related by its use of flexible tensioned substrate. The design is that of a hinged articulated system of rigid frame panels stowed flat in a 56- x 53- x 7-in. (142- x 135- x 17.8-cm) envelope against the North and South panels of the spacecraft, this envelope including the stowed array retention fittings. The package is configured in a way such that the array can be stowed on panels having their outboard surface 27.6 in. (70.1 cm) away from the spacecraft thrust axis, for a Delta launch vehicle. The design primary objective was to establish a weight-competitive "rigid panel" deployable array system for lower-power regimes (0.75 - 1.5 kW at the end of a 7-yr mission life), in contradistinction to the CTS design conceived as a forerunner of future, much higher-power systems.

Tradeoff Studies

Tradeoff studies were conducted with respect to the design criteria of: 1) 1.3-kW power at beginning of life for two array wings, 2) available stowage stack height, 3) adequate separation of adjacent substrates to prevent cell damage during stowage, 4) deployment control; 5) deployed array stiffness, and 6) minimum weight. As a consequence, a five-panel configuration was adopted, with a V-shaped elevation arm and a rectangular framework providing peripheral and intermediate support to the flexible substrates. The material of construction for the metallic parts is mainly magnesium alloy in rectangular or round tube form, whereas the substrate material is fiberglass-reinforced Kapton. Figure 2 shows the array both in the deployed and stowed configuration.

Table 1 Integrated CTS Solar Array Tests

Phase of array operation	Breadboard model	Development model	Qualification model	Protoflight model
Stowed phase	Sinusoidal vibration, Z axis only; fold behavior, stowage	Sinusoidal vibration, two axes (Y and Z), acoustic excitation; depressurization; stowed thermal vacuum cycling	Sinusoidal and random vibration, three axes	Sinusoidal and random vibration, three axes, on flight spacecraft
Deployment phase	Manual deployment	Repeated jettisons and deployments	Repeated jettisons and deployments; thermal vacuum jettison and deployment	Limited jettisons and deployment; thermal vacuum jettison and deployment; double deployment off spacecraft
Deployed phase			Deployed array thermal vacuum testing; pseudo-thermal balance test; deflection sensor alignment	Deployed array thermal vacuum cycling; pseudo-thermal balance test

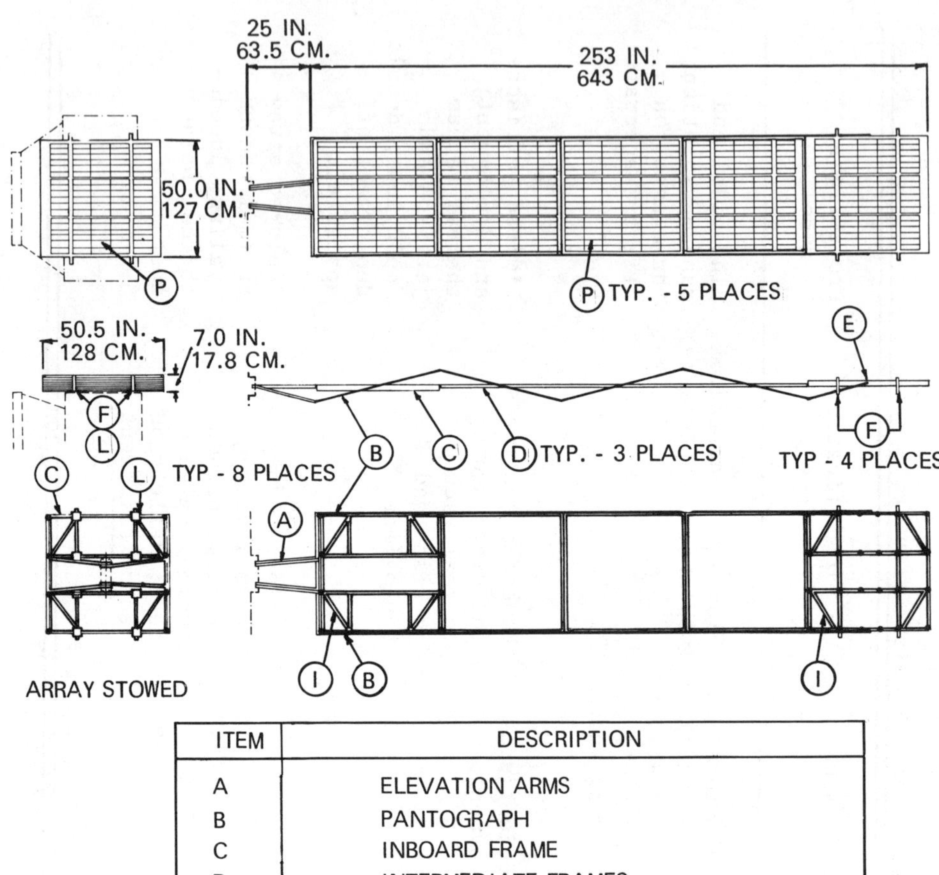

ITEM	DESCRIPTION
A	ELEVATION ARMS
B	PANTOGRAPH
C	INBOARD FRAME
D	INTERMEDIATE FRAMES
E	OUTBOARD FRAME
F	TIE-DOWN FITTINGS
I	LATERAL BRACING
L	STOWED ARRAY SUPPORTS ON SPACECRAFT
P	SOLAR PANELS

Fig. 2 Rigid frame/flexible substrate array: deployed and stowed configuration.

Design for the Stowed Phase

The array was designed to fit a CTS-type spacecraft and the Delta launch vehicle constraints. The array is attached at four hard-points to the spacecraft panel and released by pyrotechnic means upon ground command. Four additional hard-points are provided for purposes of array preloading. Because the array overhangs the supporting spacecraft panels, dynamic response during vibration testing and launch is an important design criterion. This requirement was satisfied by designing more rigid inboard and outboard panel frame members and by preloading the stowed array so that no individual response of frame members is possible. Preloading is by means of a system of set-screws that are adjusted to produce an out-of-plane distortion of the inboard and outboard frame members. Because the substrates are flexible, potential cell damage during vibration testing and launch was another important criterion. This was satisfied by spacing the substrates apart by the frame member depth of 3/4 in. (1.9 cm) and by providing edge clamping, as well as intermediate supports to the substrate at points of frame preloading. It also was expected that aerodynamic damping, as a consequence of the air cushions trapped between adjacent panels, would play some part in attenuating substrate responses.

Design for Deployment

Array deployment is by means of articulation, energized by springs located at each panel hinge and controlled by a pantograph system and an electric motor. Upon completion of deployment, latches at the hinges provide a locking feature.

Design for the Deployed Phase

In the deployed phase, the framework provides peripheral support for the flexible substrates, which are tensioned bidirectionally to a force level of approximately 4 lb (17.8 N) per side by low-stiffness peripheral springs. With all natural frequencies in excess of 0.15 Hz, no dynamic interaction problems with a CTS-type attitude control system are expected. No thermal gradient problems affecting cell performance are expected, provided that a reasonable distance of the order of 3/4 in. (1.9 cm) between cells and frame members is maintained.

Present Status of the Program

A breadboard model of the array has been built to study the responses of the stowed array during vibration and acoustic testing. Although the array configuration is flight-representative, materials of construction have been adapted to those available on short notice, aluminum alloy square tubing of greater than required thickness was used for frame members, and nylon-reinforced mylar instead of fiberglass-reinforced Kapton was used for the substrates. Cells have been simulated by glass chips bonded in two layers to represent cell weight and fragility.

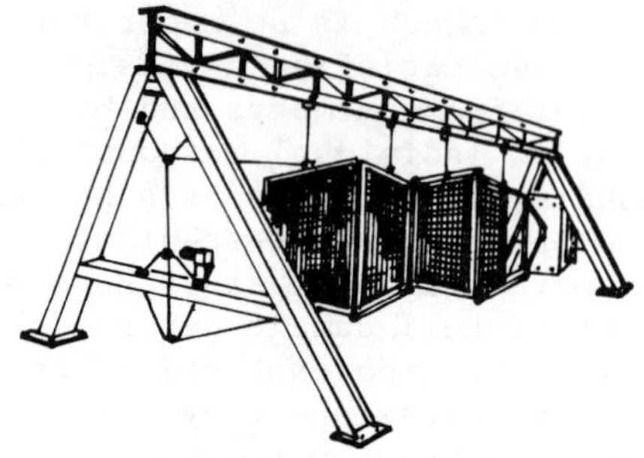

Fig. 3 Ground deployment of rigid frame/semiflexible substrate array.

A computer/math model of the stowed array has been prepared, test predictions have been made, and the necessary levels of array preloading have been established. A complete system vibration and acoustic test has been carried out successfully.

VI. Semiflexible Rigid-Frame-Supported Substrate System

Developed by TRW Systems Inc.,[5] the design consists of a frame within which is supported a semiflexible substrate constructed of aluminum honeycomb core and Kapton face sheets. This design relies on the substrate responding with a membrane effect to quasistatic, vibratory, and acoustic inputs, maintaining both Kapton skins in tension throughout most of the panel. Peripheral and intermediate frame members react with the primary, normal-to-plane components, as well as the induced in-plane components of membrane tension, and have to be designed accordingly. A 3-1/2 panel articulated mechanical system has been designed by Spar, using this principle of substrate design combined with a simple aluminum alloy frame and producing 0.75 kW at the end of a 7-yr life in geostationary orbit.

Design for Stowage

The array is stowed by being held down at four hardpoints against the spacecraft panel, with the cells on the outermost panel exposed and thus available for power generation before array deployment. Two additional in-plane reaction points are required on the spacecraft panel for this particular 3-1/2 panel configuration for restraining lateral array vibration response during test and launch. The array, when stowed, occupies a volume of 51.5-x 51-x 6.2-in. (131-x 155-x 15.8-cm).

The stowed array is preloaded to prevent relative motion between the frame members under launch environment. The preload is applied between the outboard two and inboard two panels by introducing a predetermined normal-to-panel displacement at each end. Substrates have been arranged within the stack to maximize the distance between adjacent panels, required for the substrate membrane action to take place, without sacrificing available cell area. Two intermediate pads per panel are required on the cell side of the two intermediate panels to prevent cell-to-cell contact in the worst-case out-of-plane vibratory response of the substrates.

Design for Deployment

Deployment is initiated by firing two sets of redundant bolt cutters in a predetermined sequence, whereupon the strain energy stored in the torsion bars at the hinges unfolds the panels. Coordination of deployment is achieved by a system of cables and pulleys. Passive control of deployment is provided

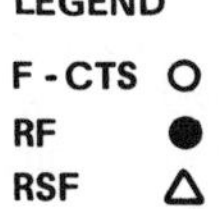

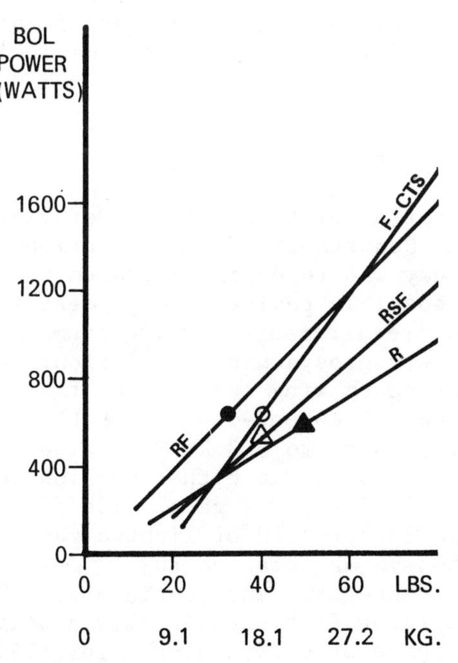

Fig. 4 Variation of BOL power per array wing with total weight of stowed array.

by a fluid damper at the base of the array elevation arms which keeps the deployment velocities at an acceptable level. Figure 3 shows the array during ground deployment with a deployment velocity augmenting device required to overcome aerodynamic drag on the deploying array and produce qualification level latch-up velocities.

Design for the Deployed Phase

The end of deployment is marked by locking of the hinges by means of latches. A residual spring torque in the deployment springs prevents any free-play from developing at the hinges under moderate spacecraft maneuver loads. The deployed array, under no-free-play conditions, has an estimated first natural frequency of about 0.16 Hz for an aluminum frame version and 0.22 Hz for an optional graphite epoxy composite one of about the same weight.

Testing

TRW carried out development tests on an array of this type [5]. Static loading tests, followed by a sinusoidal vibration resonance survey and acoustic tests, were performed. A preliminary development test program was carried out in Canada. This involved the manufacture of one representative panel, using laminated glass slides to simulate solar cells. This panel was mounted representatively to the outboard frame of the rigid frame array described in Section V and subjected to sinusoidal and acoustic excitation.

VII. Rigid Substrate System

This system was designed for a relatively low-power general-purpose spacecraft for the Department of Communications. The approach adopted in the study was to design a system using existing technology. Several possible payloads were identified for this spacecraft, with power requirements ranging from 0.5 to 0.8 kW at the end of a 7-yr geosynchronous mission. A modular design was adopted where each wing consists of four rigid honeycomb panels producing 0.4 kW end-of-life (EOL). Removal of one panel reduces the power to 0.3 kW EOL. The cells on the outboard panel are arranged in either version to be facing outward and be available for power generation in the stowed phase. The substrate is constructed of lightweight 1-lb/ft^3 (16 kg/m^3) aluminum honeycomb core and thin 0.004-in. (0.10 mm) aluminum facesheets. The panels are hinged together and stowed accordion fashion. In this configuration, the array occupies a space of 60.5-x 51-x 7-in. (154-

Table 2 Array Characteristics

Array Characteristics	Type of Array			
	Flexible substrate (F-CTS)	Rigid frame flexible substrate (RF)	Rigid frame semiflexible substrate (RSF)	Rigid substrate (R)
Stowed phase:				
Overall dimensions [length (Z) x width (X) x height]	11x52x 2 3/4 in. (28x132x7 cm)	50.5x50x7 in. (128x127x17.8 cm) (not including pantographs & retention fittings)	61x51.5x6.2 in. (155x131x15.8 cm)	60.5x51x6 3/4 in. (154x130x17.1 cm)
Face-to-face distance between cell and adjacent hard surface	0.006 in. (0.015 cm)	0.75 in. (1.9 cm)	1.25 in. (3.2 cm) for intermediate panels; 0.65 in. (1.65 cm) for inboard panel and spacecraft	0.25 in. (0.64 cm)
Cell protection	0.625 in. (1.6 cm) thick foam pads bonded to pallet and pressure plate; foam interleaves 0.06 in. (0.15 cm) thick	None	None	None

Table 2 (continued)

Stowage volume	0.91 ft^3 (0.026 m^3)	10.2 ft^3 (0.29 m^3)	11.3 ft^3 (0.32 m^3)	12.1 ft^3 (0.34 m^3)
Stowage pressure	2 psi (13,800 N/m^2)	Frame preload	Frame preload	Panel preload
Transfer orbit power (average at 40 V)	0	80 W	94 W	94 W
Deployed phase:				
Locking	Elevation arm lock only	Locked elevation arm and panel hinges	Locked elevation arm and panel hinges	Locked elevation arm and panel hinges
Substrate tension	8 lb (35.6 N) along length of array only, kept constant by tip tensioning mechanism	4 lb (17.8 N) applied bidirectionally to each panel substrate using fiberglass rods	None	None
Elevation from spacecraft	29 in. (72.5 cm)	25 in (64 cm)	70 in (178 cm)	60 in (152 cm)

Table 2 (continued)

Frame member material	No frames	Magnesium alloy	Aluminum alloy	No frames
Number of panels	30 (26 with cells)	5	4 (3.4 with cells)	4
Panel size	8.6x51.6 in. (21.8x131 cm)	50x50.5 in. (127x128 cm)	51.5x60 in. (131x152 cm)	51x60 in. (130x152 cm)
Total deployed length	286 in. (7.26 m)	278 in.(7.06 m)	285 in (7.23 m)	301 in. (7.65 m)
Total weight per wing	40 lb (18.1 kg)	33 lb (15.0 kg)	40 lb (18.1 kg)	50 lb (22.7 kg)
Lowest natural frequency	0.10 Hz	0.14 Hz	0.16 Hz	0.31 Hz
Available power BOL per wing	630 W	630 W	520 W	580 W
BOL power/weight	16 W/lb (35 W/kg)	19 W/lb (42 W/kg)	13 W/lb (29 W/kg)	11.5 W/lb (25 W/kg)
Δpower/Δweight	28ΔW/Δlb (62ΔW/Δkg)	21 ΔW/Δlb 46 ΔW/Δkg)	18 ΔW/Δlb (39 ΔW/Δkg)	13 ΔW/Δlb (29 ΔW/Δkg)
BOL power/stack height	229 W/in. (90 W/cm)	90 W/in. (35 W/cm)	84 W/in. (33 W/cm)	86 W/in. (34 W/cm)

Table 2 (continued)

Δ power/Δ stack height	700 Δ W/Δ in. (274 Δ W/Δ cm)	104 Δ W/Δ in. (41 Δ W/Δ cm)	Odd number of panels added: 66 Δ W/Δ in. (26 Δ W/Δ cm) Even number added: 125 Δ W/Δ in. (49 Δ W/Δ cm)	116 Δ W/Δ in. (46 Δ W/Δ cm)
BOL power/ stowage volume	690 W/ft^3 (24,500 W/m^3)	63.5 W/ft^3 (2240 W/m^3)	46 W/ft^3 (1625 W/m^3)	48 W/ft^3 (1710 W/m^3)
Δ power/Δ stowage volume	2110 Δ W/Δ ft^3 (74,700 Δ W/Δ m^3)	71 Δ W/Δ ft^3 (2510 Δ W/Δ m^3)	Odd number of panels added: 36 Δ W/Δ ft^3 (1290 Δ W/Δ m^3) Even number added: 69 Δ W/Δ ft^3 (2430 Δ W/Δ m^3)	65 Δ W/Δ ft^3 (2300 Δ W/Δ m^3)

x 130-x 19-cm). Spring energy in the hinges provides deployment power, and control is effected by a pantograph linkage and an electric motor.

VIII. Comparison of the Four Systems

Each of the four systems described has been designed to solve the same basic set of design problems but each with slightly different objectives in mind. These designs are compared in Table 2. Figures 4-6 show the variation of beginning of life (BOL) power with weight, stowage volume, and stack height for these four specific arrays.

The rigid frame/flexible tensioned substrate (RF) array is the most weight-effective at 19 W/lb, but, extrapolating the present designs to larger arrays, the CTS-flexible substrate (F-CTS) array design becomes more efficient at BOL powers exceeding 1200 W per wing. The rigid frame/semiflexible substrate (RSF) array and the rigid substrate (R) array are not weight-competitive. These conclusions must be tempered by the following remarks: 1) The weight used to evaluate the F-CTS array excluded the weight of the body array structure, solar cells and wiring required for transfer orbit power; if this extra weight were included, the F-CTS superiority would appear at higher power levels. 2) The lack of competitiveness of the RSF Array is strongly determined by the particular features of the design reported here. If semiflexible substrates were added to the frames of the RF array, this new array, delivering 605 W BOL power, would weigh 39 lb (17.7 kg), for 16 W/lb instead of the 13 W/lb of the present RSF design. The growth rating would remain 18 Δ W/Δlb.

The F-CTS array has outstanding stowage volume and stack height efficiencies compared with the other arrays. This is obviously the type of array to use when the required power is high, particularly when the available stowage space is limited, as is the case with any of the expendable launch vehicles. Again the RF array appears more efficient than the RSF array, but the latter could be made to have the same ratings as the RF array by installing a semiflexible substrate to the framework of the RF array.

All the quoted array performance figures are based on using 2-x 2-cm 200-μm-thick, N on P silicon solar cells of 1-Ω-cm base sensitivity with 100-μm-thick ceria-doped quartz cover glass. These cells, used on the CTS solar array, produce, at their maximum power point, typically 65 mW at 25°C (air mass zero). The beginning of life (BOL) performance figures are based, however, on array power output at the

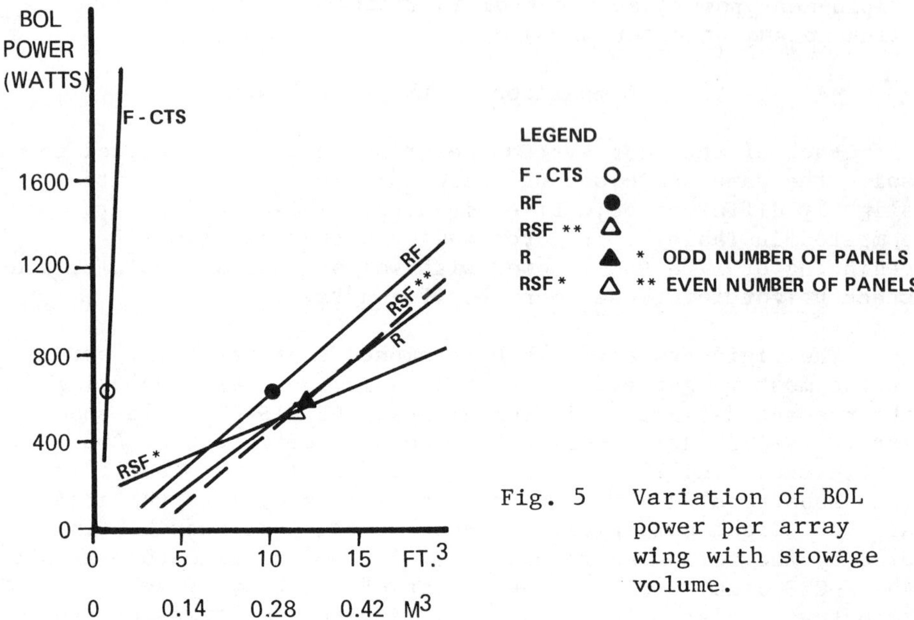

Fig. 5 Variation of BOL power per array wing with stowage volume.

appropriate predicted array operating temperature, at an assumed summer solstice sun intensity and inclination.

The following guidelines on the choice of array type for a given application can be drawn: 1) Rigid frame/flexible substrate arrays are the best choice for medium power systems where weight is critical. 2) Rigid frame/semiflexible arrays should be considered for medium - power systems with a more limited available stack height. The RSF array allows adjacent substrates to be stowed closer to each other, thus allowing a greater number of panels in the array. Semiflexible substrates also need less care in handling and are easier to install than are flexible substrates. 3) Flexible substrate arrays of the CTS type constitute the only efficient design when the required BOL power is greater than about 1500 W per wing. They also offer the best option for applications with limited stowage heights and volumes, regardless of the required power. As demonstrated by ground-testing and flight experience on the CTS program, flexible substrate arrays provide excellent protection to the solar cells and interconnects during vibration. 4) The rigid substrate array is never competitive and should only be considered in a program where weight is not critical. Its greatest asset is its relatively lower cost.

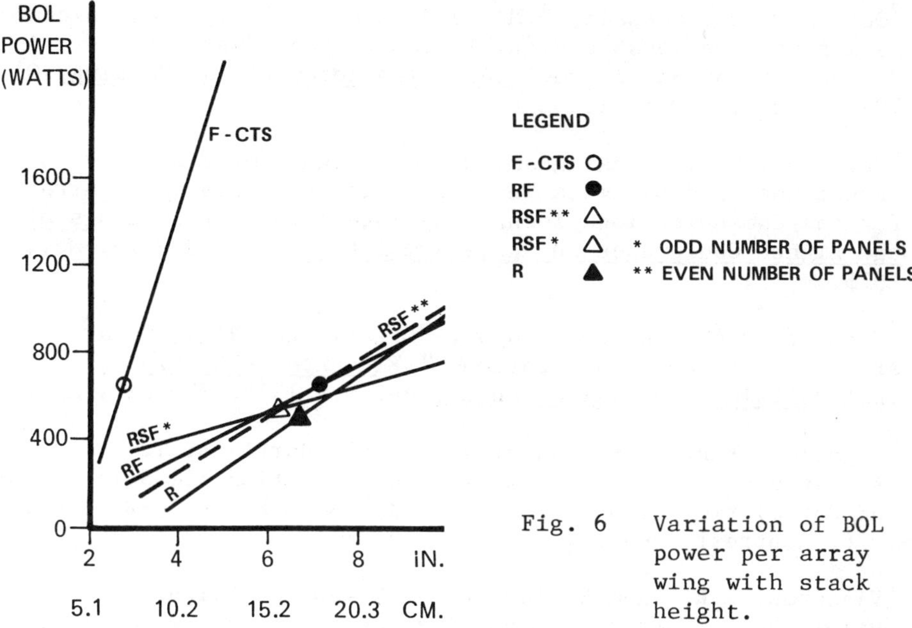

Fig. 6 Variation of BOL power per array wing with stack height.

Any of the four types of array design are applicable to future communications spacecrafts whether their mission be based on an expendable launch vehicle or the space-shuttle. The particular demands of each mission and program will influence the choice of array design. Total spacecraft weight will remain a critical item, as weight saved in the power subsystem can be utilized to provide additional communications capability. Weight at reasonable cost will probably be the prime factor influencing the design of future deployable array systems.

References

[1] Sachdev, S.S., Quittner, E., and Graham, J.D., "The Communications Technology Satellite Deployable Solar Array Subsystem," International Photovoltaic Power Generation Conference, Sept. 1974, Hamburg, Germany.

[2] Smith, J.S., Sachdev, S.S., and Hunter, J.A., "Testing of the Communications Technology Satellite Deployable Solar Array Subsystem," Eleventh IEEE 1975 Photovoltaic Specialists Conference, May 1975, Phoenix, Ariz.

[3]Quittner, E., Gossain, D.M., and Sachdev, S.S., "Mechanical Design of a Deployable Flexible Solar Array System for a Communications Satellite," CASI/AIAA Aeronautical Meeting, Oct. 1974, Toronto, Ontario.

[4]Harrison, T.D., Buckingham, R., and Vigneron, F.R., "Functional and Dynamics Testing of the Flexible Solar Array for the Communications Technology Satellite," Proceedings of the Eighth Conference on Space Simulation, NASA SP-379, Nov. 1975.

[5]Corbett, D.A. and Dean, W.J., "Lightweight Rigid Solar Array Structural Considerations," Eleventh IEEE 1975 Photovoltaic Specialists Conference, May 1975, Phoenix, Ariz.

[6]Vigneron, F.R., Parthasarathy, A., and Harrison, T., "Analysis of the Structural Dynamics of a Flexible Solar Array and Correlation with Ground Testing," AIAA Paper 76-242, April 1976, Montreal, Quebec.

[7]Vigneron, F.R. and Millar, R.A., "Plan for Flight Evaluations of Attitude Stabilization and Flexible Solar Array Dynamics for the Communications Technology Satellite," 26th International Astronautical Federation Congress, Sept. 1975, Lisbon, Portugal.

ULTRALIGHT SOLAR ARRAY (ULP)
FOR FUTURE COMMUNICATION SATELLITES

D.E. Koelle, and H. v. Bassewitz

Messerschmitt-Bölkow-Blohm GmbH (MBB), Munich, Germany

Abstract

The ultralight solar array ULP is conceived for three-axis-stabilized spacecraft with sun-oriented solar arrays and a power requirement of 1 to 10 kW. Its modular concept allows easy adaption to each specified power. The array consists of a number of identical panels, each having a size of 1×3 m^2 and consisting of a rigid carbon fibre frame with a pretensioned solar cell blanket. Up to this time, the design of the solar array and the required tooling and ground support equipment for fabrication and test have been designed completely; aspects such as stability against space radiation, thermal gradients, and thermal cycling have been investigated by numerous tests. One full-size panel (1×3 m^2) has been exposed successfully to sinusoidal and random vibration levels that correspond to amplified loads of the Atlas - Centaur launcher. Presently the development of anticharging layers, blanket repair techniques, and nondestructive test techniques of the electrical bonds is being performed. A complete ULP array (2×6 panels) is being fabricated and will be subjected to final qualification tests in 1977.

Presented as Paper 76-286 at the AIAA/CASI 6th Communications Satellite Systems Conference, Montreal, Canada, April 5-8, 1976.

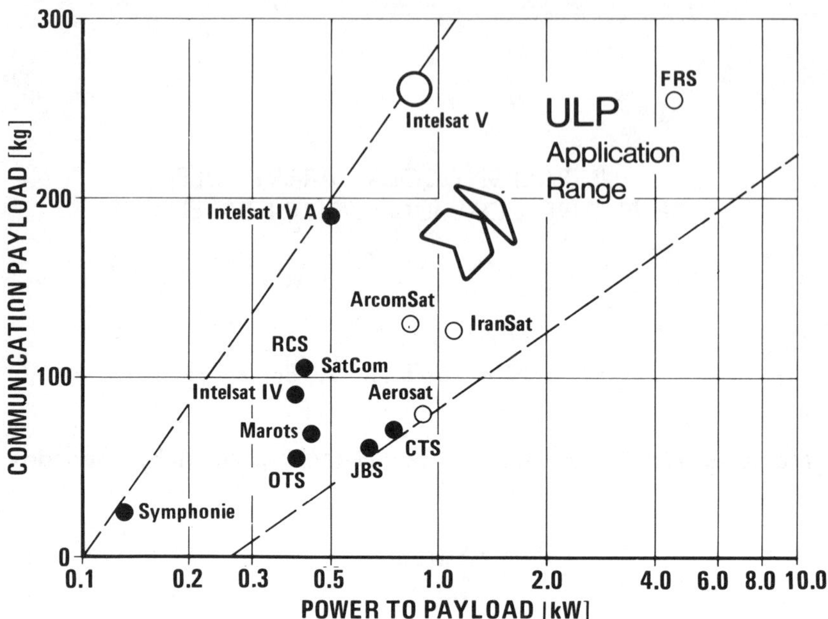

Fig. 1 Power range of communication payloads.

1. ULP Array Program

The ULP (Ultra-Lightweight Panel) solar array is a semirigid (hybrid) flat-pack advanced solar array. It is under development in Germany by MBB and AEG and funded out of the German national space technology program with support by Intelsat via Comsat Corporation. Intelsat decided, after an request for proposal action in December 1975, to participate in this development, which is dedicated to future high-power satellites with more than 1-kW payload power requirements. The general trend in the area of communication satellites is shown in Fig. 1. In Germany, for example, a high-power direct TV broadcasting satellite (FRS) with 6 kW/EOL (end-of-life) has been studied for several years, and supporting technology projects have been initiated.

The general configuration of the ULP is shown in Fig. 2; it consists of a number of identical panels, 1,1 x 3,2 m^2 each, interconnected and deployed by a special spring-actuated mechanism with a closed cable-loop (MBB patent). This relatively simple redundant system has been developed and qualified for the OTS/MAROTS solar arrays under fabrication at MBB. They are shown in Fig. 2, too, and will be flown in 1977.

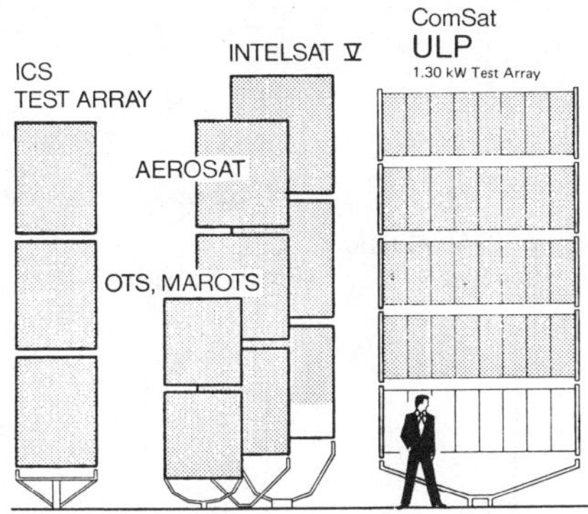

Fig. 2 MBB's solar arrays.

In comparison to this smaller array, which has a rigid CFC-aluminium honeycomb sandwich structure, the ULP-concept is characterized by a CFRP (carbon-fibre-reinforced plastic) frame structure with a flexible substrate suspended under pretension within this frame. This feature allows one to reduce the specific structure mass to 0.7 kg/m^2, compared with the 1.4 kg/m^2 mass of the CFC sandwich panel or the conventional aluminium panels with more than 5 kg/m^2 (Fig. 3).

The ULP development has entered its third phase since December 1975, and a full four-panel wing with yoke will be assembled early 1977. (Fig. 4). The qualification of a single panel was completed successfully at the end of 1975, after the initial design and concept verification phase in 1973/1974. An extensive test series is planned in 1977 to qualify the ULP array.

It has been proposed to fly a ULP prototype on a test satellite.

2. ULP Applications and Performance

The original requirements for ULP were formulated from the direct TV-broadcasting satellite project studied in Germany. Figure 5 illustrates the spacecraft with four (redundant) high-power transmitter tubes, 450 W each. The power requirement was 6 kW EOM, resulting in a solar array power of about 9 kW BOM.

The spacecraft was designed to be launched by an Atlas Centaur. In the meantime, a redesign was performed for adaption to the European Ariane launch vehicle and a liquid-propellant apogee injection system (Symphonie technology). A simplified platform

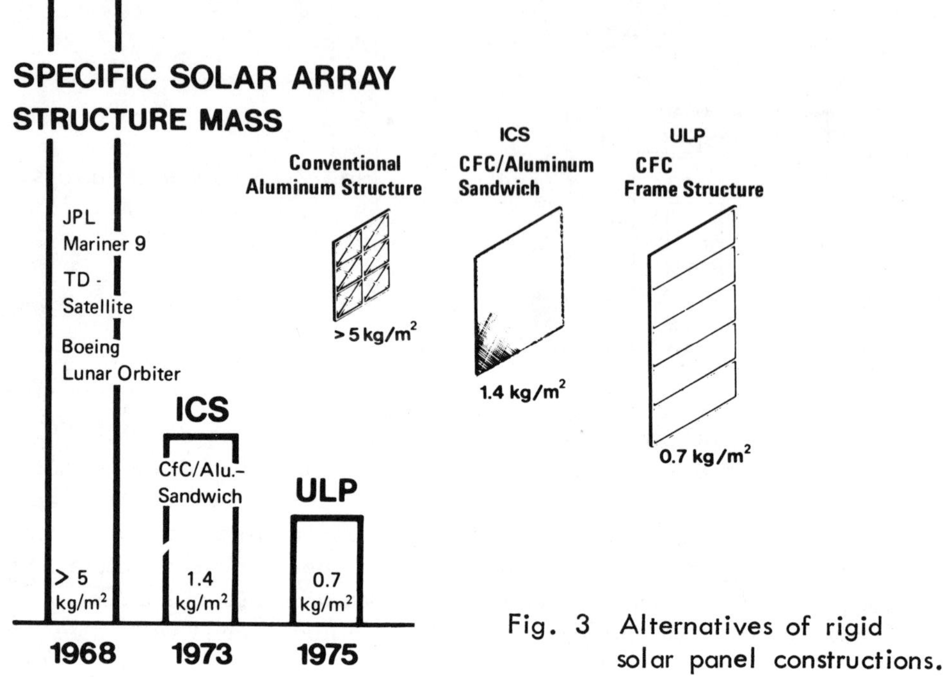

Fig. 3 Alternatives of rigid solar panel constructions.

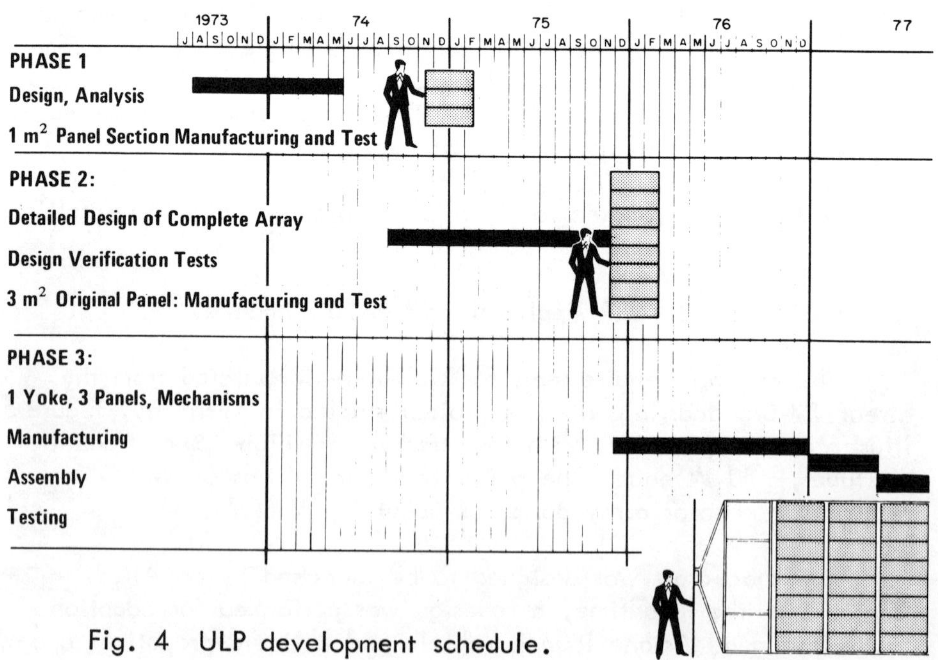

Fig. 4 ULP development schedule.

called ARPAS (Ariane Platform for Application Satellites) has been conceived and proposed as an Ariane "passenger experiment." The launch could be performed with Ariane test flight L.04 in 1980.

Figure 6 is a conceptual drawing of the ARPAS test platform with the ULP solar array. The modular concept allows easy adaption of the array to any power requirement. The inherent advantage of the concept is the superior stiffness in addition to the low mass.

The ULP specific performance is depicted in Fig. 7. The major factor is the single solar cell power output. The parameter used in Fig. 7 is the effective power per cell under the desired conditions (BOM, EOM). The 25°C values given by the cell manufacturers are as follows: AEG HEC (high-efficiency cell), 126 mW, and AEG Blackcell, 154 mW; Spectrolab Hybrid, 132 mW, and Spectrolab Helios, 138 mW; and ComSat Blackcell, 160 mW. In comparison, for the BOM performance some 20% must be deducted for matching losses and the higher operation temperature in orbit. For EOM power, the actual power delivered will be some 40% less. All values just mentioned relate to a 2 x 4-cm^2 cell of 200-μ thickness and a 100-μ cover glass.

The ULP array application range lies within a 2 x 2 panel array with 1-kW output to a 2 x 15 panel array with some 10-kW power output. A smaller version of the ULP presently is under study for such cases where the 1 x 3-m size is difficult to accommodate.

3. ULP Development Status

3.1 Design and Analysis

Deployed Configuration. The solar array consists of two identical deployable wings, each with a number of identical panels and an interconnecting frame structure (yoke) mounted to the spacecraft which allows sun tracking of the array and which avoids shadowing of the solar cells. The panel construction consists of a rigid carbon fibre composite (CFC) supporting framework to support a pretensioned flexible substrate on which the solar cells are bonded. The substrate is pretensioned in one direction only, parallel to the 3-m-long edge of the panel. The value of pretension is selected with regard to the natural frequencies of the substrate and in order to keep vibration amplitudes within the panel frame during launch. The rectangular panel frame is subdivided by thin CFC rods, which

Fig. 5 ULP on a candidate high-power spacecraft.

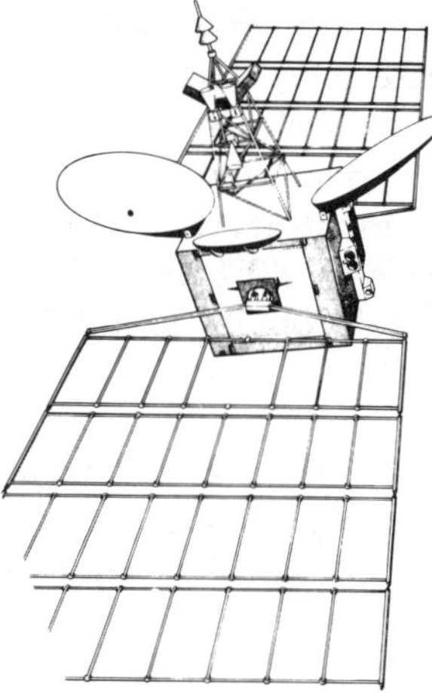

Fig. 6 ULP on an Ariane platform for application satellites (ARPAS).

clamp the substrate. Thereby vibration nodes are forced to the substrate and limit its amplitudes additionally. The 1-m-long beams of each panel form the main beams of the deployed array. They are responsible for its stiffness. Panels and yoke are interhinged rigidly, by form-closed latched deployment units having the same stiffness as the main panel beams; thus, the wing has an equal stiffness along its whole length.

The INTELSAT-ULP test array geometry is shown in Fig. 8. The first panel built and tested can be seen in Fig. 9.

Stowed Configuration. Panels and yoke are zigzag folded without spacing to a very compact package, which is pressed to the spacecraft sidewall at six points. Panels and yoke touch each

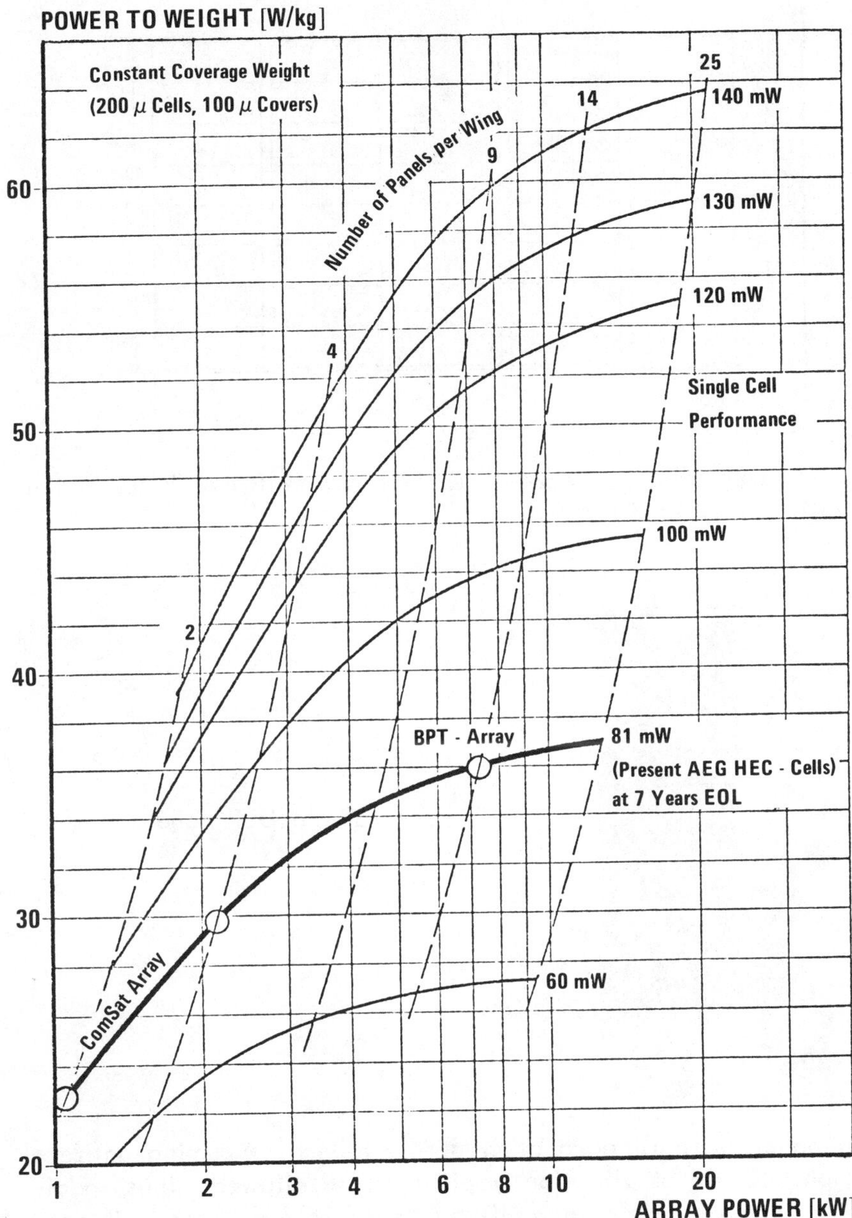

Fig. 7 ULP performance range.

other along their framework; the preloaded hold-down mechanism prevents relative vibration of the single-panel frames. Since the substrate with solar cells is much thinner than the frame (substrate 0.5 mm, frame 25 mm), there is little risk that the cells of ad-

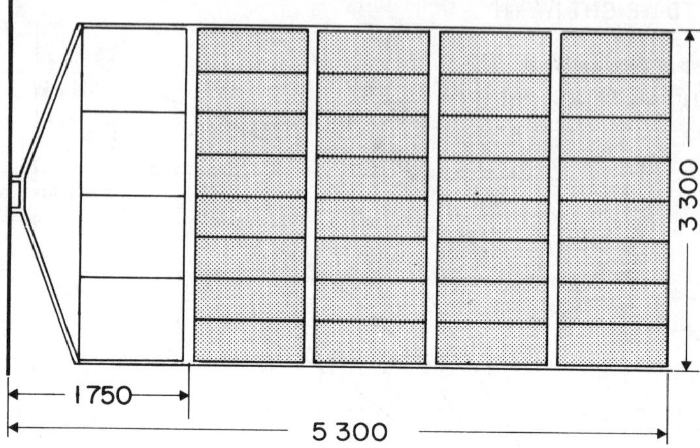

Fig. 8 ULP solar array for INTELSAT with 2.8 kW (BOL) and a specific power of 40 W/kg.

Fig. 9 Single ULP panel.

jacent panels will hit each other during launch. Assuming antiface vibration, 12 mm of vibration amplitudes are allowed. Thus, additional covering of the solar cells by layers of foam, etc., is not required for the array stowage.

<u>Mechanisms.</u> During launch, each wing package is held down to the spacecraft on six points by thin metal bands. After release by one central pyrotechnique, the hold-down bands roll up to spirals and remain on the outermost panel. Subsequently the

array deploys, driven by pretensioned spiral springs that are mounted to each interpanel hinge. The deployment is controlled by MBB's so-called closed cable loop (CCL), which enforces simultaneous movement of all panels. Form-closed latching of the hinges guarantees a homogeneous rigidity of the whole wing.

Power Analysis. The array power is produced by an AEG 10-Ω-cm, n-on-p silicon high-efficiency solar cell of 200-μ thickness and 2 x 4 cm dimensions, with a 100-μ-thick cerium-stabilized microsheet cover slide. This cell features a shallow diffused junction and multifinger grid configuration and is characterized by a high EOL performance of 95.7 mW at 25°C after 2×10^{15} meV electron particle radiation equivalent doses.

The cells are interconnected in series and parallel by automated parallel gap resistance welding using an advanced interconnect design of silver-plated molybdenum. The current program uses a cell layout with 144 cells in series and 25 in parallel per panel (total 3600 cells). For a 7-year mission in geostationary orbit, this layout yields a worst-case, EOL, equinox (t = 58°C) power per panel of 268 W, taking into account particle degradation of the cells as well as losses due to micrometeorites, random cell failure, and cell mismatch at EOL. At the EOL operation voltage, the two outer panels of both stowed panel packs provide a transfer orbit power of 218 W spin-average at 25°C and sun normal to spin axis.

Thermal Design. A 43 node model had been established in order to predict the temperature along the CFC bars, the hinge, and on the blanket. It could be shown that no critical temperatures exist. During the eclipse phase, the minimum temperature is -184°C. Lower temperatures are not expected, since the infrared radiation of the Earth provides sufficient heat input. A variation of the heat capacity of the blanket was shown to have no effect on this minimum temperature.

Structural Analysis. The Atlas-Centaur levels are lateral 1 g and axial 1.5 g over the complete frequency range from 5 to 200 Hz. An amplification of the launcher level in the area of spacecraft natural frequencies is taken into account (Fig. 10). The pretension of the solar cell blanket is selected such that 1) natural frequencies of blanket and spacecraft are separated; 2) blanket vibration amplitudes do not exceed panel thickness; and 3) thermal shrinking (eclipse period) does not result in frame overloading.

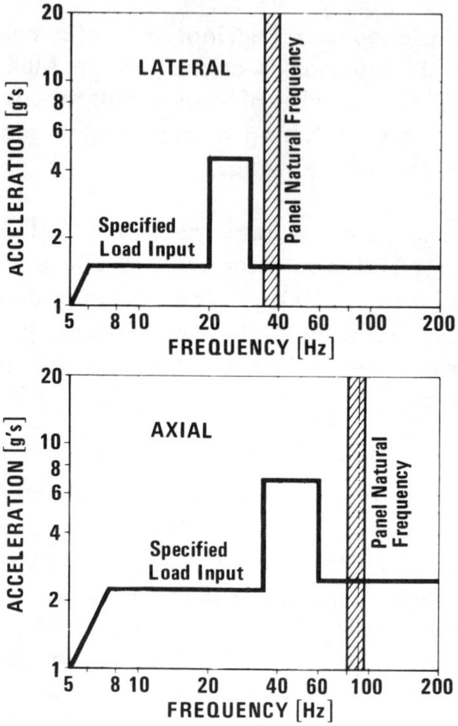

Fig. 10 Decoupling of resonance frequencies during launch.

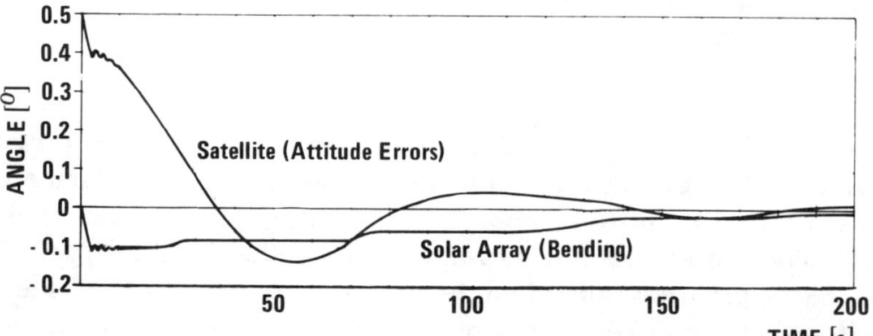

Fig. 11 Compatibility of ULP with AOCS.

The stiffness of the deployed wings has been calculated using a multinode computer program under the following assumptions: 1) rigid attachment of the array to the spacecraft (cantilever), and 2) mass of the cell blanket included in the CFC frame.

The first natural frequencies are as follows: bending, 0.048 Hz; torsion, 0.15 Hz; and lateral displacement, 0.36 Hz. Additional

Table 1 Mass breakdown of 2-and 7-kW class ULP arrays

Part description	Unit weight, kg	INTELSAT array (2.15 kW EOL) Units	Total weight kg	TVBS array (7.12 kW EOL) Units	Total weight kg
Panel frame	1.85	10	18.5	30	55.5
Solar cell blanket	3.75	8	30.0	28	105.0
Yoke	1.85	2	3.7	2	3.7
Yoke deployment unit	0.55	2	1.1	2	1.1
Panel deployment unit	0.11	20	2.0	60	6.6
Closed cable loop	0.02	10	0.2	30	0.6
Stowage and release mechanism	2.9/3.2	2	5.8	2	6.4
Cabling	0.8/5.7	2	1.6	2	11.4
BAPTA including electronics	3.50	2	7.0	2	7.0
Total weight			69.9		197.3
Power/weight ratio			30.8 W/kg		36.1 W/kg

investigations taking into account the effects of possible gaps at the interpanel hinges showed that especially the bending frequency cannot be expressed as a fixed value. The dynamic behavior depends on the excitation level. For free vibration, an eigenfrequency will not exist. With increasing excitation, the behavior is approaching the idealized case quoted previously. However, the feasibility of operation of such a system in combination with the attitude control of the spacecraft has, in principle, been demonstrated by computer analyses (Fig. 11). Compatibility of the dynamic behavior of large solar arrays with the attitude control is a dominant requirement. A frequency specification can be understood only as a simplifying requirement.

Deployment Analysis. Because of the closed cable loop, the wings will deploy as one-degree-of-freedom systems. The pretension of the deployment units is adapted to the bending torques of the electrical panel connections. To obtain the lowest bending torques, specially developed flat cable spirals are used. The main deployment data of the large 7-kW array are duration, 40 - 60 sec; final angular velocity, 45 - 90 deg/sec; and latch-up shock, 50 - 90 Nm.

Interfaces to Spacecraft. A main interface problem is the compatibility of the dynamic behavior of the deployed array with the spacecraft attitude control system. The required attitude accuracy of a "high-power satellite" is 0.1°. Control maneuvers of the spacecraft must not excite the array to amplitudes resulting in attitude errors that exceed the 0.1° limit.

MBB performed an extended analysis using a typical AOCS system with gimballed momentum wheel. Even under the worst-case condition of 0.5° initial error, the spacecraft attitude will be re-established within less than 3 min. During this maneuver, the solar array is bent only to one side and moves back without oscillation to its nominal position.

Mass Breakdown. Table 1 shows a mass breakdown of two typical ULP arrays with an electrical performance closed to both of the ULP design limits: the INTELSAT array with 2 kW, and a TVBS array with 7.1 kW (both after 7-y lifetime. As can be seen by Table 1, both arrays have been modular assembled, and, of course, the power weight ratio increases for the larger array type.

3.2 Current Development

The flexible blanket is being developed by AEG-Telefunken. The blanket substrate consists of a carbon fibre reinforced kapton foil. The carbon fibre cloth is bonded to the blanket rear side and since the fibres are electrically conducting the complete blanket rear side may be grounded to the spacecraft in order to prevent charge build up under magnetic substorm conditions. The carbon fibre layer is capable of handling 10 mA/cm^2 without significant voltage rise.

3.3 Tests

The main tests are listed in Table 2. To investigate the stability of the carbon fiber structure under space environment, especially Young's modulus and strength, a series of irradiation tests has been performed with CFRP samples. To save time, accelerated tests with

Table 2 Main ULP test activities

Test	Sample	Test conditions	Date of completion
Stability of CFC against space radiation First test program Second test program	CFC bars	Aequivalent to 10 yr life in synchronous orbit	 1974 May 1977
Thermal bending of ULP structure	Main CFC beams of ULP panel	Temperature gradient front/backside up to 40°C	1974
Thermomechanical behavior of blanket	Blanket with life cells	- 170°C up to + 100°C	1976
Thermal cycling First program Second program Third program	Panel with reduced size of ca. 20 x 30 cm^2	- 175°C up to + 70°C; 500 cycles 3000 cycles	 1975 April 1976 Nov. 1976 Dec. 1976
Launch vibration Sinusoidal Random	1 original panel (1 x 3 m^2)	 See Fig. 12 See Fig. 13	1975
Launch vibration Sinusoidal Acoustic noise	INTELSAT array package	 See Fig. 12 Atlas-Centaur noise	 July 1977 July 1977
Deployment	INTELSAT array	Thermal vacuum	Aug. 1977
Thermal cycling	1 original panel (1 x 3 m^2)	- 175°C up to + 70°C 25 cycles	June 1977

corresponding higher irradiation intensities up to a factor of 10 have been used. Partial degradations, as well as the following improvements of the mechanical material properties, have been observed. The tests showed that accelerated tests might be too unrealistic and too severe. Therefore, a second test program was started which will be completed in May 1977.

The high advantage of the carbon fiber structure "minimum thermal bending" has been demonstrated with original ULP beams of 2-m length. A temperature gradient up to 40°C between front and backside of the structure has been applied. (The calculated gradient for ULP in orbit is some 10°C.) The test results showed that the outermost tip of an 18-m-long ULP wing will deflect not more than 20 mm during operation in orbit. 3000 thermal cycle tests have been performed on sample panels and a thermal cycling test of a full size panel is presently subject of a proposal to the German National Space Agency.

As a milestone of the ULP development, the full-size panel in Figs. 9 and 12 was exposed to sinusoidal and random vibration tests at the end of 1975. The panel is covered with Al cell dummies and with some 700 life cells, which are placed on the expected critical areas. A special test adapter allows fixation of the test panel on its original hold-down points, but nevertheless the test conditions of the single panel are more severe than for a complete array package, which has higher stiffness and strength. The vibration behavior of structure and blanket has been observed via accelerometers and two high-speed cameras. The blanket pretension has been measured simultaneously during all tests. The applied test

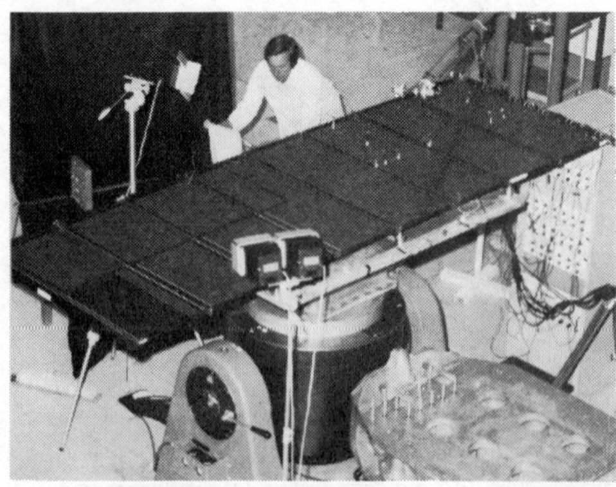

Fig. 12 Vibration test setup of single panel.

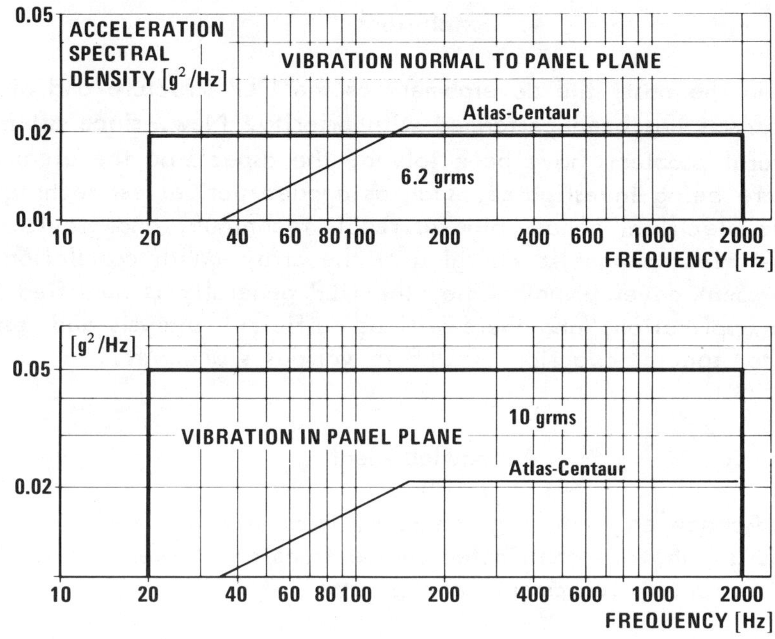

Fig. 13 Random vibration test levels.

levels are shown in Fig. 13. The random vibration levels have been increased as much as possible to demonstrate the application range of ULP for different launchers. The panel is capable of withstanding a much lower frequency range than required for the Atlas-Centaur, and the low-frequency ranges are the most critical ones. The panel did survive all tests successfully, although the summarized test duration is about 1/2 hr. The natural frequencies are closed to the predicated values; occurring loads and amplitudes were within allowable limits. Observed problems, such as the vibration of the free blanket edges have been solved by small design improvements during the test period. The mechanical compatibility of structure, blanket, and solar cells has been demonstrated, since no one cell was broken by the tests.

The successfull completion of these vibration tests was the go-ahead for further ULP development. Now a four-panel wing for Comsat is being fabricated, and the complete package will be exposed to sinusoidal vibration tests, acoustic noice tests, and deployment tests under thermal vacuum. Thermal cycling tests will be performed with small samples.

4. Conclusions

In the past, the development of the ULP structure and of the mechanisms has been performed with priority. Now, since all major structural problems have been solved, the aspects on the electrical side are being investigated, such as nondestructive test technique of the electrical bonds, blanket repair technique, and provisions against electromagnetic charging of the array. With completion of the present development phase, the ULP generally is qualified for space application, and there will be sufficient analysis and test data for special adaption of ULP to various spacecraft.

Acknowledgment

Acknowledgment is given to R. Buhs of AEG-Telefunken, Hamburg, who has contributed the sections on "Power Analysis" and "Current Development" of this paper.

DEVELOPMENT OF SPACECRAFT POWER SYSTEMS USING NICKEL-HYDROGEN BATTERY CELLS

R. E. Patterson* and W. Luft+
TRW Defense and Space Systems Group, Redondo Beach, Calif.
and
J. D. Dunlop≠
Communication Satellite Corporation, Clarksburg, Md.

Abstract

A 10-cell nickel-hydrogen battery system was designed, fabricated, and tested. The battery system contains cylindrical cells, 3.50 in. diam, with a maximum capacity of 50 A-h. The design objective was 50 W-h/kg at 100% depth-of-discharge. Basically cells are all girth-supported by an electrically insulated mounting bracket. The cells feed through the mounting panel such that one end of each cell extends above the panel, whereas the other end extends below. Heat transfer is by a combination of radiation directly from the vessel wall and conduction through the girth band to the panel from which radiative transfer also occurs, both to a secondary radiator. Two-cell test samples successfully survived vibration and pressure fatigue tests. Thermal performance was as predicted. Energy density for the 35-A-h design was measured to be 39.2 W-h/kg and for the 50-A-h design was projected to be 51.4 W-h/kg at 100% depth-of-discharge.

I. Introduction

The objective of the work described was to develop a flight-qualified nickel-hydrogen battery system that can reduce the total energy storage mass in a typical communication satellite by 50% or more and can increase mission life and reli-

---

Presented as Paper 76-288 at the AIAA/CASI 6th Communications Satellite Systems Conference, April 5-8, 1976, Montreal, Canada.

*Now Director of Special Projects, Intermedics, Inc., Freeport, Texas.
+Assistant Manager, Power Sources Engineering Department.
≠Manager, Electrophysical Devices Department.

ability. The nickel-cadmium battery system used in conventional communication satellites represents approximately 15 to 20% of the total satellite dry weight exclusive of payload, or typically about 90 kg for a satellite having a 1-kW power requirement. A nickel-hydrogen battery system for the same size satellite would have a mass of 45 kg or less and also would have greater cycle life expectancy.

This paper describes the design, fabrication, and testing of a 10-cell 50-A-h nickel-hydrogen battery system. The cells used in the battery are cylindrical in configuration and were designed at Comsat and built by Eagle-Picher Industries. The battery was designed for a maximum capacity of 50 A-h. Because of cell availability, 35-A-h cells were used to fabricate the 10-cell battery. No modification of the battery mechanical design was required to accommodate 35-A-h cells rather than

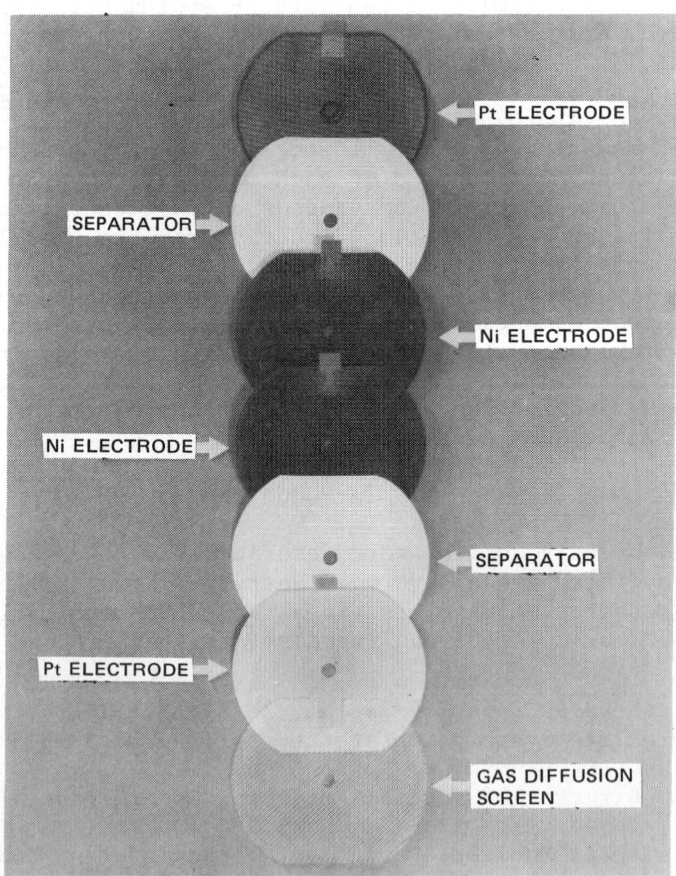

Fig. 1 Elements of a nickel-hydrogen electrode stack.

50 A-h cells, since both types have the same external diameter. Because 35-A-h cells generate less heat then 50-A-h cells, the battery effective radiator area was reduced for the thermal vacuum test so that the thermal behavior of the battery with 35-A-h cells would be similar in response to the unmodified design with 50-A-h cells.

II. Nickel-Hydrogen Cell Design

The nickel-hydrogen cells described herein use a design developed at Comsat Laboratories. In its construction, these cells resemble the Ni-Cd cell, except that the cadmium electrode is replaced by a platinum electrode, which consumes hydrogen gas on discharge and evolves it on charge. The components used to make up the electrode stack are shown in Fig. 1. The two positive plates, positioned back to back, are electrochemically impregnated nickel electrodes. These electrodes were fabricated by Eagle-Picher using the Bell Laboratories process.[1] The hydrogen electrode structure consists of Teflon-bonded platinum black supported within a thin, fine-mesh nickel screen with a Teflon backing. A plastic gas diffusion screen facilitates hydrogen diffusion to the back of this platinum electrode. Reconstituted fuel cell grade asbestos is the separator material used because it is more stable than nylon in the aqueous KOH electrolyte. The electrode-electrolyte-separator stack is surrounded by an atmosphere of hydrogen under pressure.

Fig. 2 Components of a 35-A-h nickel-hydrogen cell.

Components of the nickel-hydrogen cell are shown in Fig. 2, with the stack assembled before the pressure vessel shells. The rated capacity for this cell is 35 A-h. During normal operation, the hydrogen pressure within the cell varies from 600 psi in the fully charged state to 100 psi in the discharged state to 1.0 V. The pressure vessel is fabricated from Inconel-718 material, 18 mils thick. Plastic compression seals are used at both ends to insulate the positive and negative feed-through terminals from the pressure vessel.

Rated capacity for this cell can be increased from 35 to 50 A-h by heavier loading of the positive electrodes. The pressure vessel would remain the same; however, the hydrogen pressure within the cell would vary from 815 psi in the fully charged state to 100 psi in the fully discharged state. Since the same pressure vessel design can be used for both 35- and 50-A-h cell capacities, the same battery design for 50-A-h cells also can be used for 35-A-h cells.

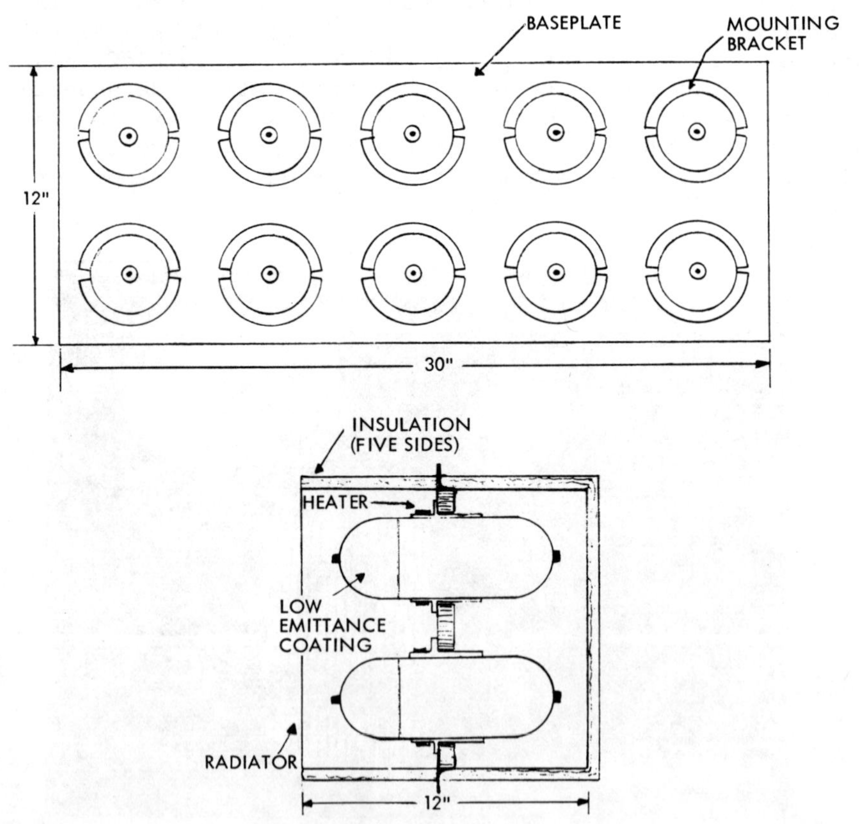

Fig. 3 Nickel-hydrogen battery design.

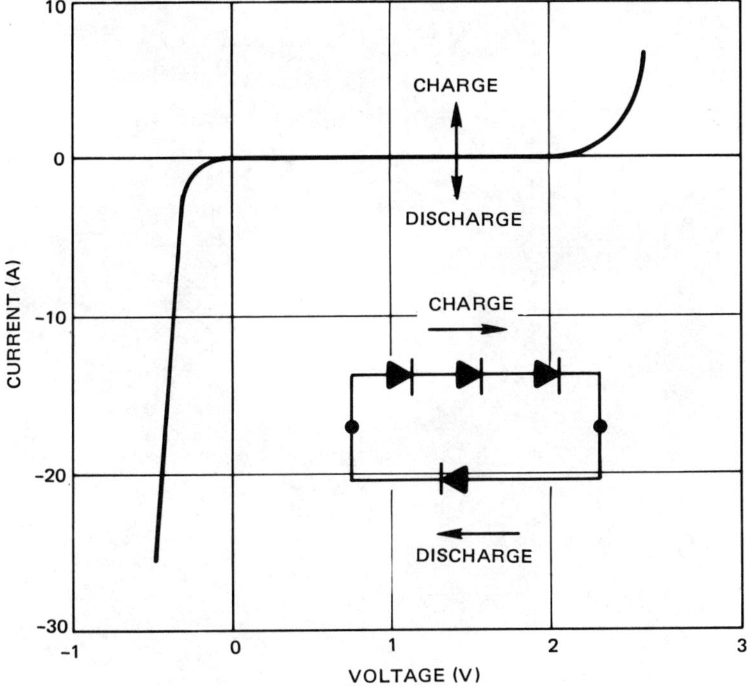

Fig. 4 Nickel-hydrogen cell bypass circuit and characteristics.

III. Battery System Design

In assembling groups of cells into a battery for integration into a spacecraft, the following must be considered: 1) minimum overall weight, 2) sufficient support and protection of cells in shock and vibration environments, 3) heat transfer from cells, 4) spin-stabilized or body-stabilized spacecraft, 5) required spacecraft platform area, and 6) electrical insulation of cell vessels from their mounting. A fundamental consideration in the battery design is the means of heat dissipation from the cells when mounted in the spacecraft. Radiation to space is the primary means of maintaining the battery cell temperature within desired limits. This radiation may be direct, where the surface in conductive contact with the heat-producing cell is exposed to space, as from an equipment platform to which the cells are coupled conductively, or the radiation may be indirect, where the conductive surface radiates to a surface, which, in turn, radiates to space. Although the latter requires about twice as much area to dissipate a given quantity of energy compared to the former, it is easier to maintain the battery within a reasonably small tem-

Fig. 5 Cell-and-bracket assembly attached to mounting panel.

perature range during periodic direct sun impingement caused by seasonal variation in the sun angle.

For a body-stabilized synchronous spacecraft, the battery should be mounted to a panel that receives a minimum of solar radiation (i.e., north or south panel). Here a cover with proper thermal coatings is required to which the battery radiates and which, in turn, radiates to deep space (indirect radiation), to avoid any direct sun impingement on the cells and the resulting unacceptably wide temperature excursions. For a spin-stabilized spacecraft, the battery would be mounted to an internal panel, with the heat-dissipating elements facing either forward or aft to their best view of a colder body for radiative transfer. A protective cover over the cells may be desired for safety during handling and installation of the battery but would not be desirable for flight unless there was a possibility of direct sun impingement on the cells.

Interface Definition

The battery consists of 10 series-connected 50-A-h cells as shown in Fig. 3. The cells are all girth-supported by an electrically insulated clamp. Heat transfer is by a combination of radiation directly from the vessel wall and conduction through the girth band to the support plate, from which radiative transfer also occurs, both to the secondary radiator. The cells are arranged such that their internal stack is on the outboard side of the mounting plate; thus the path from the heat-generating elements to the radiation surface is minimized.

The spacecraft installation concept for the preferred battery design is to prepare a cutout in the spacecraft equipment platform consistent with the battery mounting panel. The mounting panel, which also supports the individual cells, can be load-bearing if required to reinforce the platform cutout; this would require accurately located platform mounting hardware.

Electrical Design

A key design problem in using any battery in an unattended, rechargeable, energy storage application is that of proper recharge control and protection against loss of the battery resulting from defective cells. Performance characteristics determined in work reported by Giner and Dunlop,[2] Patterson and Sparks,[3] Harsch et al.,[4] and Levy et al.[5] were analyzed, and the following observations were made:

1) Nickel-hydrogen batteries require protection electronics to prevent loss of a battery resulting from open-circuited cells.

2) Protection against overcharge and overdischarge with bypass electronics is not required.

3) Absolute cell pressure is not a straightforward indicator of cell state of charge, although the time differential of the pressure during charge is a reliable indicator of maximum state of charge.

4) Recharge ratio is an effective charge control method for maintaining nickel-hydrogen cell energy balance for synchronous orbit duty cycles.

Requirements for electronic bypass circuitry are that it bypass an open-circuited cell during all operational modes (charge or discharge during eclipse seasons and trickle charge during noneclipse seasons) and that it must not interfere in any way with the operation of an unfailed cell. An example of the second requirement is that the bypass circuit must allow a reversed (but not open-circuited) cell during discharge to be charged during the subsequent charge period (i.e., reversed cells are not to be bypassed permanently).

The bypass circuit selected for cell protection for the nickel-hydrogen battery is shown in Fig. 4. Also shown in the figure are the current voltage characteristics of the circuit which are discussed in Sec. V. Protection in the forward (charge) direction is provided by three silicon diodes in series, whereas protection in the reverse (discharge) direction is provided by one Schottky barrier diode.

Thermal Design

Thermal control of the battery involves a number of interfaces. First, design of the cell must be such that heat dissipated in the cell stack is transferred to the containing vessel wall with a minimum of temperature difference. Dissipated heat then must be transferred to an appropriate radiator surface for ultimate rejection to space while maintaining the cell within allowable temperature limits. Thermal design of the battery system was based on a 24-hr synchronous orbit with a maximum discharge period of 1.2 hr. A body-stabilized spacecraft was assumed with a north- or south-facing surface available for the battery system, which has 10 50-A-h cells.

Radiator Area. Heat-rejection capability of the radiator is dependent on the effective sink temperature, which the radiator sees. The effective sink temperature is a convenient way to characterize the radiator thermal environment. It is calculated by summing all external heat inputs to the radiator and converting them to a corresponding radiation boundary temperature. For a body-stabilized spacecraft in synchronous orbit, the sink temperature will vary between about -65°C to a value approaching -273°C for a north- or south-facing panel with second surface mirrors. The minimum sink temperature value depends on the radiator view to appendages, such as the solar array, whereas the upper sink temperature represents, in addition, solar input at an angle of 23.5° from horizontal.

Assuming a 1.3-W trickle charge dissipation, a 25°C baseplate temperature, and a -65°C sink temperature, the required radiator area for a 10-cell battery is 0.096 m^2.

Mounting Bracket Design. To aid in design of the cell mounting bracket, an analytical thermal model was constructed which included the cell stack, mounting bracket, and baseplate. The model was solved for various mounting bracket thicknesses and hydrogen gap widths from the electrode stack edge to the pressure vessel inner diameter. A 10-W steady-state dissipa-

Fig. 6 Ten-cell nickel-hydrogen battery.

tion rate for the stack was assumed. Actual heat rates may approach this value for significant time periods in application, depending on the depth of discharge and charge control method. Results of this analysis indicated the following:

1) The hydrogen gap width should be 0.040 in. or less to give a 5.5°C or lower internal thermal gradient in the cell.

2) There is little heat-transfer advantage in making the mounting bracket thicker than about 0.025 in.

Thermal Control. Fig. 3 shows the battery thermal design. The mounting plate is a honeycomb panel with aluminum facesheets and core; inserts are provided to accept cell mounts. The outboard facesheet is thicker to distribute heat in the panel. A silvered Teflon radiator panel is utilized, with a black paint and vacuum-deposited aluminum coating pattern provided on the inward surface to give the proper thermal coupling to space.

The outward-facing domed ends of the cells have a low-emittance coating to reduce internal cell gradients. All other internal surfaces are painted black to maximize radiation coupling. Heaters are provided on the back side of the mounting panel to prevent the cell temperature from dropping below 0°C during active charge periods where internal dissipation is negative or small or during periods where the battery is inoperative.

Mechanical Design

The cell mounting bracket has two functions: to support the cell on the platform under launch loads, and to transfer the heat generated in the cell to the mounting platform. The launch vibration and shock requirements are estimated to be approximately 50 Earth accelerations, giving a total load of 150 lb in any direction for a nominal cell mass of 3 lb. Thus, the applied structural loads are relatively small, and the major design requirement is the transfer of the internally generated heat. The mounting bracket covers the 3-in. length of the cell stack and has an attachment flange at its center. The total area of the sleeve is 33 in.^{2}, resulting in an average shear load of only 4.5 psi due to the maximum applied load. Thus, the adhesive bond between the bracket and the cell must support only an insignificant external load. The critical strength requirement is the ability of the bond to sustain the shear stress that develops as a result of the growth of the pressure vessel under pressure. A two-part flexible silicone rubber adhesive system, which cures uniformly over the required large area of contact having a minimum bond thickness of 10 mils, is used. The low modulus adhesive minimizes the effect of the mount on the stresses in the pressure vessel, reduces the shear stress peaks at the edges of the bonded area, and provides a more uniform distribution of the shear stress. These advantages are gained at the expense of increased shear deformation of the adhesive. The honeycomb mounting panel

Fig. 7 Two-cell vibration test setup.

design was found by stress analysis to be more adequate for meeting mechanical launch load requirements.

IV. Battery Fabrication

Battery fabrication consisted of 1) fabricating cell mounting brackets and the honeycomb mounting panel; 2) attaching the brackets to the cells and then attaching the cell-and-bracket assemblies to the mounting panel; 3) attaching the wiring harness, which includes power cables, voltage sensing cables, bypass electronic cables, and heater cables; 4) installing heaters to the bottom of the battery; and 5) assembling the bypass circuits. Fabrication of thermal control hardware is included as part of the thermal vacuum test discussion.

Mechanical Hardware Fabrication

The cell mounting brackets were machined from aluminum bar stock. Following machining, all outer surfaces, with the exception of the bottom surface of the mounting flange, were painted flat black for high thermal emittance. The honeycomb mounting panel was fabricated by bonding aluminum facesheets to both sides of a 0.5-in.-thick sheet of aluminum honeycomb core. Ten cutouts were made, each surrounded by eight bonded threaded inserts for attachment to the 10 cell-and-bracket assemblies. Both sides of the panel were painted flat black for high thermal emittance.

Mechanical Assembly

Mechanical assembly consisted of bonding the mounting brackets to the cells and then attaching the cell-and-bracket assemblies to the mounting panel. The brackets were attached to the cells using a silicone rubber adhesive. Three layers of fiberglass scrim were used for insulation. A clamping tool was used to hold the cell-and-bracket assembly together while the silicone rubber adhesive cured.

Each cell-and-bracket assembly was attached to the mounting panel using eight screws. A thin layer of silicone adhesive was used between the panel and the cell bracket attachment flange to insure thermal contact. Figure 5 shows a cell-and-bracket assembly attached to the mounting panel.

Electrical Assembly

Following mechanical assembly, the wiring harness was attached to the battery. The harness included power cables, voltage sensing cables, bypass electronics cables, and heater cables. Cables were soldered to cell terminal tabs, which, in turn, were attached mechanically to the threaded cell terminals with two bolts. The wiring harness terminates at four connectors attached to the mounting panel. Figure 6 shows the 10-cell battery exclusive of cell bypass circuits and thermal control hardware. The cell bypass circuits were attached mechanically to an aluminum heat sink. Circuit element electrical interconnectors were soldered directly to the diode terminals.

Thermal Control Hardware

Details of the radiator are discussed as part of the thermal vacuum test description. The eight battery heaters were attached to the bottom of the battery mounting panel with silicone adhesive.

V. Battery System Performance

Tests were performed on components, cells, two-cell test samples, and the 10-cell battery in order to evaluate critical elements of battery system performance.

Vibration Test

A vibration test was performed on a two-cell test specimen in order to demonstrate that the cell design and cell mounting design are capable of meeting flight acceptance vibration specifications. Two 35-A-h cylindrical nickel-hydrogen cells were bonded to mounting brackets. The cell-and-bracket assemblies then were bonded and bolted to a modified mounting panel. The modified panel represents one-fifth of the full mounting panel for 10 cells. Since the fundamental frequency of the modified panel is close to that of the full panel, it was expected that the vibration response would be similar to that

Table 1 Vibration spectrum

Frequency, Hz	Power spectral density
20-100	Increase at 6 dB/octave to 0.1 G^2/Hz
100-1000	0.1 G^2/Hz
1000-2000	Decrease at 6 dB/octave

of the full panel. Figure 7 shows the two-cell test specimen mounted to the vibration test fixture.

The test consisted of a one Earth acceleration (G) sinusoidal sweep from 10 to 2000 Hz at 2 octaves/min, followed by a 5-min random exposure at 12 G rms, with the vibration spectrum as shown in Table 1. These exposures were performed in the direction of the long panel axis (x) and the pressure vessel axis perpendicular to the mounting panel surface (y). The amplification factors measured at the top of the cell were 15 and 9.5 in the x and y directions, respectively, for sinusoidal excitation in the y direction. The fundamental frequency was 160 Hz.

There was no apparent physical damage to the unit following the sinusoidal and random vibration test exposures. Also, there was no apparent electrical degradation determined by electrical capacity measurements performed prior to and following the vibration test.

Fatigue Test

A pressure cycle fatigue test was performed on two pressure vessels bonded to thermal mounting brackets, which, in turn, were bonded and bolted to a mounting panel. The purpose of the test was to determine the fatigue life of the pressure vessels, as well as the fatigue life of the thermal mounting brackets and the bond between the pressure vessel and the mounting bracket.

The pressure vessels were cycled between 50 and 650 psi using oil as the pressurizing fluid. Both pressure vessels

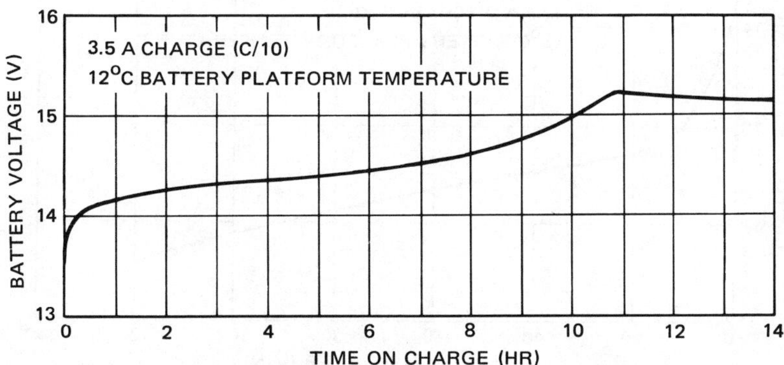

Fig. 8 Battery voltage during charge.

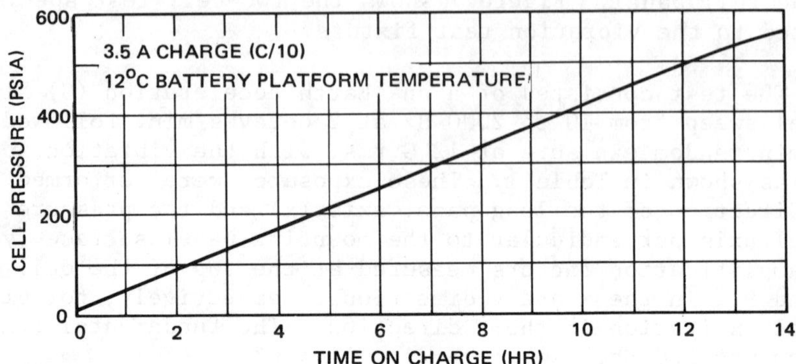

Fig. 9 Cell pressure during charge.

failed at the electrical terminal feed-through collar weld following successful completion of 26,800 and 120,000 cycles. Both failures were pinhole leaks. There was no apparent damage to the mounting brackets and no apparent delamination of the pressure vessel to mounting bracket bonds.

Electrical Performance

The 10-cell 35-A-h nickel-hydrogen battery was given an initial charge/discharge cycle for the purpose of determining battery capacity and current, voltage, and pressure characteristics on charge and discharge. All bypass circuits were tested to determine current-voltage characteristics for all anticipated modes of operation.

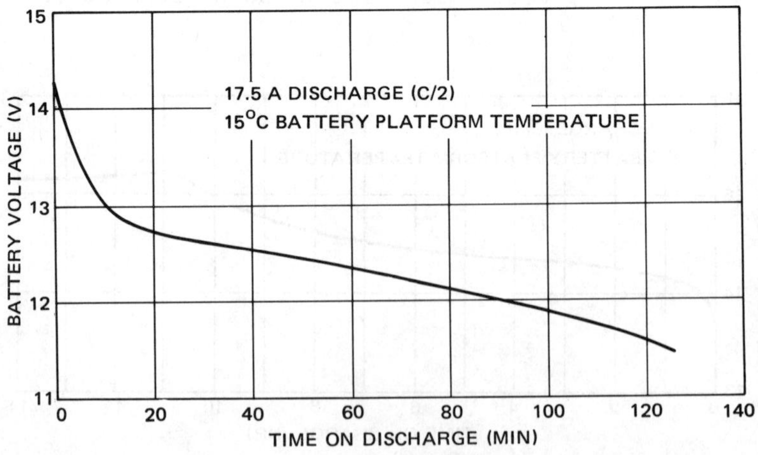

Fig. 10 Battery voltage during discharge.

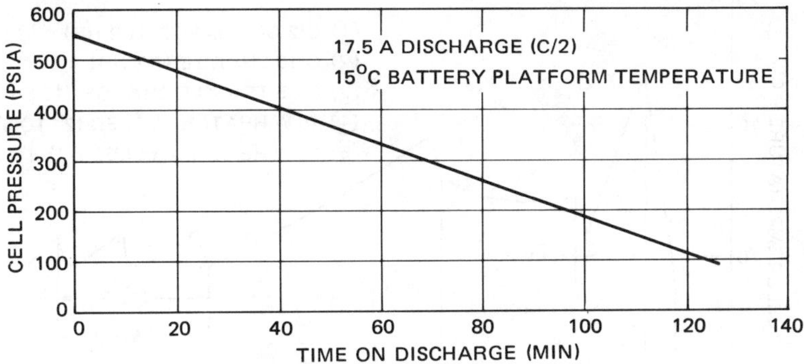

Fig. 11 Cell pressure during discharge.

Fig. 12 Nickel-hydrogen battery in spacecraft simulation chamber (radiator cover removed).

The 10-cell battery was placed on an aluminum heat sink, which contacted the bottom surface of the battery mounting platform. Cutouts were made in the heat sink to accommodate the protruding cells and wiring. The battery mounting platform was maintained between 8° and 10°C. The battery was charged at

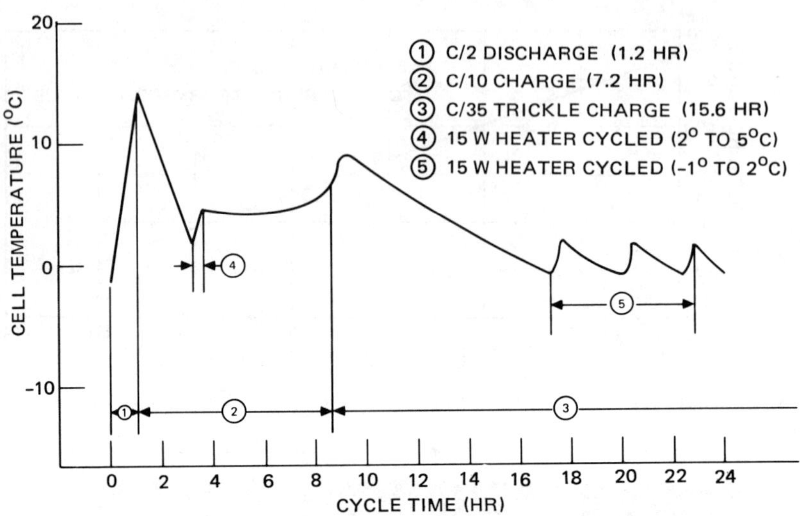

Fig. 13 Thermal vacuum test results during 60% DOD synchronous orbit cycle.

3.5 A for 14 hr and then discharged at 17.5 A until a minimum cell voltage of 1.0 V was reached by any one of the 10 cells. Figures 8 and 9 present voltage and pressure data, respectively, for the charge. Figures 10 and 11 present voltage and pressure data, respectively, for the discharge. The discharge voltage curve was integrated to obtain battery capacity that was 460 W-h.

Current-voltage characteristics of one of the bypass circuits are presented in Fig. 4. Each of the 10 bypass circuits was found to have very similar characteristics. In the event of an open-circuit cell failure, current would flow through the bypass circuit for that cell. From Fig. 4, it can be seen that the voltage drop across the bypass circuit would be 2.4, 2.2, and 0.4 V for a full rate charge of 3.5 A, a trickle charge of 1.0 A, and a discharge (negative current) of 17.5 A, respectively. The corresponding heat dissipation rates in an operative bypass circuit for the preceding conditions would be 8.4, 2.2, and 7.0 W, respectively.

Thermal Vacuum Test

The purpose of the thermal vacuum test was to provide realistic thermal response data for simulated synchronous orbit conditions. Several charge-discharge cycles were run, and a thermal equilibrium test was performed.

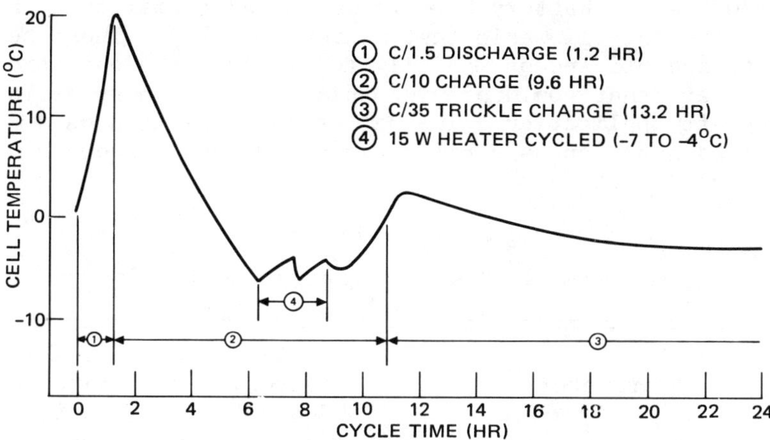

Fig. 14 Thermal vacuum test results during 80% DOD synchronous orbit cycle.

Test Setup. Several articles were fabricated to facilitate thermal-vacuum testing of the Ni-H_2 battery and are listed here: 1) an insulation and radiator support frame, 2) a chamber for spacecraft simulation, and 3) a secondary radiator. The support frame was designed to be lightweight and yet be able to support the battery. The insulation blanket consisted of 22 layers of aluminized mylar. The aluminum side of the material was positioned to face outward or away from the battery assembly. A chamber was fabricated to simulate a spacecraft environment. The insulated nickel-hydrogen battery assembly, after installation into the support frame, was placed inside the simulation chamber as shown in Fig. 12.

The thermal radiator was fabricated from 3-mil aluminized mylar. The aluminum side of the mylar was painted black. A pattern of aluminum tape was placed on both sides of the radiator to achieve the desired thermal properties. The effective radiator area was reduced in size in proportion to the relative reduction in heat generation in going from 50- to 35-A-h cells. Thermal analysis shows that the thermal behavior of the battery with 50-A-h cells and a full radiator would be very similar to the response obtained on the battery with 35-A-h cells and a reduced radiator. This radiator was placed over the opening of the spacecraft simulation chamber shown in Fig. 12.

Test Procedures. The battery test package was installed in a liquid-nitrogen-cooled vacuum chamber, with the radiator viewing the cold shroud. The test chamber heater was adjusted to keep this simulated spacecraft compartment temperature at

15° ±30°C. The battery heaters were used to maintain the battery above the allowable test minimum of -7°C. These heaters were turned off and on manually to simulate the operation of thermostats that would provide this function in an actual spacecraft application. The battery was operated on a 24-hr simulated synchronous orbit cycle at 60% and 80% depth-of-discharge.

Table 2 Mass and energy density for 35- and 50-A-h batteries

Parameter	35-A-h battery design	50-A-h battery design
Component mass, g		
Cells (10)	10,520	10,960
Mounting brackets (10)	520	520
Adhesive and insulation	107	107
Bracket screws (80)	98	98
Mounting platform	687.7	687.7
Wire, connectors, and heaters	275.9	275.9
Paint, solder, string, terminal board	62.0	62.0
Radiation cover	50	50
Cutout mass[a]	-600	-600
10-cell battery total mass, kg	11.720	12.161
Total battery energy, W-h	460	625
Energy density at 100% DOD, W-h/kg	39.2	51.4

[a]Mass removed from equipment platform.

Test Results and Discussion. Results of a 60% depth-of-discharge cycle are presented in Fig. 13. Note that maximum cell temperature of 15°C occurred at the end-of-discharge. The battery heater was used to keep the cell temperature above 2°C during the 3.5-A charge and above -1°C during 1-A trickle charge. When energized, the eight heaters provided a total constant heat dissipation of 15 W.

Results of an 80% depth-of-discharge cycle are presented in Fig. 14. A maximum cell temperature of 20°C occurred at the end-of-discharge. The steady-state battery temperature was about -2°C when the battery was charged fully and operating the trickle charge mode.

Mass and Energy Density

Detailed mass data for the 10-cell battery are presented in Table 2. Data for the 35- and 50-A-h batteries represent measurements of actual hardware, with one exception: cell mass for the 50-A-h cells was estimated in accordance with the rationale discussed previously. Since the battery is designed for 50-A-h cells, wiring and packaging masses are greater than actually required for 35-A-h cells.

Table 2 also presents energy density data for both the 35- and 50-A-h battery designs. Note that at 100% depth-of-discharge an energy density of 51.4 W-h/kg (exclusive of charge control and bypass electronics) can be achieved for the 50-A-h design.

VI. Conclusions

The battery system design discussed in this paper is well suited for synchronous orbit applications. The battery mechanical design can survive typical acceptance level vibration requirements. The thermal design is adequate for cell capacities of up to 50-A-h and depths-of-discharge of up to 80%. Energy densities at 100% depth-of-discharge are 39.2 and 51.4 W-h/kg for battery capacities of 35 and 50 A-h, respectively.

Acknowledgments

This development was supported by International Telecommunications Satellite Organization (Intelsat). Views expressed in this paper are not necessarily those of Intelsat. Work reported in this paper was supported by many individuals. Key contributors include R. Griebel (electrical design, assembly, and test), A. Kaplan (stress analysis), V. Mikol (material analysis), V. Reineking (thermal testing), E. Luedke (thermal analysis), D. Wanous (thermal analysis), J. Stockel (cell design), G. Van Ommering (cell design), and E. Burchman (material testing).

References

[1]Beauchamp, R. L., U.S. Patent 3,653,967, April 1972.

[2]Giner, J. and Dunlop, J. D., "The Sealed Nickel-Hydrogen Secondary Cell," Journal of the Electrochemical Society, Vol. 122, Jan. 1975, pp. 4-11.

[3]Patterson, R. E. and Sparks, R. H., "Nickel-Hydrogen Battery System Development," Tenth Intersociety Engineering Conference, Aug. 1975, Newark, Del.

[4]Gandel, M. G., Chang, R., and Harsch, W. C., "Nickel-Hydrogen Battery Development for Synchronous Satellites," Ninth Intersociety Energy Conversion Engineering Conference, Aug. 1974, Palo Alto, Calif.

[5]Levy, E., Jr., Rogers, H. H., McGrath, R. J., Wittman, A., "Nickel-Hydrogen Energy Storage for Satellites," AFAPL-TR-74-111, Nov. 1974, Air Force Applied Physics Lab.

INTELSAT SATELLITE ONBOARD PROPULSION SYSTEMS, PAST AND FUTURE

J. R. Owens*
Comsat Laboratories, Clarksburg, Md.

Abstract

The onboard propulsion systems utilized on the Intelsat satellites all have been catalytic monopropellant systems that are compatible with spin-stabilized configurations. The propellant mass of these long-life reliable systems has increased from 5 kg on Early Bird to 158 kg on Intelsat IV-A. The total propulsion system weight including tanks, thrusters, and propellant lines has remained about 20% of the satellite weight. The propulsion system is, in fact, the heaviest subsystem on the Intelsat IV-A satellite, weighing even more than the communications subsystem. For Intelsat V and VI, propulsion systems with higher specific impulses may be used to save substantial amounts of weight, which can be converted into increased communications capacity. Bipropellants, electrically augmented hydrazine, and ion propulsion all deserve consideration. However, the value of weight saving technology after the Space Transportation System (shuttle-tug) becomes available is not understood clearly.

Introduction

On the Intelsat satellites the onboard propulsion systems have been used to adjust the initial orbit parameters and to control the orbit velocity, orbit inclination, satellite attitude, and satellite spin speed during the operational lifetime. All systems to date have been catalytic monopropellant systems operating in a simple blow-down mode. A monopropellant is a substance requiring no auxiliary liquid (oxidizer) to decompose and release its thermochemical energy. Instead, the decomposition is initiated by passing the pro-

---

Presented as Paper 76-289 at the AIAA/CASI 6th Communications Satellite Systems Conference, April 5-8, 1976, Montreal, Canada.

*Spacecraft Specialist with the Intelsat Executive Organ, Engineering Department, Washington D.C.

pellant through a catalyst bed that is part of the thrust chamber. A blow-down operation is one in which the propellant storage tanks are not repressurized over the life of the satellite. As the propellant is expended, the ullage volume increases, and the tank pressure decreases.

To predict the systems that may be used on future Intelsat satellites, it may be helpful to review the history of propulsion systems used on the past programs. The Intelsat satellites are part of a commercial operating communications system; thus, high reliability and long life are the prime requirements. They are not experimental or technology application satellites.

Propellant Systems on Intelsat I-IV-A

Intelsat I, Early Bird, was launched in April 1965 with an initial in-orbit weight of 39.5 kg. The satellite at launch contained 5 kg of 95% concentration hydrogen peroxide, with most of the propellant used for initial orbit corrections. The satellite was built by Hughes Aircraft Corporation and the propulsion system by the Walter Kidde Corporation.

In terms of mechanical considerations and layout, the Early Bird propulsion system is very typical of all of the Intelsat satellites to date, since all have been spin-stabilized configurations. The propulsion system is divided up into two half-systems. Each half-system has two tanks, one axial and one radial thrust chamber assembly (TCA). The TCA includes the propellant valve and the catalytic thruster. The radial thruster is fired in a pulsed mode for east-west stationkeeping (keeping the satellite at a fixed longitude) in synchronism with the satellite spin rate. The axial thrusters are located parallel to the satellite spin axis, but off the spin axis so that they have a moment arm about the center of gravity. They are fired in a pulsed mode for

Table 1 Intelsat satellite onboard propulsion system hardware

Satellite	Dry weight, kg	Propellant weight kg	No. of tanks	No. of thrusters	Initial orbit satellite weight, kg
Intelsat I	3.7	5.0	4	4	39.5
Intelsat II	7.0	9.6	4	4	86.3
Intelsat III	5.6	21.8	4	4	133.0
Intelsat IV	16.8	136.4	4	6	732.0
Intelsat IV-A	16.8	158.2	4	6	828.0

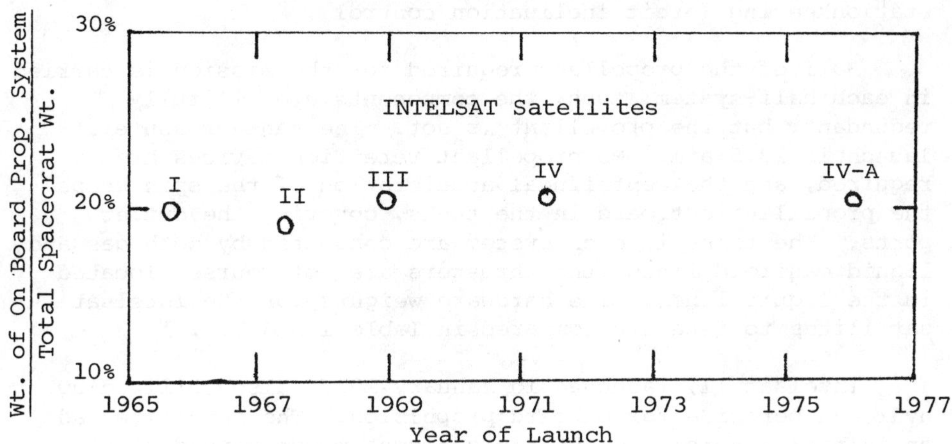

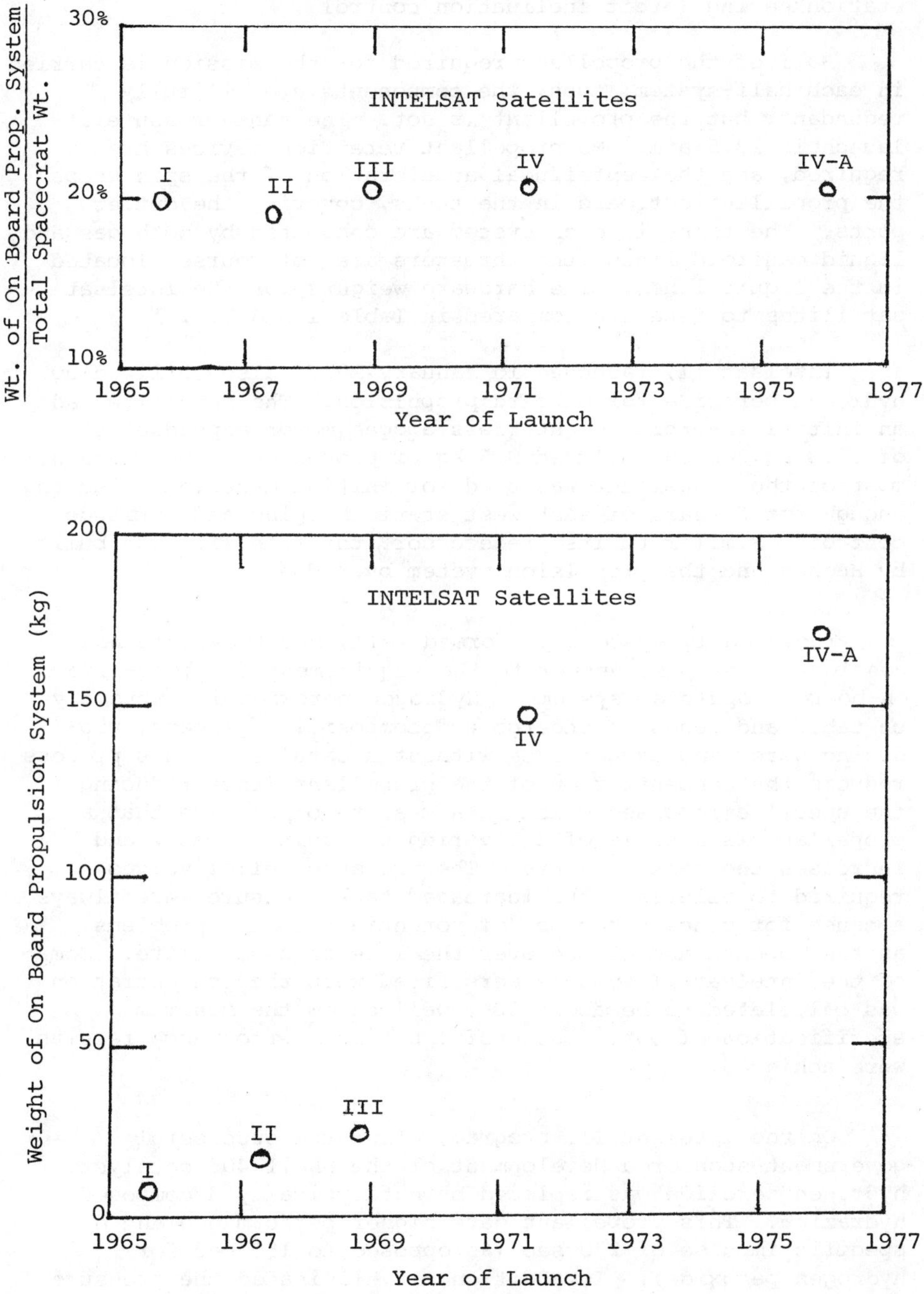

Fig. 1 Weight comparison.

attitude corrections and in a continuous mode for north-south stationkeeping (orbit inclination control).

Half of the propellant required for the mission is carried in each half-system; thus, the components are all fully redundant, but the propellant is not. The tank pressure at launch is 13.5 atm. No propellant retention devices are required, and the centrifugal acceleration of the spin keeps the propellant outboard in the tanks, covering the outlet ports. The tanks in each system are connected by both gas and liquid manifold lines; the thrusters are, of course, located in the liquid lines. The hardware weights for the Intelsat satellites to date are compared in Table 1 and Fig. 1.

Intelsat II, launched in January 1967, also utilized 90% hydrogen peroxide for onboard propulsion. The satellite had an initial in-orbit weight (less apogee motor expendables) of 86.3 kg, which included 9.5 kg of propellant. Once again, most of the propellant was used for initial maneuvers, leaving enough for 3 years of east-west stationkeeping and attitude control. Similar to its predecessor, the satellite was built by Hughes and the propulsion system by Kidde.

These early systems performed well, but they were not considered the best answer to the requirement for long-life on-board propulsion systems. Hydrogen peroxide is extremely unstable and tends to undergo a decomposition process, producing water and oxygen even without a catalyst. This process reduces the concentration of the propellant (thus reducing the specific impulse, which is a measure of impulse that a propellant is capable of delivering per unit of mass) and increases the tank pressure. The pressure relief valves required to alleviate the increased tank pressure were always a cause for concern because of potential leakage problems, as they opened and closed over the life of a satellite. Some of the Intelsat II systems were fired when the concentration was calculated to be about 50%, well below the minimum specification of 65%. Successful but low performance results were achieved.

On the Intelsat III program, after the successful government-sponsored development of the Shell 405 catalyst, hydrogen peroxide was replaced by catalytically decomposed hydrazine. This propellant gave higher performance and a specific impulse of 220 sec (as opposed to 150 sec for hydrogen peroxide). In addition, it eliminated the pressure relief valve requirements, since hydrazine is not as active as hydrogen peroxide. Thus, the system could be sealed, and truly long-life systems could be expected with the design and development of the proper TCA's. Indeed, the Intelsat III hydrazine system had no problems over the life of the satellite. It utilized double-poppet solenoid valves with soft seats. This satellite and propulsion system was built by TRW Systems Inc. The first flight model to achieve orbit was launched in late 1968.

At this time, it was decided that Intelsat satellites would utilize north-south stationkeeping to control the inclination of the orbital plane. Alternatively, the initial inclination of the orbit could have been biased so that it would pass through zero inclination midway through the operational lifetime, thus limiting the orbit inclination change to $\pm$ 6° over a 7-yr satellite life. The antenna rf beam pointing error caused by the north-south daily variation could be offset by a gimbal on the antenna reflector mount. However, this approach would result in problems of signal fading at Earth stations located either far north or far south of the equator, locations necessary for a true global system. The decision to incorporate north-south stationkeeping meant high total impulse requirements for the onboard propulsion systems, resulting in large propellant weights for propellants with moderate specific impulse.

As shown in Table 2, Intelsat III was the first of the Intelsat satellites in which the onboard propellant had to supply velocity corrections associated with north-south stationkeeping, in this case for 5 yr. The satellite mass was 133 kg (without the apogee motor), including 22 kg of hydrazine. This represented a ΔV budget of 323 m/sec, which was adequate for initial maneuvers plus 5 yr of both north-south and east-west stationkeeping. (North-south stationkeeping requires an average ΔV budget of about 50 m/sec/yr.)

Table 2 Intelsat satellite onboard propulsion system performance

Spacecraft	Propellant	ΔV, m/sec	Design life, yr	Thrust level, N
Intelsat I	Hydrogen peroxide	190	1.5	14-5.5
Intelsat II	Hydrogen peroxide	200	3	14-5.5
Intelsat III	Hydrazine	320	5	16-7
Intelsat IV	Hydrazine	432	7	26-13
Intelsat IV-A	Hydrazine	432	7	26-9

Table 3 ΔV budget for Intelsat IV and IV-A

Maneuver	ΔV, M/sec
Attitude correction	0.305
Spin-up	0.61
Spin speed control	4.17
E-W and station repositioning	14.6
Initial orbit correction	54.9
N-S stationkeeping	358.4

Intelsat IV and IV-A continued to use monopropellant catalytic hydrazine systems with similar components.[1] One more thruster was added to each half-system for initial spin-up after separation from the Atlas-Centaur launch vehicle. Another change was that the axial thrusters could be fired by onboard logic to damp large-angle nutational motion actively. These satellites had projected operational lives of 7 yr. Their ΔV budget is given in Table 3.

The total ΔV budget of 432 m/sec for Intelsat IV and IV-A resulted in a large weight penalty. Intelsat IV weighed 732 kg, of which 136.4 kg was hydrazine, and Intelsat IV-A weighed 828 kg, of which 158.2 kg was hydrazine, so that their weight penalties represented about 20% of the satellite weight (Fig.1). The total (175 kg) of propellant weight (158.2 kg) and propellant system dry weight (16.8 kg) made the onboard propulsion subsystem the heaviest subsystem on Intelsat IV-A, 9 kg heavier than the total communications subsystem. Intelsat IV and IV-A were built by Hughes, and the propulsion systems were built by Hughes and the Hamilton Standard Division of United Aircraft. The first Intelsat IV was launched early in 1971 and the first IV-A in late 1975.

Despite their weight, these subsystems have given fine performance with only a few minor problems. Because of the manner in which the propellant lines were mounted in the Intelsat IV satellite structure, each of the two half-systems had small thermal expansion loops purposely formed in the liquid lines to preclude any differential thermal expansion problems. As a result, if any bubbles were formed in the liquid line, they could get trapped in this thermal expansion loop. Since the bubble would block the liquid flow from one tank, the two propellant tanks in a given half-system would drain unequally, causing a change in mass properties and a slight shift in the satellite spin axis. This situation occurred in orbit, and a specific ground-commanded firing procedure was developed to clear the blocking bubble.

In addition, one of the axial thrusters on an early Intelsat IV probably was destroyed by a thermal soak-back problem. This problem was solved by defining very conservative firing constraints based on ground tests and detail temperature mapping of the thruster. Later models used a redesigned thruster that had no firing constraints.

Propellant Systems for Future Intelsat Satellites

Intelsat now is reviewing proposals for the Intelsat V series and conducting early systems planning for Intelsat VI. At this time, Intelsat V is planned to be launched in late 1979 on an Atlas-Centaur launch vehicle and will weigh approximately 955 kg. The use of hydrazine with comparable ΔV budgets may result in an onboard propulsion system weight in excess of 182 kg. Hence, it is appropriate to consider the alternatives that are available to reduce this weight penalty without affecting the high standards of long-life performance and reliability.

As mentioned previously, the onboard propellant mass is set by the ΔV budget for the life of the satellite. The standard rocket equation is

$$\Delta V = gI_{sp} \ \ln(\omega_f/\omega_i)^{-1}$$

where

g = gravitational constant
I_{sp} = propellant specific impulse
ω_f = final satellite weight
ω_i = initial satellite weight
$\omega_p = \omega_i - \omega_f$, the propellant weight

It is obvious that, with generally fixed ΔV budgets, the only parameter available for propellant weight saving is the specific impulse. A higher specific impulse will reduce the mass fraction, as shown in Fig. 2. This saved weight can be traded off for increased communications capacity by incorporating additional transponders and additional solar array and batteries.

To date, all Intelsat onboard propulsion systems have operated on spin-stabilized satellites, either single- or dual-body spinners. In all cases, the propulsion system has been on the spinning portion, resulting in systems of simple design.

Future Intelsat satellites may, indeed, have attitude stabilization systems different from those utilized previously. Body-stabilized satellite configurations have many attractive features for communications missions, including the capacity for greater electric power. These configurations will require different and in many cases more complex propulsion systems, i.e., a greater number of thrusters, propellant retention devices in the tanks, and thrusters with lower thrust levels. Despite these differences, large amounts of propellant will be required to support the large total impulse requirement of 7 to 10 yr of north-south stationkeeping.

Over the years, Intelsat has sponsored research and development of advanced propulsion systems to prepare for future satellite systems. A program to develop and test

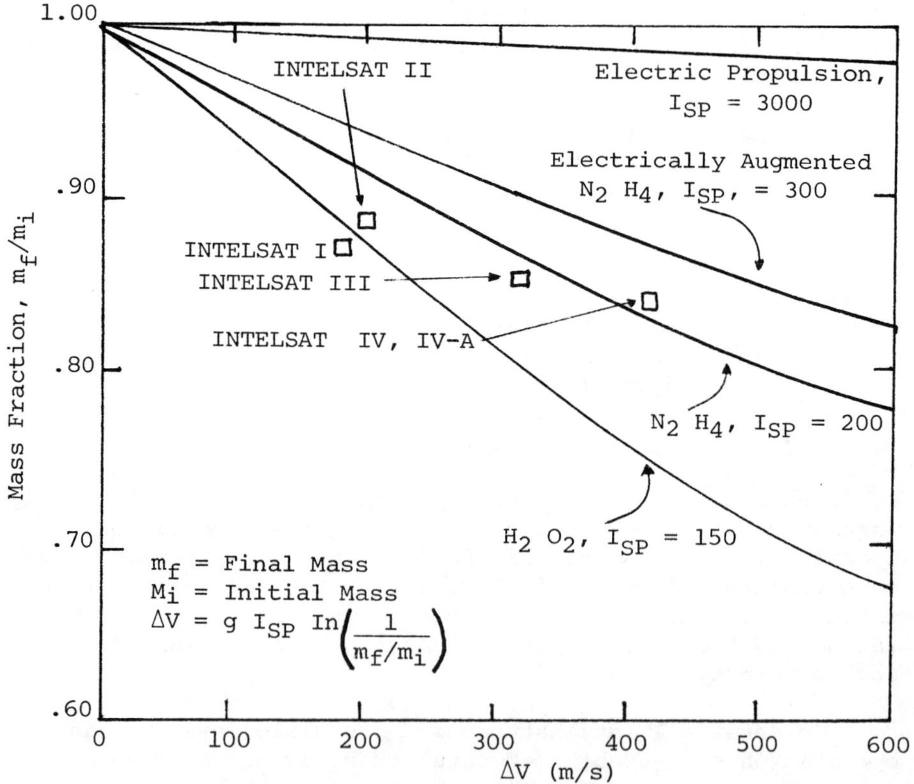

Fig. 2 Plot of ΔV vs mass fraction.

blends of hydrazine and hydrazine azide resulted in some increase in specific impulse and a decrease in freezing temperature, both attractive results. However, this mixture, when fired in a standard hydrazine thruster, resulted in earlier deterioration of the catalyst bed and retainer screens, probably because of the higher reaction temperature along with the greater amount of nitrogen and more rapid nitriding. The potential benefits were not deemed sufficiently attractive to warrant the development of a thruster compatible with this propellant.

Intelsat also has sponsored the development of electric propulsion systems such as ion-bombardment-type engines using both mercury and cesium as propellant. The chief advantage of electric propulsion over chemical propulsion is the high propellant exhaust velocity attainable or the much higher specific impulse. Specific impulse can be changed to exhaust velocity by multiplying by the gravitational constant g. A typical specific impulse range for electric propulsion is 2000 to 3000 sec, which would reduce the required propellant mass by an order of magnitude but incur a power penalty.

Work at Comsat Laboratories by Free and Huson[2] estimated that the weight of a mercury ion thruster system would be equivalent to only 27% of the weight of a comparable hydrazine system for 7 yr of north-south stationkeeping. This included the weight penalties associated with the storage tank, feed system, and power conditioning and controls. However, the weight of an additional solar array was excluded, since it was assumed that the thrusters would be fired when the batteries were not being charged. Thus, additional power would be available. This estimate did include the penalty resulting from canting the thrust axis away from the north-south axis at 45° to prevent the ion beam from impinging on a north or south solar array. Thrusters of this type would be low-thrust devices of approximately 5 mN to minimize the power requirement. They would have to be fired for many hours during almost every orbit, with the thrusting time centered at the ascending and descending nodes to impart the required velocity correction of ∿50 m/sec/yr. To convince a commercial user of the design adequacy and long-term reliability of such a device, a long-time in-orbit or at least a long-term ∿20,000-hr thermal vacuum life test would be required.

One method of reducing the life requirements and hence the life demonstration requirements of electric thrusters is to increase the thrust level. This does require more power. A method that has been investigated by Intelsat for the past few years ties the larger thruster application to an advanced electrical storage device, the nickel-hydrogen battery.[3] This advanced battery shows promise of three to five times higher energy storage efficiency than the standard nickel-cadmium batteries, with the added capability of supporting many more charge-discharge cycles. Thus, thrusters of 17 mN may be used to reduce the life requirement to about 4500 hr over 7 yr. This type of life requirement seems much more practical to prove in life test programs but requires the timely development of a second technology, the nickel-hydrogen battery.

Although none of the electric propulsion devices seem to be in the proper phase of development for Intelsat V, there is every reason to think that electric propulsion may be a reasonable contender for the Intelsat IV onboard propulsion system. A large unknown at this time is the effect of shuttle launchings in the mid-1980's (the expected Intelsat VI time frame) on the desire for weight saving technologies.

Two other candidate propulsion systems for future Intelsat satellites are bipropellants and electrically augmented hydrazine. Both of these systems offer weight saving advantages due to higher specific impulses.

An advanced bipropellant system designed and built by MBB of Germany was flown successfully on the Symphonie program. This system used monomethyl-hydrazine (MMH) as the fuel and nitrogen tetroxide (N_2O_4) as the oxidizer, with a specific impulse of 285 sec. If a similar system were used for north-south stationkeeping on a satellite of the Intelsat V class, approximately 30 kg of propellant mass could be saved. Of course, part of this gain would be offset by additional valves, propellant lines, and tankage complexity.

The successful launch and operation of the Symphonie satellite with its bipropellant propulsion system gives more confidence in this class of system. Once bipropellants are accepted for operational use, it may be possible to combine this system with the liquid apogee motor system, i.e., to use a liquid apogee motor with common propellants and a common storage system. Since this type of apogee boost system is capable of multiple firings, a liquid apogee motor would be capable of a more precise final orbit injection, and less propellant would have to be budgeted to correct the initial orbit errors. This would represent an additional weight saving.

Electrically augmented hydrazine systems add electrothermal energy to the hydrazine flow, hence increasing the gas temperature. The specific impulse (or exhaust velocity) increases with the square root of the absolute temperature of the gas; thus, the specific impulse can be increased to 300 s with a gas temperature of about 1900 K. This can be compared to the gas temperature associated with catalytic decomposition of about 1250°K. This higher temperature causes concern for long-life operation, but two companies, TRW Systems and Avco Corporation, are pursuing this technology actively and realistic life tests are underway.

Once again, if this technology were used for north-south stationkeeping in a satellite of the Intelsat V class, propellant weight savings of about 32 kg would be possible, but in this case there would be a power penalty. The power requirement of this type of thruster is 1 to 1.3 W/mN; thus, a 0.5-N-thrust thruster would require from 500 to 600 W of power during the stationkeeping maneuver. This type of system would be more attractive if it could be operated off the spacecraft batteries rather than requiring an additional solar array.

Conclusion

The primary reason for switching from catalytic hydrazine to some more advanced type of onboard propulsion system on future Intelsat satellites is weight savings. North-south stationkeeping for 7 yr requires a large total impulse, about 337,500 N-sec. Thus, any increase in propellant specific impulse, a measure of the efficiency of the propellant, results in appreciable weight savings. However, any such gains must be considered in terms of the fine performance and reliability history of the catalytic hydrazine system over the past 7 yr of Intelsat operational life.

Acknowledgment

This paper is based upon work performed in Comsat Laboratories under the sponsorship of the International Telecommunications Satellite Organization (Intelsat). Views expressed in this paper are not necessarily those of Intelsat.

References

[1]Vonnegut, M. A., "INTELSAT IV Positioning and Orientation System," _COMSAT Technical Review_, Vol. 2, Fall 1972, pp. 360-365.

[2]Free, B. A., and Huson, G., "Selected Comparisons Among Propulsion Systems for Communication Satellites," AIAA Paper 72-517, April 24-26, 1972, Washington, D.C.; also, _Progress in_

Astronautics and Aeronautics: Communications Satellite Technology, Vol. 33, edited by P. L. Bargellini, The MIT Press, Cambridge, Mass., 1974, pp. 245-259.

[3]Free, B. A., and Dunlop, J. D., "Battery Powered Electric Propulsion for North-South Stationkeeping," COMSAT Technical Review, Vol. 3, Spring 1973, pp. 211-214.

ION PROPULSION FOR COMMUNICATIONS SATELLITES

D. H. Mitchell* and M. N. Huberman+

TRW Defense and Space Systems, Redondo Beach, Calif.

Abstract

This paper examines the feasibility of integrating electrostatic ion thrusters into a 1000-kg communications satellite for North-South stationkeeping. This use of ion propulsion can save about 100 kg in propellant and tankage mass over conventional hydrazine systems, for a 7-yr mission. Some of the practical spacecraft interfaces considered in this paper are stationkeeping schedules, contamination, electrical power sources, power processing, thermal control, attitude control, stowage, and reliability. Engine types examined (all in the 5-to 25-mN thrust force range) include Kaufman-type mercury bombardment, magnetoelectrostatic containment cesium, and ionized mercury. Both spinning and body-stabilized satellites are considered. Major considerations in the study include the amount of life-test data available on the thrusters, current system reliability estimates, spacecraft contamination due to thruster material transport, component redundancy requirements, yaw-axis sensor requirements, and operational interfaces with the communications payload. No substantial obstacles were found. A preliminary development and demonstration plan shows that an ion propulsion stationkeeping subsystem could be operational in the early 1980's.

I. Introduction

Ion propulsion offers a significant weight-savings payoff when used for North-South stationkeeping on satellites in geo-

---

Presented as Paper 76-290 at the AIAA/CASI 6th Communications Satellite Systems Conference, April 5-8, 1976, Montreal, Canada.

*Senior Staff Engineer.

+Section Head.

synchronous orbit. TRW Systems recently conducted the Electric Propulsion/Spacecraft Integration Study[1] under contract to Intelsat. The purpose of this study was to identify any obstacles to the use of electrostatic ion thrusters for North-South stationkeeping on 7-yr operational communication satellites, and to suggest methods of eliminating or circumventing such obstacles.

Stationkeeping corrects for the significant out-of-orbit-plane gravitational forces exerted on a synchronous equatorial Earth satellite by the Earth, sun, and moon. These forces cause the satellite orbit to precess, and, if not corrected, the satellite antenna patterns will trace out everlengthening figure-8's on the Earth. Ground-based tracking antennas would have to be programmed to compensate for this daily motion. In order to be able to use simple, fixed-base ground antennas, a stationkeeping requirement often is imposed on the spacecraft.

North-South stationkeeping is an expensive operation, in terms of propulsion impulse, because it requires rotating the orbit plane. Since the natural disturbance forces produce orbital inclination drifts in the range from 0.749 to 0.946 deg/yr, a velocity increment of 40 to 51 m/sec/yr is required for correction. Using conventional hydrazine, with an I_{sp} of 219 sec, a 1000-kg spacecraft with a 7-yr lifetime requires about 180 kg of propellant just for North-South stationkeeping. The tenfold or greater increase in specific impulse provided by ion propulsion over hydrazine can result in net mass savings on the order of 100 kg. For communications satellites launched by Atlas-Centaur, this is an increase of about 50% in payload capability.

Ideally, stationkeeping maneuvers are accomplished by applying an impulse each time the spacecraft crosses the line of nodes between the actual spacecraft orbit and the geosynchronous orbit. The thrust levels obtainable with ion thrusters, however, are not large enough to provide an impulsive force. Therefore, they must fire for a finite period of time before and after the node is reached. The firing time depends upon the thrust level, and a minimum thrust level is established when the firing time reaches 12 hr. At that point, the thrust force must change direction. If this is not possible (for instance, if all of the thrusters point in the same direction), thrusting can be done only once a day. Ideally, thruster forces will be applied in one direction at one node and in the opposite direction at the other node. If thrusting can be done only once a day, the alternate node is used occasionally to remove off-axis torques that produce orbit eccentricity.

Table 1 Study ground rules

Orbit	Geosynchronous
Spacecraft mass at start of operational life	1000 kg
Mission duration	7 yr
Total N/S stationkeeping ΔV requirement	350 m/sec
N/S stationkeeping accuracy	±0.1°
Spacecraft ACS mode	1) Spinner
	2) Body-stabilized, momentum bias
Spacecraft power level	600 to 900 W, end of life
Power sources	1) Violet cells, rigid lightweight arrays
	2) Nickel-hydrogen batteries
Thruster types	1) Modified Kaufman-type mercury
	2) Magnetoelectrostatic containment (MESC) cesium
	3) RF ionized mercury
Thruster force range	5 to 25 mN
Body-stabilized spacecraft thruster locations	1) Array-mounted
	2) Body-mounted

Table 2 Candidate thrusters

NASA Lewis/Hughes	SIT-8	5 mN Hg
NASA Goddard/EOS	ATS-6	5 mN Cs
GFW/Giessen	RIT 10	10 mN rf ionized Hg
UKRAE/Culham	T 4/A	10 mN Hg
Intelsat/EOS	12 CM MESC	9-25 mN Cs
NASA Lewis	SERT II	25 mN Hg

II. Design Constraints

Ground Rules

The most significant ground rules of the study, based on communications satellites of the 1980-1985 time period, are summarized in Table 1.

Candidate Thrusters

Specific electrostatic ion thrusters considered for the study are listed in Table 2. Detailed data on these thrusters were obtained from the sponsors and manufacturers. There are performance differences between the various thrusters, but no fundamental obstacles to the ultimate operational application of any candidate were identified.

Mission and Operational Considerations

The stationkeeping subsystem is required to provide a total effective impulse of 97,220 mN-hr (350,000 N-sec) over the 7-yr mission. This means, neglecting thrusting inefficiencies, that a minimum thruster operating life of 9722 hr would be required for each of two 5-mN engines firing alternately at opposite orbital nodes. Two 25-mN thrusters each would require, to first order, only 1944 hr of life. These considerations thus roughly scope the range of thrust-vs-life tradeoffs to be considered.

Inefficiencies due to off-nodal thrusting and thruster canting to avoid solar array impingement increase the total required thruster operational life requirements. The required thruster operating life vs thrust and approximate power level for daily two-node thrusting, 275 days/yr, is shown in Fig. 1. These data assume that no stationkeeping thrusting is performed during the two 45-day eclipse seasons per year.

The electric propulsion subsystem consists of thrusters, a propellant storage and feed system, and the power processing unit (PPU). There are several ways to generate and distribute the electrical energy required to operate the subsystem. The PPU's may be powered 1) directly from the solar array, 2) directly from the battery, or 3) from a load bus that takes power from both the array and battery.

If only array power is to be used, low-power, low-thrust systems would be favored to minimize the added array. These thrusters are not as efficient as the larger ones and require many more hours of firing to meet the total impulse require-

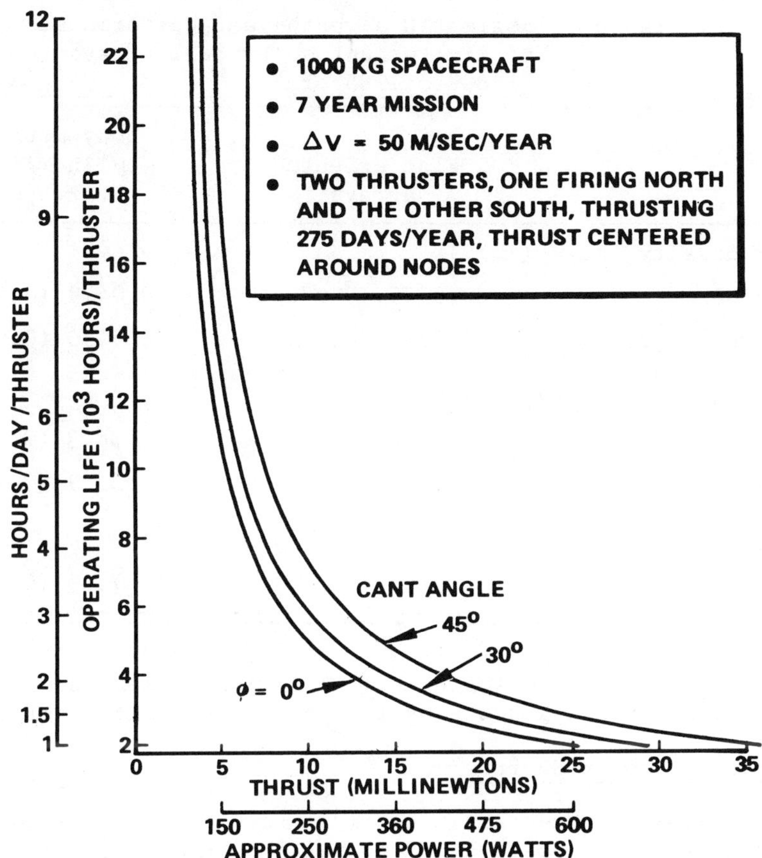

Fig. 1 North-South electric propulsion stationkeeping requirements.

ments. Using battery-augmented power allows the use of higher force, more efficient thrusters, and greatly reduces the total firing time.

The onboard battery normally is not used by the payload or any housekeeping subsystem during sunlight operations. This means that it is essentially free to use for stationkeeping at least 275 days/yr, discounting only the 45-day eclipse seasons near each equinox. Since the battery can be recharged in 10 to 11 hr, battery-powered stationkeeping also can be phased in with other operations during the eclipse season. This will provide more days per year for thrusting, reduce the required firing time per day, and be slightly more efficient.

The 11-hr recharging time for the battery also allows for two thrusting periods per day for certain battery-powered

Table 3 Estimated seventh-year performance of lightweight solar cell arrays (1980 to 1985 era)

	Flat oriented arrays	Body-mounted cylindrical arrays
Power density, kg/W (lb/W)		
Equinox	0.0355 (0.0783)	0.0419 (0.0925)
Summer solstice	0.0372 (0.0821)	0.0450 (0.0992)
Power/area ratio, W/m^2 (W/ft^2)		
Equinox	83.0 (7.71)	36.3 (3.37)
Summer solstice	79.1 (7.35)	33.9 (3.15)
Areal density, kg/m^2 (psf)		
	14.3 (0.603)	7.39 (0.312)

modes. This adds to the number of firing cycles but reduces the total thrusting hours by the increased orbital efficiency.

Electrical Power Sources

Standard spacecraft electrical power sources of the 1980's have been postulated for use in the study. Solar arrays are assumed to have high-efficiency, violet response cells[2] mounted to rigid, lightweight arrays.[3] The estimated performance of this combination is shown in Table 3. Flat oriented arrays are used on the body-stabilized spacecraft configurations and body-mounted cylindrical arrays on the spinners.

Nickel-hydrogen batteries[4] are considered as the primary energy storage system. In addition to greatly increased energy density compared to current conventional nickel-cadium batteries (30 to 33 W-h/kg vs 13 to 15), nickel-hydrogen cells have a greatly superior cycle-life performance. This feature, shown in Fig. 2, allows consideration of battery-operated ion propulsion strategies for little added battery weight. A 7-yr mission could require as many as 1700 to 4000 high-depth-of-discharge battery cycles for stationkeeping, depending on the thruster strategy. The main spacecraft load requires only 630 cycles. Only nickel-hydrogen cells can provide this dramatic increase.

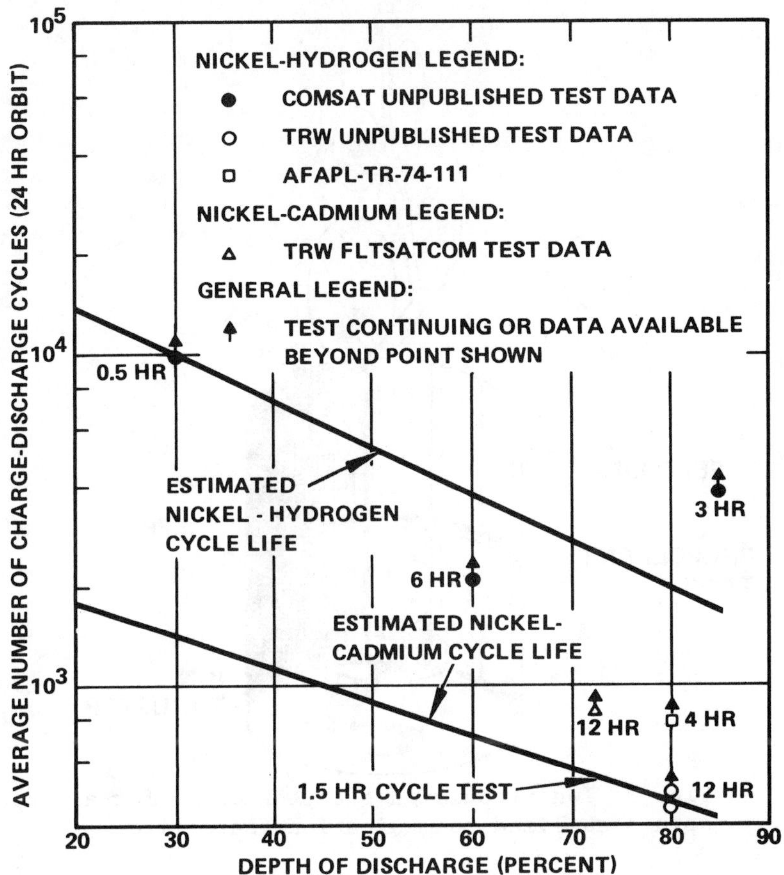

Fig. 2 Cycle life of nickel hydrogen cells.

III. Spacecraft/Thruster Configurations

Electric propulsion thruster configurations were developed for each of the prescribed spacecraft types: spin-stabilized, body-stabilized with array-mounted thrusters, and body-stabilized with body-mounted thrusters. Emphasis was on adding ion propulsion stationkeeping to spacecraft configurations developed for other considerations.

The very low thrust levels introduced by the use of electric propulsion for North/South stationkeeping dramatically increase the time required to perform stationkeeping. Ion propulsion required roughly 800 to 4000 hr of yearly thrusting, compared with roughly 1.6 hr of yearly thrusting required with the nominal hydrazine propulsion system. Stationkeeping

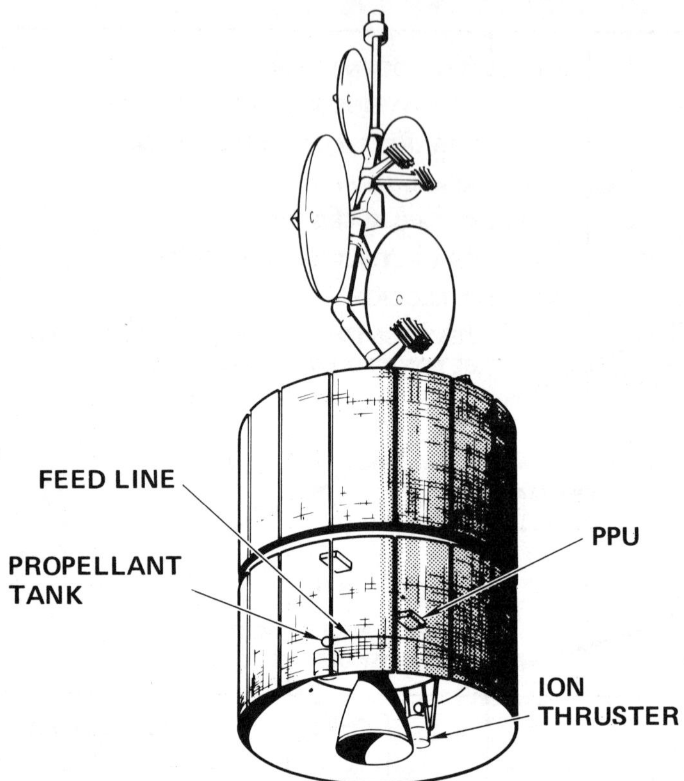

Fig. 3 Ion propulsion stationkeeping for a spin stabilized communications satellite.

thrusting scheduling therefore becomes a major systems problem that is strongly seasonal and configuration-dependent. All configurations are designed to complete the mission with one thruster failure.

Spin-Stabilized Spacecraft

An ion propulsion subsystem for a spin-stabilized communication satellite of the Intelsat IV-A configuration is shown in Fig. 3. Each ion thruster and its supporting propellant tank and power processing unit (PPU) are mounted on a special truss near the "lower" or south-oriented end of the spacecraft. This location minimizes potential contamination problems. No thrusters can be mounted on the upper end because of interference with the antenna farm. The truss structure is required for support from the main platform and central cylinder, since there is no solid structure near the arrays.

The location of the apogee kick motor (AKM) along the spacecraft centerline prevents mounting an ion thruster on the board location, it is not necessary to cant the ion thruster nozzles to point through the spacecraft center of mass, since the induced wobble is extremely small because of the large spacecraft kinetic energy from spinning.

A more serious limitation on the spinner is the need to keep the spacecraft principal axis closely aligned with the spin axis. This means that flow from the propellant tanks must be near-symmetric. If each thruster has its own tank, at least four independent propulsion systems are required, since a single failure prohibits use of propellant from its symmetric tank. This can be avoided by having a self-balancing tank or an interconnecting propellant line between opposing tanks which will equalize the levels.

Thruster operational life requirements are more severe for a spin-stabilized satellite than for body-stabilized satellites. This results from the fact that the selected spinner configuration allows for thrusting at one orbital node only. Thus the effective total impulse required for each thrust period is doubled. This leads to corresponding increases in the required thruster "on" time. Furthermore, the daily thrusting schedule is pushed into periods where the orbital cosine loss is a major factor. For example, at 5-mN thrust, the daily thrust period is 11.3 hr for a 365-day/yr thrust schedule. If this mission were to be attempted with only one thruster, the total required operational life would be 28,878 hr for the 7-yr mission. This is beyond the demonstrated capability of ion thrusters.

The preceding considerations would appear to be a driving motivation to use higher thrust levels on the spinner. Unfortunately, additional power is harder to obtain on an existing spinner than on a body-stabilized spacecraft. For this reason, an attractive option for the spinner appears to be the sequential use of several low-level thrusters to achieve a high total firing time, for example, three 5-mN thrusters, which are required to operate for 9626 hr each. Additional redundant standby systems also can be included on the total spacecraft design.

For high-thrust, battery-assisted thruster system configurations, the spinner does not impose unique difficulties, since this mode would employ daily single node thrusting even for the body-stabilized configurations. Thrust vectoring is desired for spin-rate control. Maintaining the design spin rate is important for operating dual-spin satellites with unfavorable inertia ratios, such as Intelsat IV-A. In order to minimize the hydrazine used for control of spin variations due

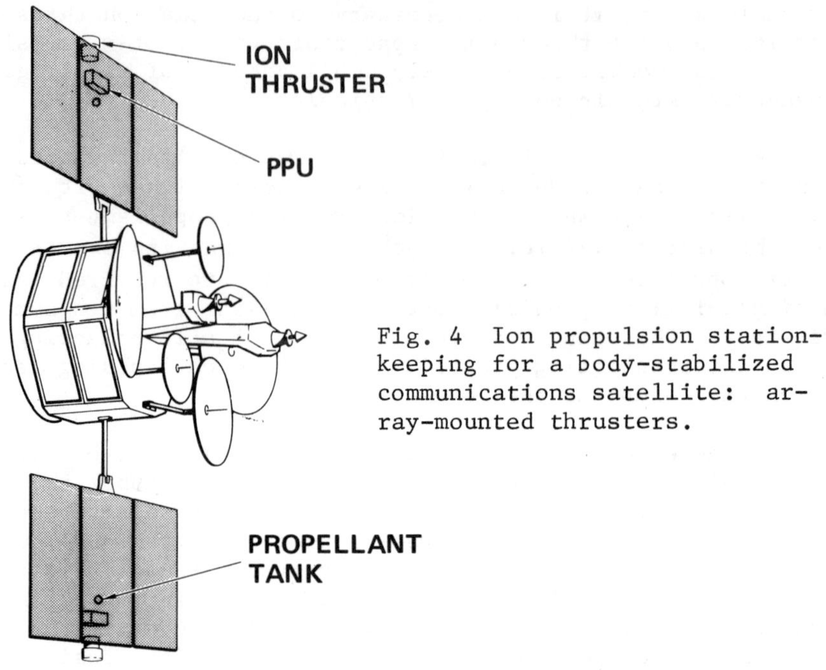

Fig. 4 Ion propulsion stationkeeping for a body-stabilized communications satellite: array-mounted thrusters.

to ion thrusting, resolution of such vectoring should be in the order of 0.1°. A single-axis gimbal is adequate, since the only concern is for misalignment about a radius from the spacecraft centerline.

Except for the foregoing, North/South stationkeeping has little effect on a spin-stabilized spacecraft. The precession errors due to stationkeeping are small, and attitude control maneuvers are not required immediately. There are no obscured sensor fields of view. Depending on battery charging strategy, the desired approach may be to fire every day or to avoid the eclipse or solstice seasons. Firing every day is the preferred strategy.

Body-Stabilized Spacecraft with Array-Mounted Thrusters

Array-mounted thrusters on a body-stabilized spacecraft (Fig. 4) minimize the contamination problem and provide a great deal of flexibility in stationkeeping scheduling and thruster selection. North and/or South firings can be used, and so redundancy can be achieved with only one thruster per array. However, since it is not desirable to run propellant across the rotating joint between the arrays and the main body, there would have to be adequate propellant for a complete 7-yr mission on each array. This is still lighter than a two-thruster arrangement, since each thruster requires a separate power processing unit, and these are relatively massive.

Dynamic coupling with the spacecraft attitude control system for this configuration was examined and found to be within acceptable limits. The spacecraft mass unbalance created by propellant depletion if only one thruster is used is not a problem on a body-stabilized spacecraft.

Thrust vectoring capability for the engines can counteract both steady-state misalignment and transient effects such as thermal warpage. Otherwise, it is estimated that up to 45 kg of hydrazine might be required to correct stationkeeping misalignment torques. Thrust vectoring is inherent in several cesium engines, and simple gimbals can be added easily to the mercury engines.

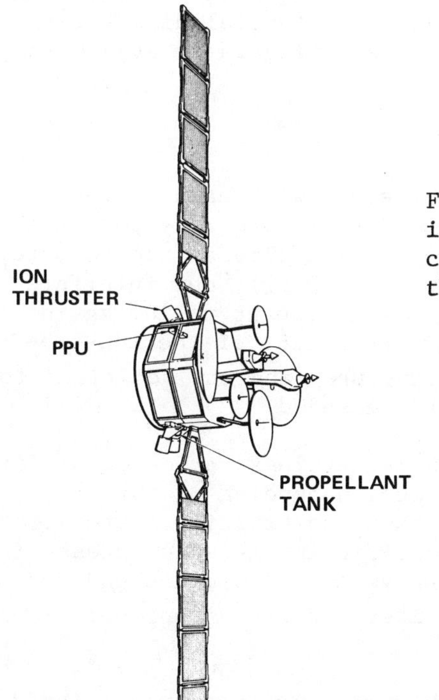

Fig. 5 Ion propulsion stationkeeping for a body-stabilized communications satellite: body-mounted thrusters.

Body-Stabilized Spacecraft with Body-Mounted Thrusters

Mounting the thrusters on the main body of a body-stabilized spacecraft, as shown in Fig. 5, eliminates direct mechanical interfaces with the solar array but requires that a high-aspect-ratio array be used and that the thrusters be canted outward to minimize contamination. A thorough examination of the material transport problem as the arrays rotate near the thruster stream shows that this configuration is feasible with a cant angle of at least 30° and the use of a small shield at the thruster exit. New engine modifications that reduce contamination also may be desirable. The thruster cant angle introduces radial orbital thrust components whose net effect can be cancelled by equal thrusting at opposite nodes. Therefore, both North-and South-pointing thrusters are needed. To allow for failure of one, the requirement is for two redundant pair of thrusters (4 total).

The body-mounted thrusters also must be vectored to guarantee alignment with the spacecraft mass center. Otherwise, about 25 kg of hydrazine will be required to correct for misalignment errors. Although not as critical as for the array-mounted thrusters, it still is worth adding a thrust vectoring capability. With this capability, the potential also exists to unload reaction wheel momentum, resulting in an even more efficient RCS design.

System Performance

Evaluation of the various thrusters with each of the spacecraft configurations shows that there are many potential combinations that will keep the operating life of the thruster under 12,000 hr (sometimes as low as 2000 hr), not interfere with normal spacecraft operations, and save up to 100 kg in mass. This mass savings is after the weight of the ion propulsion subsystem, its support structure, and any additions to the electrical power subsystem are included.

In principle, the lowest power-to-thrust ratios occur for specific impulses as low as 1000 sec, depending on thrust level.[5] This study, with its emphasis on existing systems, confined its considerations to specific impulses in excess of 2000 sec. Electric propulsion subsystem performance and added mass requirements of some of the most promising configurations are shown in Table 4.

It is interesting to note that mass savings are achieved going to higher specific impulse for most of the configura-

Table 4 Typical stationkeeping subsystem performance (at 2500 sec I_{sp})

Satellite configuration	Thrust mN	Days/hr/ nodes	Life, hr	Thrusters/ tanks/ propellant %	Added mass, kg	Subsystem power, W	Added array power, W
	5	365/11.3/1	3x9,626	6/2/100	93.2	148	89
Spin-stabilized	10	275/5.5/1	10,595	2/2/100	47.6	257	...
	17	275/3.0/1	5,860	2/2/100	55.3	391	...
Body-stabilized	5	365/4.0/2	2x10,157	6/2/100	64.2	148	47
Array mount	17	275/3.0/1	2x2,930	2/2/200	77.9	391	...
Body-stabilized	5	365/4.7/2	2x11,932	4/2/100	67.3	148	47
Body mount	17	275/3.6/1	2x3,416	4/2/100	92.9	391	...

tions. For shared propellant tank cases, this mass advantage is small, and it pays to design for low specific impulse in order to reduce power. This is because the propellant mass for redundant systems, with nonredundant shared propellant, is a small fraction of the total mass over the entire specific impulse range of interest. This reduces the physical problem of designing larger arrays and, for battery-assisted propulsion, reduces the total required discharge per thrusting period. On the other hand, systems with redundant propellant tanks and double propellant do benefit significantly (as much as 10 kg) from higher specific impulse. The cesium systems have been estimated conservatively to fall into this latter category, since a propellant feed isolator is not yet available to allow propellant sharing between thrusters. The cesium system added masses can be approximately equal to those of the mercury systems if propellant feed isolators are used.

IV. Thruster/Spacecraft Integration Considerations

Material Deposition

A major portion of the study was devoted to examining potential spacecraft contamination effects due to material transport from the thruster plume. Documented in detail elsewhere,[6] the conclusions are as follows.

Body-stabilized spacecraft with array-mounted thrusters and spin-stabilized spacecraft with end-mounted thrusters are inherently insensitive to contamination, since critical equipment surfaces are located well behind the exhaust plane. However, there is a major concern for body-stabilized spacecraft with body-mounted thrusters when their nominal ion thrust vector is parallel to the solar array boom. In order to minimize the exposure of the solar array panels as they rotate daily through the exhaust plume, long, high-aspect-ratio arrays are necessary. It also is necessary to cant the thruster about 30° relative to the solar array axis and to use exhaust shields to reduce thrust beam ion interception to acceptably low levels.

Five types of particles were modeled: 1) thrust ions, 2) sputtered metal, 3) un-ionized propellant, 4) propellant charge exchange ions, and 5) charge exchange metal ions. Of these, the major concern is the propellant charge exchange ions. These ions form downstream and escape through a wide range of exit angles, migrating to regions unscreened by the thruster shield. At present, there is not sufficient experimental data to define accurately the charge exchange fluxes for candidate thrusters to the degree needed to guarantee freedom from contamination. In this regard, cesium, because of its higher chemical reactivity, causes greater concern than mercury. Also, the potential danger is greater at high thrust levels, since the charge exchange collision rate varies approximately as the thrust squared. Hence, the total mission deposition will, to first order, be proportional to the thrust level.

A major study conclusion is that more measurements of thruster effluent fluxes at wide angles are needed. Additional preventive measures, such as more extensive shielding, the use of a third decel grid, or incorporation of the small hole accel grid concept into the MESC engine, may be required if unexpected problems are uncovered.

Design Life and Reliability Demonstration

A major deterrent to the acceptance of electric propulsion for operational spacecraft has been concern over design life and reliability. The predicted design life of a thruster often has been the extrapolated theoretical operational life based on wearout rates observed in shorter-duration tests. Historically, as the life-testing time has been increased, thruster failures have occurred because of the onset of previously unrealized wearout modes. Thus, with the benefit of hindsight, it could be recognized that the originally claimed design life was incorrect. From this, it follows that predicted design life must be demonstrated before electric propulsion can be used in operational spacecraft.

Many long-term electron bombardment thruster and system ground tests have been performed in the past decade. Most of these tests have been thruster-only and indicate that design lives of 10,000 hr or more are feasible as far as thruster wearout mechanisms are concerned. Unfortunately, most of the thrusters that have been tested do not include the latest technological developments. For example, the longest-running test is on the SIT-8 thruster. The ultimate SIT-8 design, however, probably will incorporate at least three design changes from the life-test thruster: 1) a small hole accel grid, 2) a tantalum baffle, and 3) plasma-sprayed heater assemblies. There is still a need for hands-off, mission-equivalent ground testing of the finalized total system design. An additional requirement is to include an adequate cycle life demonstration, at least on the order of several thousand cycles.

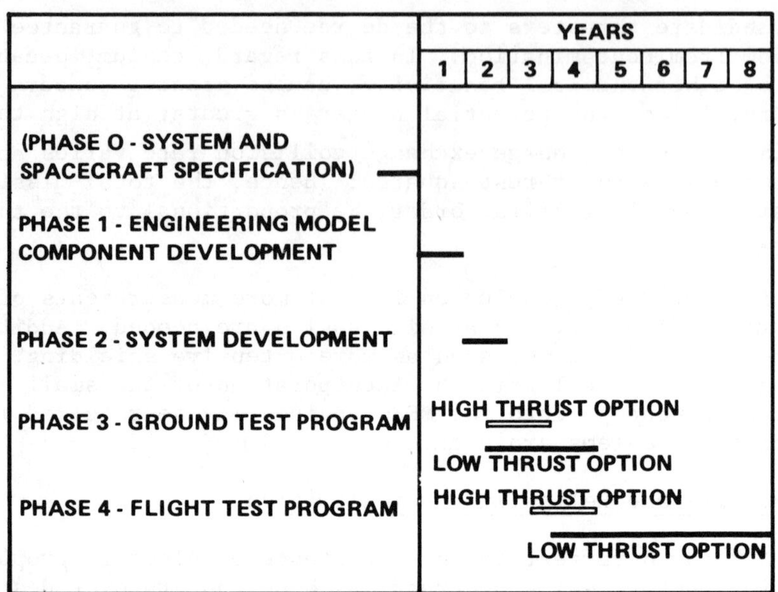

Fig. 6 Electric propulsion subsystem development and test plan.

There have been three electron bombardment engine flight tests: SERT I and SERT II (Kaufman mercury engines) and ATS-6 (MESC cesium engine). In general, these tests duplicated ground test experience. The one major exception was the inability of each of the ATS-6 engines to restart after a successful initial on-cycle. It is theorized,[7] and partially corroborated by later laboratory tests, that this problem was due to a design deficiency in the cesium capillary feed system which manifested itself under zero-gravity flight conditions.

All three flight tests demonstrated the ability to operate with no observable communications interference. The ability to charge-neutralize the spacecraft and the space charge within the thruster exhausts also was demonstrated. The most general conclusion to be drawn was that no fundamental operational difficulties with ion bombardment thrusters were discovered. Furthermore, the SERT II test achieved in-flight steady-state operational lifetimes of 2011 and 3781 hr before failure. This demonstrated flight lives on the low end of the scale of stationkeeping operational requirements. Results from ground test programs indicated that the SERT II flight failures were due to localized ion bombardment erosion of the accel grid in the vicinity of the neutralizer. All subsequent ion engine designs have been able to eliminate the localized erosion problem by repositioning the neutralizer. Unfortunately, the SERT II ground tests results were obtained too late in the program to change the flight design. The subsequent flight failure demonstrated the importance of having an adequate ground test mission simulation prior to the final flight system design freeze.

The major ion engine demonstration test failures to date have not been classical reliability-connected random failures but rather the result of unforeseen wear-out mechanisms that appeared as the operational life was extended. As previously discussed, these developmental endurance test programs have proceeded to the point where thruster wear-out mechanisms have been identified, understood, and reduced to a tolerable level for operational periods of the order of 10,000 hr or more. Since each succeeding thruster design incorporated changes intended to eliminate the previously identified failure mechanisms, the preceding failures no longer could be incorporated into the reliability data base. Furthermore, standardized manufacturing techniques have not been developed yet. Consequently, there is not an adequate data base for predicting ion engine failure rates.

The approach for this study has been to develop failure rate estimates by analogy with similar flight-qualified components. Estimates by TRW and Hughes[8] indicate individual stationkeeping thruster system failure rates of the order of 10,000 to 30,000 failures/10^9 hr. Failure rates of this order or lower are a necessity when viewed in the light of mission constraints. Even under the most pessimistic of assumptions, at least 19 candidate electric propulsion subsystem designs examined in the study were predicted to exceed the reliability allocation of 0.95. This gives the designer a wide selection of acceptable configuration. The effect on reliability of other spacecraft subsystems caused by electrical propulsion is relatively minor, causing at most a 0.006 decrease in a typical overall spacecraft reliability because of increased utilization of the attitude and velocity control subsystem.

High confidence level statistical confirmation of failure rates would require on the order of 10^5 hr of total testing for finalized systems with well-established manufacturing processes. The costs for such tests may be prohibitive. A more palatable approach would be to test only key critical components and subassemblies, while using analytical techniques and related historical data for straightforward mechanical features. Such an approach currently is being followed by the NASA Lewis Research Center. A reliability demonstration procedure testing 10 units each of critical thruster components (i.e., 10 cathodes, 10 neutralizers, etc.), each unit to undergo 10,000 hr of cycled test, is recommended. This procedure should ensure that the thruster failure rate upper limit of about $20{,}000/10^9$ hr is not exceeded.

Power Processing

Power processor unit (PPU) designs for electric propulsion applications were addressed in the study. Although PPU technology lags ion engine development, it appears that techniques developed for other applications, such as traveling wave tubes and colloid propulsion engines, can minimize the development process.

Electrical Power Subsystem

One of the interesting results of the study is that the impact on the spacecraft electrical power subsystem can be kept to a minimum.[9] Use of the charge array greatly reduces the magnitude of the required solar array increase. Most higher-power thruster systems can obtain energy from the spacecraft batteries, augmented by the existing load and battery charge arrays, with no increase in battery or array size.

Thermal Control

Interface thermal control requirements were examined for each of the configurations. The significant item was the PPU, and simple passive thermal control designs were found to be adequate.

Electromagnetic Interference

Major considerations for electromagnetic interference are plasma noise/receiver and power processor/spacecraft interactions. The first interaction appears minimal, since the plasma power in the rf range appears very small. The transmission paths appear to be minimal, but this must be verified on an actual flight vehicle. Power processor/spacecraft interactions most likely would be caused by transient radio frequency interference during current onset. Rate, not amplitude, effects are most important and will have to be considered during

PPU design. Electromagnetic interference system verification tests will be required on any development.

V. Development and Demonstration Program

There have been difficulties encountered with every electric propulsion flight test to date. Therefore, it is essential that a development and demonstration program culminate in a successful flight demonstration. Detailed design analysis and thorough testing are needed at every level of the process. This takes time. The plan developed in the study is based on thorough ground testing of the complete system before flight tests. For this reason, four to eight years are required to complete the entire program. It may be possible to compress this schedule for practical considerations, at increased risk.

The complete development and demonstration program, shown in Fig. 6, is divided into four major phases: phase 1) engineering model component development; phase 2) system development; phase 3) ground test program; and phase 4) flight test program. A "phase 0" period also is required prior to initiation of the development program to select the system configuration, prepare and update specifications, issue a request for proposal, evaluate proposals, and award a contract. Phase 0 could add a year more to the total program.

Phase 1 lasts for 1 yr and includes component design, fabrication, and performance test tasks needed to clean up all remaining loose ends in the way of final flight system design. Specific tasks in phase 1 will depend to a great extent on the operational readiness of the configuration selected. Phase 2, also 1 yr, includes the integration of phase 1 components into a system engineering model and subsequent performance verification, accelerated life testing, and diagnostic wear testing. Phase 2 also includes manufacturing all hardware required for ground endurance testing, ground reliability test programs, the flight test and spares.

The ground test program, phase 3, overlaps the last six months of phase 2 to allow for facility preparation and then extends for an additional 2-yr maximum total system hands-off cycled life test until the end of year 4. This is based on an assumed 12,000-hr mission operating life plus 20% margin of safety, plus 1/6 cycle off time for the low-thrust option. The use of battery-assisted higher thrust levels could cut more than 1 yr from this time. Phase 3 also includes interchangeability demonstrations of major system subunits and statistical failure rate testing of critical thruster components and subsystems.

Phase 4 is the flight test. The launch date is 1 to 2 yr after the initiation of phase 4. This is based on a flight system design freeze at the end of phase 3 and the initiation of long-lead spacecraft integration tasks 1 yr prior to completion of the ground test. Launch date also depends on availability of the spacecraft, test range, and a host of other considerations.

A key problem at this point for the ion propulsion system is the lengthy flight time required to achieve total mission simulation. For piggyback experiments onboard operational spacecraft, other mission constraints will make it difficult to reduce significantly the time required, especially if battery recharge cycles are the pacing item. For array-powered thrusters, it may be possible to build up operational time more rapidly by deliberately thrusting inefficiently far from the nodes. A more detailed analysis can be made when the mission is better specified.

VI. Conclusions

The major conclusions of the study are as follows:

1) There are no fundamental obstacles to the use of electric propulsion for North-South stationkeeping of communication satellites. There can be a significant mass savings, and the spacecraft impacts can be minimized.

2) A development/implementation program of four to eight years, including flight testing, is required for conclusive demonstration.

Specific technical conclusions include the following:

1) Solar array power normally dedicated to battery charging can be made available to the electric propulsion subsystem in order to reduce the amount of additional array required.

2) 5-mN thrusters can operate directly off the solar array power for their required 10,000 to 12,000 hr of operation.

3) With proper power budgeting, nickel-hydrogen batteries can be used to allow higher-thrust, shorter-operating-life thrusters without added battery weight.

4) Thrust vectoring is desired for all configuration in order to minimize a) disturbance torques on body-stabilized spacecraft, and b) spin rate variations on spin-stabilized spacecraft.

5) The use of a high-aspect-ratio solar array, plus exterior thruster shields and outward canting of approximately 30°, is required for body-mounted thrusters on body-stabilized spacecraft.

6) For spin-stabilized spacecraft, it is necessary to store the propellant symmetrically with respect to the principal axis of inertia. There is no similar requirement for body-stabilized satellites.

7) More reliability test data would be extremely desirable, especially on thrusters.

Acknowledgement

This paper is based in part upon work performed under the sponsorship of the International Telecommunications Satellite Organization (Intelsat). Any views expressed in this paper are not necessarily those of Intelsat.

References

[1]"Electric Propulsion/Spacecraft Integration Study - Final Report," Rept. 26951-6001-TU-00, March 1976, TRW.

[2]Arndt, R. A., "Effects of Radiation on the Violet Solar Cell," COMSAT Technical Review, Vol. 4, Spring 1974, pp 41-52.

[3]Luft, W. and Patterson, R. E., "Lightweight Rigid Solar Array Development," 11th IEEE Photo-Voltaic Specialists Conference, May 1975, Scottsdale, Ariz.

[4]Patterson, R. E. and Sparks, R. H., "Nickel-Hydrogen Battery System Development," 10th Intersociety Energy Conversion Engineering Conference, Aug. 1975, Newark, Del.

[5]Free, B. A., "Chemical and Electric Propulsion Tradeoffs for Communications Satellites," COMSAT Technical Review, Vol. 2, Spring 1972, pp 23-145.

[6]Sellen, J. M., Jr., Cole, R. K., and Komatsu, G. K., "Material Deposition Processes for North-South Stationkeeping Ion

Thrusters on Three-Axis Stabilized Spacecraft," JANNAF Propulsion Meeting, 1975, Anaheim, Calif.

[7]Worlock, R. A., James, E. L., Hunter, R. E., and Bartlett, R. O., "The Cesium Bombardment Engine North-South Stationkeeping Experiment on ATS-6," AIAA Paper 75-363, March 1975, New Orleans, La.

[8]Molitor, J., "Ion Propulsion Flight Experience, Life Tests and Reliability Estimates," Journal of Spacecraft and Rockets, Vol. 11, Oct. 1974, pp 677-685.

[9]Rusta, D. W., "Power Source Requirements of Electric Propulsion Systems Used for North-South Stationkeeping of Communication Satellites," 11th Intersociety Energy Conversion Engieering Conference, Sept. 1976, Las Vegas, Nev.

Chapter IV – Spacecraft Antennas

DUAL-POLARIZATION MULTIPLE-BEAM ANTENNA

J. W. Duncan,* S. J. Hamada,+
and P. G. Ingerson+

TRW Defense and Space Systems Group,
Redondo Beach, Calif.

Abstract

A dual-polarization multiple-beam antenna has been developed for frequency-reuse satellites that require amplitude isolation and polarization orthogonality between regional coverage antenna beams. The antenna consists of an offset paraboloidal reflector with a multielement array feed. The feed elements are dual polarized, so that each element can radiate simultaneously both right-hand and left-hand senses of circular polarization. Selected subsets of elements are used to produce low-sidelobe regional coverage beams. Test data are presented for an experimental model antenna that radiates orthogonally polarized beams that are shaped for hemispherical coverage in the Atlantic Ocean.

I. Introduction

The design and implementation of a single-aperture multiple-beam antenna that provides at least 27 dB isolation between regional coverage beams by means of amplitude isolation and/or polarization orthogonality is a challenge to antenna technology. The basis for this statement is that a dual-polarization multiple-beam antenna presents many more design problems than does a single-polarization antenna. The antenna design involves a great many considerations, all of which impact the

---

Presented as Paper 76-248 at the AIAA/CASI 6th Communications Satellite Systems Conference, April 5-8, 1976, Montreal, Canada.

*Special Assistant for Antennas, Communication and Antenna Laboratory.

+Senior Staff Engineer, Communication and Antenna Laboratory.

antenna performance in varying degrees. Initially one is faced with the selection of the basic antenna configuration, that is, a lens with an array feed or a reflector with an array feed. The antenna design process continues with a tradeoff between aperture diameter, focal length, number of feed elements, and the size, weight, and complexity of the antenna system. The beam synthesis procedure that is implemented in the design of the array feed system is of particular significance, since this will determine sidelobe characteristics and the amplitude isolation between copolarized beams. The type of element used in the array feed system has a major impact on the achievable polarization isolation between orthogonally polarized beams. The dual-polarized element must exhibit a very low axial ratio over the element radiation pattern for the entire frequency bandwidth. Mutual coupling effects in the array feed can degrade the axial ratio of elements seriously, so that the element design must account for these effects if good polarization isolation is to be realized.

The design considerations just outlined are treated in greater detail in the following material, which leads to a selected design approach for a dual-polarization multiple-beam antenna for frequency-reuse satellites. The design approach has been validated with test data from an experimental model antenna.

II. Basis of Antenna Design

Offset Paraboloidal Reflector

The principal technique for radiating multiple-zone coverage beams from a single antenna aperture is to utilize a multielement feed system to illuminate a lens or a reflector. In the case of a reflector, the feed system must be offset from the aperture to avoid blockage and scattering by the feed which would degrade the radiation pattern of the antenna seriously. Figure 1 illustrates an offset-fed paraboloidal reflector, which is illuminated by a multielement array feed. The offset paraboloid is defined by the focal length F, the projected aperture diameter D, angles θ_1 and θ_2 that define the extremities of the reflector in the xz plane, the angle of feed offset θ_o, and the coordinate x_c, which locates the center of the reflector. The array feed usually is comprised of small waveguide horns or similar types of elements. Typically, the element apertures are 0.75λ to 1.5λ in dimension, where λ is the wavelength.

A zone coverage beam is obtained by overlaying the specified region on the Earth with a collection of contiguous spot

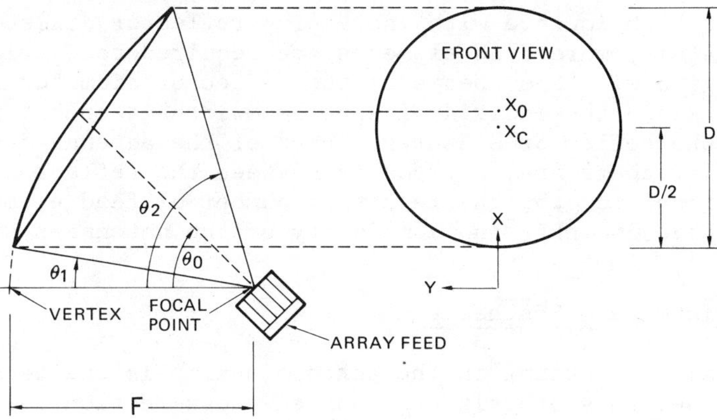

Fig. 1 Offset paraboloidal reflector with multielement array feed.

beams. Each element of the feed system radiates one of the spot beams. The signals of the set of elements that produce the overlaying spot beams are combined to achieve a composite beam whose -3 dB contour approximates the perimeter of the specified region. Multiple regional coverage beams are obtained from a single reflector aperture by exciting selected subsets of feed elements.

The shape of the beam produced by a set of feed elements is determined by the location of the elements in the focal region of the reflector, by the reflector parameters (aperture diameter, focal length, feed offset angle), and by the complex excitation coefficients of the array feed elements. When the array feed is located in the reflector as shown in Fig. 1, only one element can be placed at the focal point; all other feed elements necessarily are defocused. A defocused feed produces a spot beam with very high-amplitude sidelobes. If this sidelobe radiation extends to an adjacent copolarized beam, it will degrade or nullify frequency spectrum reuse seriously between the two beams. One method of alleviating this problem is to use a large-diameter reflector and a large focal length. The beamwidth of individual spot beams decreases with increasing reflector diameter, and, correspondingly, the sidelobes extend for a smaller solid angle from the main beam. If the reflector can be made sufficiently large, the distribution of sidelobe energy into an adjacent copolarized zone can be reduced substantially. Because sidelobe degradation from feed defocusing decreases with increasing F/D (focal length/diameter) ratio, the focal length should be selected as large as is practicable. It is clear from the foregoing that the effectiveness of beam contouring and the quantitative levels of beam isolation both improve with increasing reflector diameter. Unfortunately, more feed elements are required to overlay a given region with spot beams as the reflector diameter increases, since the individual spot beamwidths decrease inversely with the reflector diameter. Part of the antenna design process is, therefore, a tradeoff between the reflector diameter and focal length, the resulting number of feed elements, and the size, weight, and complexity of the antenna system.

Low-Sidelobe Beam Synthesis

Equally important in the antenna design is the technique that is used to synthesize the shaped-cross-section regional coverage beams. The antenna designer has the option to utilize a beam synthesis technique that does or does not include sidelobe control (sidelobe reduction or cancellation). If sidelobe cancellation is a characteristic of the beam synthesis procedure, then a smaller-diameter reflector can be used to achieve values of beam isolation which are greater than can be achieved using a larger reflector with a feed system that does not implement sidelobe cancellation.[1] For a given aperture diameter, the low-sidelobe beam synthesis technique requires a larger number of feed elements to produce a prescribed beam coverage than occurs for beam synthesis without sidelobe control. In either case, the number of feeds required to accomplish shaped-beam coverage increases with the reflector diameter.

A low-sidelobe array feed system has been developed[2] for lens and reflector antennas which has specific application to frequency-reuse satellites. The beam synthesis procedure incorporates sidelobe cancellation. As a result, the array feed can generate multiple regional coverage beams with high-amplitude isolation between adjacent copolarized beams. This development has been implemented in the experimental model antenna described in Sec. III.

The low-sidelobe beam synthesis procedure will be illustrated with three examples of regional beams designed for Intelsat station coverage as viewed from an Atlantic Ocean satellite at 335.5° E long. Figure 2 shows a beam designed to cover Western hemisphere stations in the Atlantic. The beam is indicated by constant gain contours superimposed on a map of the Earth which includes the Intelsat stations. The beam is produced by a 20-element feed in a 1.52-m (60-in.-diam) offset reflector at a frequency of 3.95 GHz. The focal length is 45 in. The boresight axis of the reflector is directed to nadir. The array feed has been displaced from the focal point to scan the beam to the position shown. All stations in the Western

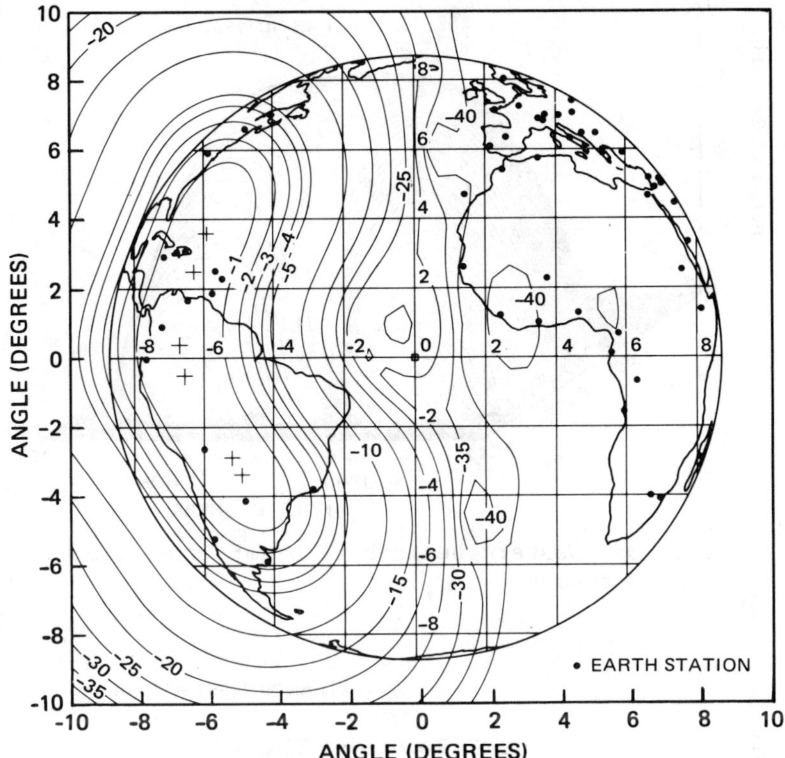

Fig. 2 Western hemisphere beam for 1.52-m reflector with 20 feeds.

hemisphere are within the -3 dB gain contour. All stations except two in the Eastern hemisphere lie beyond the -40 dB contour. The calculated peak directivity of the Western hemisphere beam is 28.3 dB. This particular beam was selected for implementation in the experimental model antenna. Two isometric perspectives of the West hemisphere beam are presented in Figs. 3a (view from the southeast) and 3b (view from the northwest). These two figures clearly show the effectiveness of sidelobe cancellation which is achieved with the beam synthesis technique.

A second example of shaped-beam synthesis is presented in Fig. 4, which shows an Eastern hemisphere beam for the satellite at 335.5° E long. The reflector antenna configuration is the same as the foregoing (1.52-m diam), but the feed now is comprised of 26 elements. The calculated peak directivity of the beam at 3.95 GHz is 27.25 dB. All stations in the Eastern hemisphere lie within the -3 dB gain contour. Note that Mill

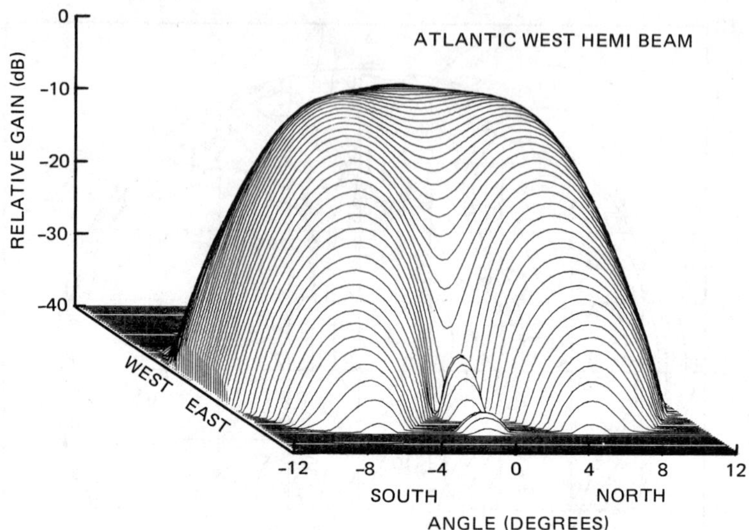

Fig. 3a Western hemisphere beam: view from the southeast.

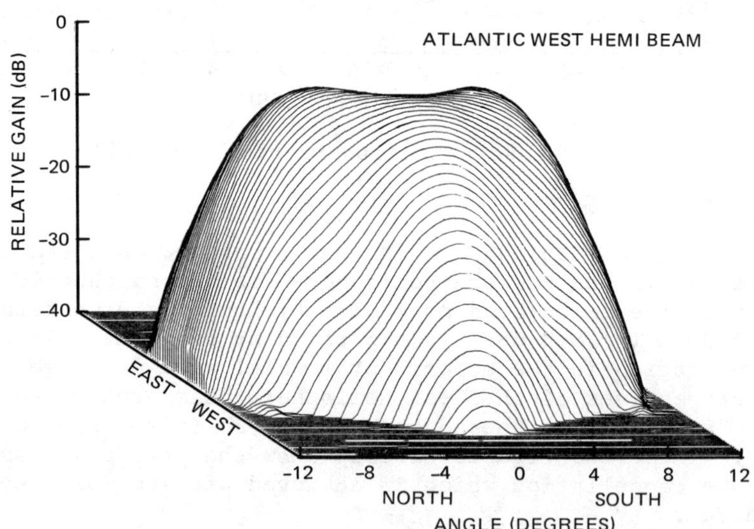

Fig. 3b Western hemisphere beam: view from the northwest.

Village, Canada (station MV) lies on the -35 dB gain contour, whereas most other Western hemisphere stations appear beyond the -40 dB contour.

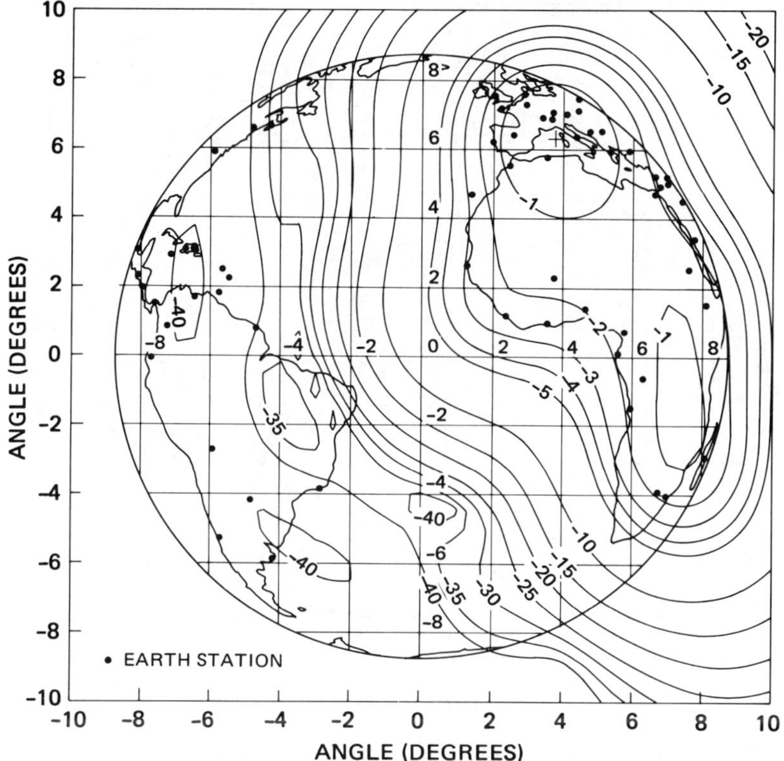

Fig. 4 Eastern hemisphere beam for 1.52-m reflector with 26 feeds.

A third example of low-sidelobe beam synthesis is the Eastern hemisphere beam shown in Fig. 5. The beam is produced by a 36-element feed in a 1.98-m (78-in.-diam) offset reflector at 3.95 GHz. The focal length is 58.5 in. The peak directivity of the beam is 28.1 dB. All stations are covered within the -3.5 dB contour. Comparing Figs. 4 and 5, one can see the improved beam contouring, gain, and beam isolation to Western hemisphere stations which results with the 78-in. reflector as compared to the 60-in. reflector. In this case, a 30% increase in reflector diameter required almost a 40% increase in the number of feed elements.

Element Design Considerations

A third major consideration in the antenna design is the type of element used in the array feed system. The radiation characteristics of the element are of paramount importance when dual-polarized frequency-reuse beams are to be provided by the antenna. In the following, we shall consider that certain of

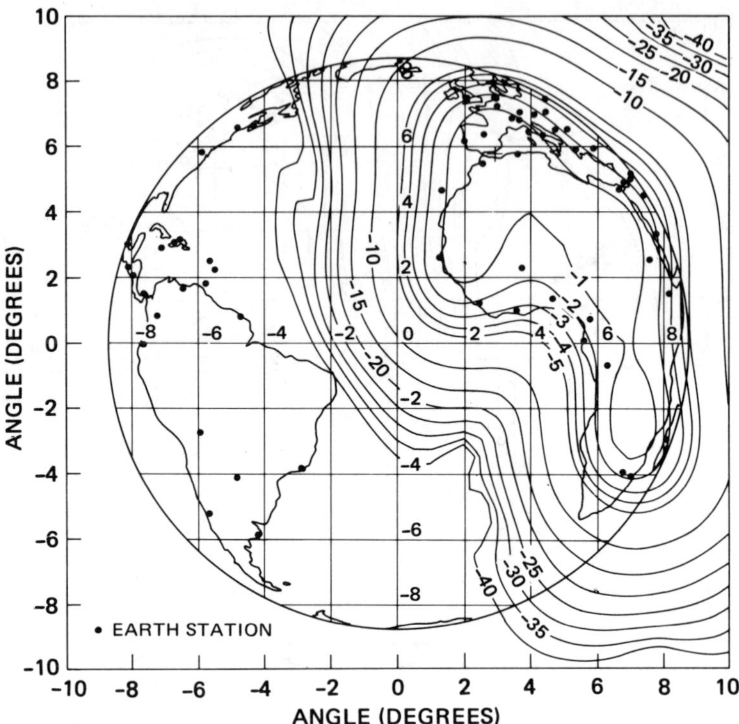

Fig. 5 Eastern hemisphere beam for 1.98-m reflector with 36 feeds.

the feed elements must radiate both senses of circular polarization. Thus, a dual-polarized element has two input ports. Excitation of one port causes the element to radiate left-hand circular polarization (LHCP). Excitation of the other port causes the element to radiate right-hand circular polarization (RHCP). The dual-polarized element can be a member of two subsets of elements in the total feed array, in which case it is used to produce two regional coverage beams that are polarized orthogonally.

Consider that the offset reflector radiates two overlapping beams that are polarized orthogonally to permit frequency reuse between the beams. Cross-polarized radiation determines the realizable polarization isolation between the beams. An offset paraboloid does not generate a cross-polarized (cross-pol) field when a perfect circularly polarized feed is located at the focal point.[3] However, cross-pol is generated when the feeds are displaced from the focal point, as is the case with a multielement array feed. For example, the maximum cross-pol amplitude of the experimental model antenna is about -47 dB (calculated) assuming perfect CP feed elements. The principal contributor to cross-pol radiation is the feed element itself, which must radiate both senses of polarization with a very low axial ratio. Thus, the experimental model antenna maximum cross-pol amplitude is -32 dB (calculated) assuming that the 20 feed elements are identical, each with an axial ratio of 0.4 dB. These calculations, along with measurements on the experimental model antenna, indicate that an element axial ratio (in the feed array environment) on the order of 0.4 dB is necessary to assure 27 dB polarization isolation between beams.

In order to exhibit a low axial ratio, the feed element radiation pattern must be rotationally symmetric (equal E-plane and H-plane beamwidths), and the element phase center must be unique. When the elements are arranged in close proximity in the array feed system (element spacings on the order of one wavelength), mutual coupling can degrade the axial ratio of the individual elements seriously. This occurs because mutual coupling is different for the two orthogonal polarization components of a circularly polarized element. A further complication is that mutual coupling, which is a function of frequency, element spacing, and the array lattice, is different for different types of elements. Thus, square aperture horns are less satisfactory than circular aperture horns in a circularly polarized array.[4] Because mutual coupling is a certain effect with multielement array feeds, the feed element design should include a means for "adjusting" the element in the array environment to compensate for mutual coupling. Alternatively, the elements can be "predesigned" to account for the mutual coupling perturbation, which is a function of element location and which can be measured experimentally by means of near-field probing techniques.

The considerations just outlined led to the development of a cup-dipole element for the dual-polarized array feed system. The element, which is described in detail in the following section, provides an axial ratio less than 0.4 dB over the frequency band 3.70 to 4.075 GHz. Moreover, the design of the element is such that it can be adjusted for the mutual coupling effects of the array environment.

III. Experimental Model Antenna Development

Dual-Polarized Cup-Dipole Element

The dual-polarized feed element that radiates both senses of circular polarization is a circular-cross-section cup excited by orthogonal coplanar dipoles. The cup is approximately 3.0 in. in diameter and 1.5 in. deep. The two dipoles excite

Fig. 6 Array feed for the Western hemisphere beam.

the cylindrical cavity in orthogonal TE_{11} circular waveguide modes. The cup-dipole radiates a pattern with equal E-plane and H-plane beamwidths when either dipole is excited. This is necessary, of course, to obtain a low axial ratio over the pattern beamwidth. The two orthogonal dipoles are driven by a 90° stripline hybrid to produce a circular polarization from the element. Excitation of one port of the hybrid produces RHCP; excitation of the other port yields LHCP. Figure 6 shows 20 of the cup-dipole elements assembled in an array feed that is designed to produce the Western hemisphere beam of Fig. 2. The cup-dipole elements are spaced 3.0 in. apart in an equilateral triangle lattice of contiguous apertures.

The excellent axial ratio characteristics of a single cup-dipole element are shown in Figs. 7a-7c. First, referring to Fig. 1, we point out that the reflector of the experimental antenna subtends a solid angle of ±30° about the feed tilt axis (angle θ_o). Figure 7a shows the radiation pattern of the element as measured at the LHCP port using a rotating linear source antenna. The frequency is 3.70 GHz. The axial ratio is less than 0.25 dB within ±30° of the beam maximum. At 3.88 GHz, the axial ratio is less than 0.4 dB over ±30°, as is shown in Fig. 7b. Figure 7c shows that the axial ratio is less than 0.4 dB over ±30° at 4.075 GHz. These patterns were measured in a plane (ϕ=0) containing one of the dipoles of the element. Measurements performed in the orthogonal plane ($\phi=\pi/2$) are equally good for both RHCP and LHCP ports. These data show the excellent pattern symmetry and axial ratio characteristics of the cup-dipole element. The design of the cup-dipole incorporates a balun and a two-step coaxial transformer section that matches the dipole impedance to a 50-Ω line with an input voltage standing wave ratio (VSWR) less than 1.06:1 over the frequency range 3.70 to 4.10 GHz.

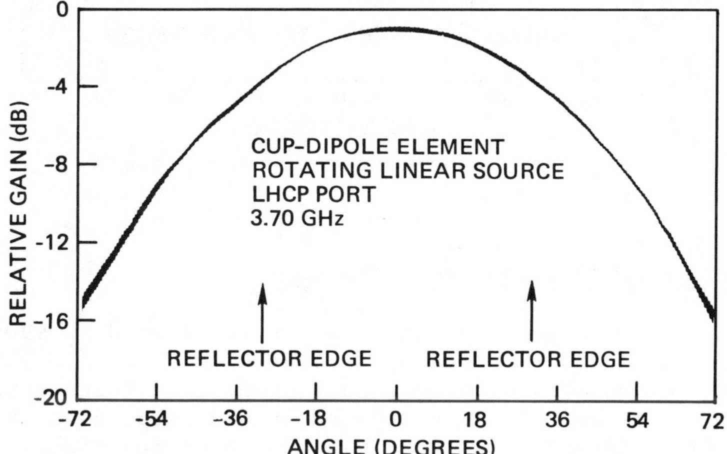

Fig. 7a Cup-dipole radiation pattern at 3.70 GHz.

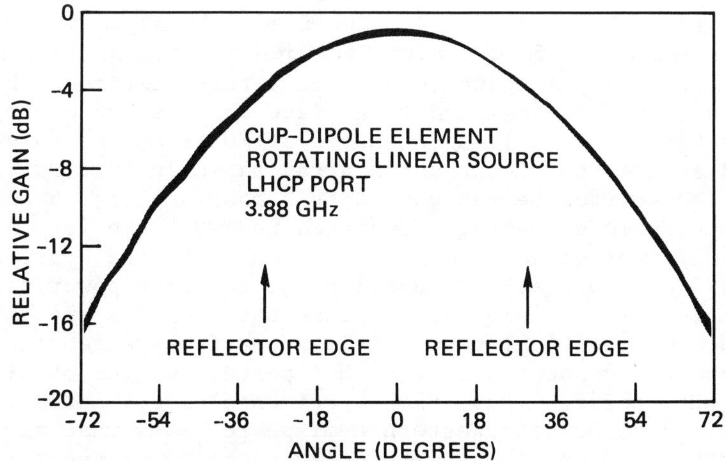

Fig. 7b Cup-dipole radiation pattern at 3.88 GHz.

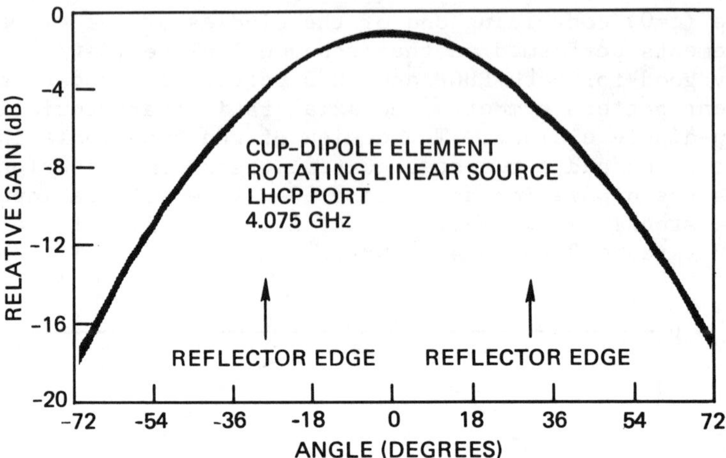

Fig. 7c Cup-dipole radiation pattern at 4.075 GHz.

Antenna Description and Test Results

An experimental model antenna was fabricated to substantiate the low-sidelobe beam synthesis technique; to evaluate the performance of the cup-dipole element in a large array feed system; and to assess the effect of mutual coupling on beam contouring, sidelobe control, axial ratio of individual elements, cross-polarized radiation, and polarization isolation. The experimental model antenna is shown in Fig. 8. The antenna is a 60-in.-diam offset reflector with a 45-in. focal length. The reflector is an off-axis sector of a 150-in.-diam parent paraboloid, where $F/D_p = 0.3$. The feed tilt angle θ_o is 49.25°. The array feed, which includes a total of 45 cup-dipole elements, is shown mounted on a test fixture in Fig. 9. The feed system is designed to radiate the Western hemisphere beam of Fig. 2 by exciting a subset of 20 elements, which are located at the top of the total array shown in Fig. 9. The 20-element subset can be recognized by comparing Figs. 6 and 9. Each cup-dipole element is connected to a 90° hybrid. A 20-way stripline power divider board is connected to the hybrids by semirigid coaxial cable to provide the required power distribution to the array elements. Because the elements are dual polarized, two identical power divider boards were fabricated; one board was connected to the LHCP ports, and the other board to the RHCP ports. Thus, the experimental model antenna can radiate two coincident Western hemisphere beams that are polarized orthogonally. The remaining 25 elements of the array feed are not excited. Instead, the input ports of the cup-dipole elements are terminated in matched loads in order to evaluate mutual coupling effects that would exist in a complete East and West beam feed system. Thus, the 25 loaded elements simulate an Eastern zone feed system.

A near-field test facility has been implemented at TRW Defense and Space Systems Group to aid in the development and adjustment of complex array feeds. Mutual coupling between feed elements perturbs the excitation coefficients and axial ratios of the array radiators. It is virtually impossible to diagnose the performance of a large array feed once it is installed in the reflector and the antenna is under test on a far-field pattern range. This major problem is solved by using the near-field test facility to measure and adjust the array feed prior to installing the feed in the reflector. Thus, the near-field test facility is used as a diagnostic tool to measure and adjust for mutual coupling effects and, thereby, guarantee the performance of the feed system when it is mounted in the reflector.

Fig. 8 Experimental model antenna: 60-in.-diam reflector, 45-element feed.

Fig. 9 45-element array feed on near-field test fixture.

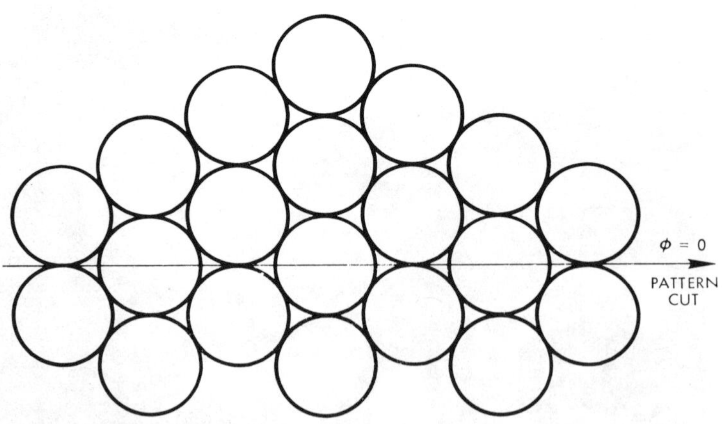

Fig. 10 Layout of the 20-element Western hemisphere array feed.

A probe is scanned over the array feed aperture in the near field, and the complex (amplitude and phase) vector field components are measured and recorded. Computer software processes the near-field data and calculates the far-field radiation pattern of the array feed and, also, the far-field radiation pattern of the reflector antenna. Particularly important is that the near-field probing can detect the amplitude and phase differential of the orthogonal field components of a circularly polarized element. Suffice it to say that we consider near-field testing as absolutely necessary for the successful development of dual-polarized array feeds.

The 45-element array feed is shown in the near-field test facility in Fig. 9. A layout of the 20-element feed that produces the Western hemisphere beam is presented in Fig. 10 to identify the plane $\phi=0$. The correlation between near-field measurements and calculated far-field radiation patterns is illustrated with measurements performed at 3.83 GHz. Figure 11a shows the calculated (from near-field data) far-field radiation pattern of the array feed in the plane $\phi=0$ for linear polarization. The location of the edge of the reflector relative to the array feed pattern is indicated on the figure. Figure 11b shows the far-field pattern of the array feed as measured in an anechoic chamber using a rotating linear source antenna. As can be seen, the correlation between the measured and calculated far-field feed pattern is quite good.

The correlation between the measured and calculated far-field pattern of the offset reflector is illustrated in Figs. 12a and 12b. The patterns are two-dimensional constant gain contours over a 20° x 20° field of view about the boresight axis of the reflector. Figure 12a shows the reflector radiation pattern that was calculated from near-field measurements of the array feed at 3.83 GHz. This pattern was obtained before any adjustments were made on the array feed to compensate for mutual coupling perturbations. After the array feed was mounted in the reflector, the far-field pattern of the antenna was measured at 3.83 GHz. The measured pattern is shown in Fig. 12b. Notice the remarkable correlation between the traces of the -30 dB contours on the right-hand side of Figs. 12a and 12b. These data clearly show the accuracy and the utility of the near-field test facility for predicting reflector antenna performance from near-field feed measurements.

Near-field probing of the feed aperture verified that mutual coupling had, indeed, perturbed the axial ratios of the array feed elements. The effects were sufficiently great that

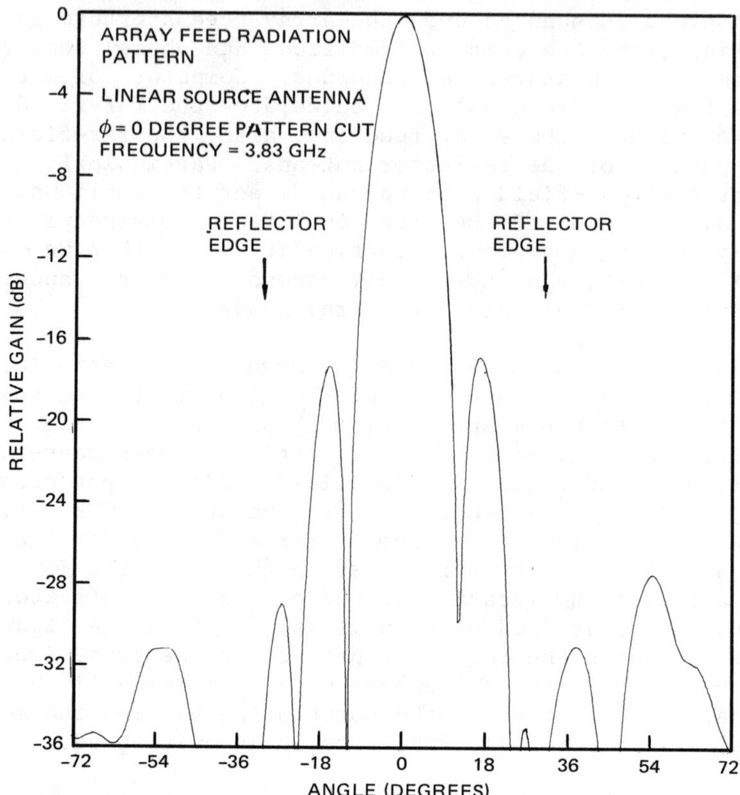

Fig. 11a Array feed radiation pattern calculated from near-field feed measurements.

small adjustments in amplitude and phase were necessary to restore the low axial ratio of the cup-dipole elements. Following these adjustments, the array feed was mounted again in the offset reflector, and pattern measurements were performed at frequencies of 3.70, 3.88, and 4.075 GHz. Data will be presented here for the midband frequency 3.88 GHz. First, the voltage coefficients at the output ports of the 20-way power divider were measured in the laboratory. These quantities represent the input coefficients to the array elements in the absence of mutual coupling. Using these coefficients, the radiation pattern of the reflector was calculated at 3.88 GHz. The pattern, which is idealized in the sense that mutual coupling is neglected, is shown in Fig. 13a. Basically, this pattern shows that the design of the stripline power divider network was implemented correctly to produce the Western hemisphere beam of Fig. 2.

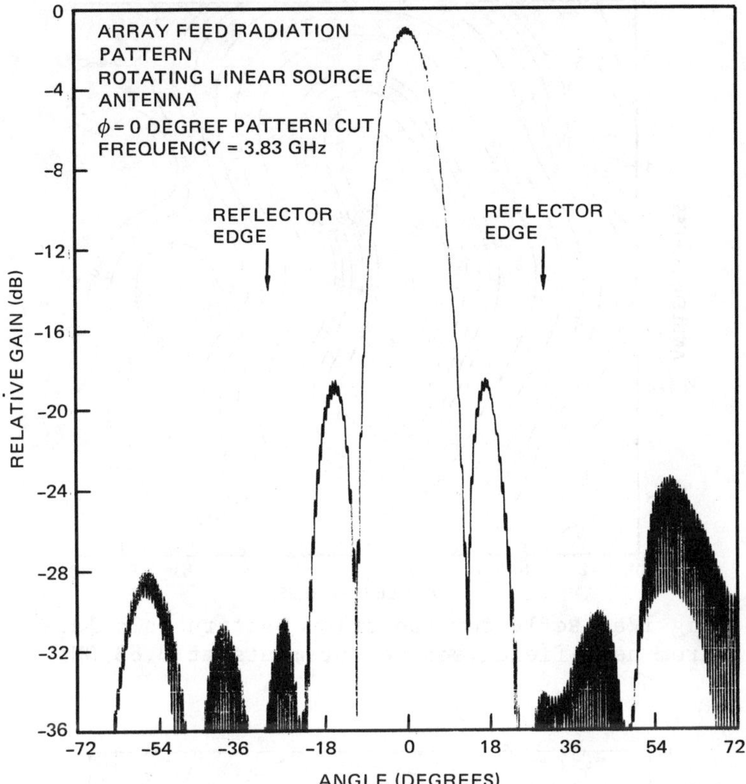

Fig. 11b Measured far-field pattern of the array feed.

The measured principal polarization pattern (LHCP) of the experimental model antenna is shown in Fig. 13b. In contrast to Fig. 12b, Fig. 13b is a measured pattern obtained after the array feed adjustments were carried out. As can be seen, there is excellent agreement between the design goal that is Fig. 13a and the measured results given in Fig. 13b.

The polarization isolation that results between two orthogonally polarized beams is determined by the cross-polarized radiation associated with each beam. Assuming that regional coverage in the principal polarization component is achieved within the -3 dB gain contour, then the cross-polarization gain level must be -30 dB or less to insure -27 dB polarization isolation between beams. Recognizing this requirement, the cross-polarization pattern (RHCP) of the experimental model antenna as measured over the 20° x 20° field of view is shown in Fig.

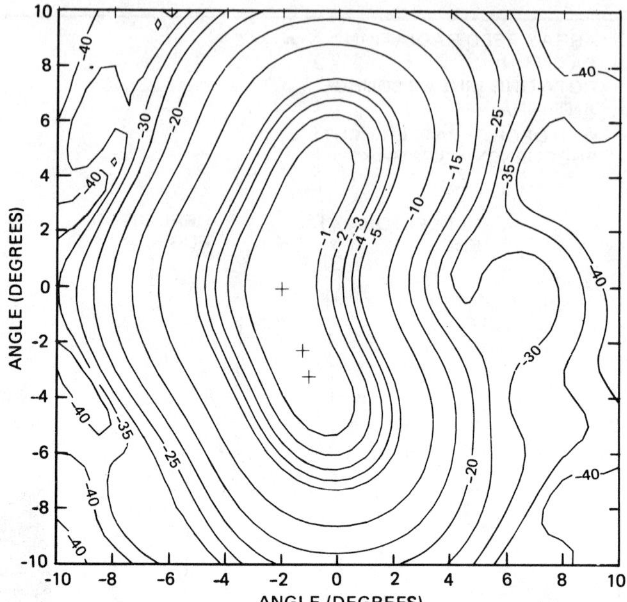

Fig 12a Reflector radiation pattern calculated from near-field feed measurements at 3.83 GHz.

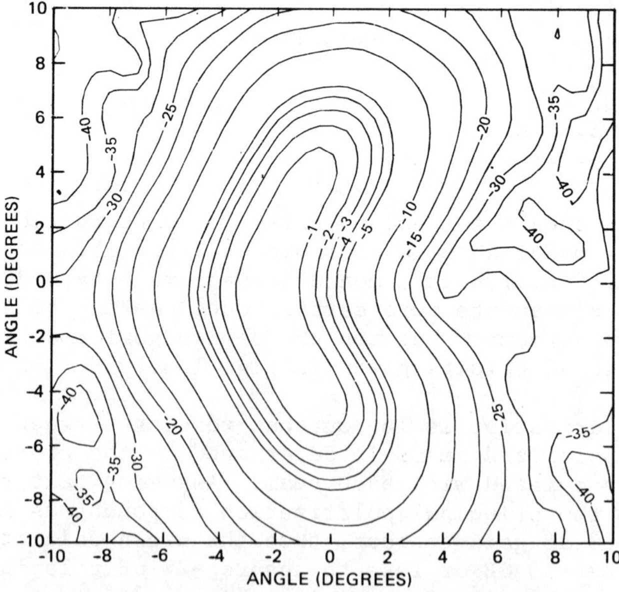

Fig. 12b Measured far-field pattern of reflector antenna at 3.83 GHz.

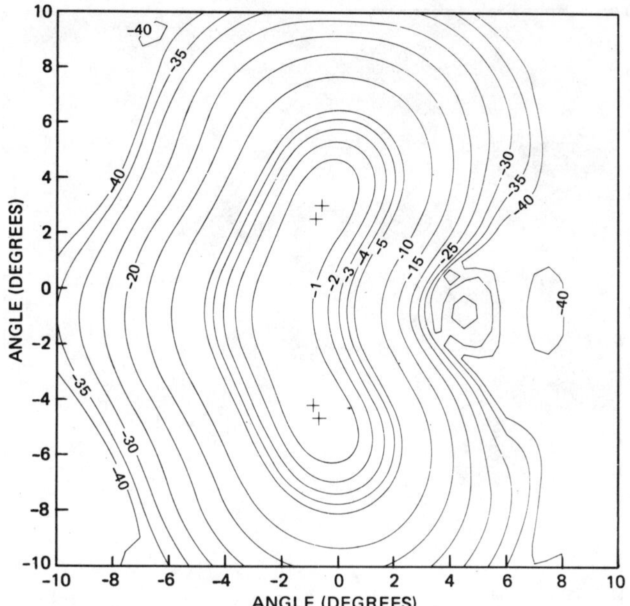

Fig. 13a Calculated radiation pattern (LHCP) of experimental antenna at 3.88 GHz.

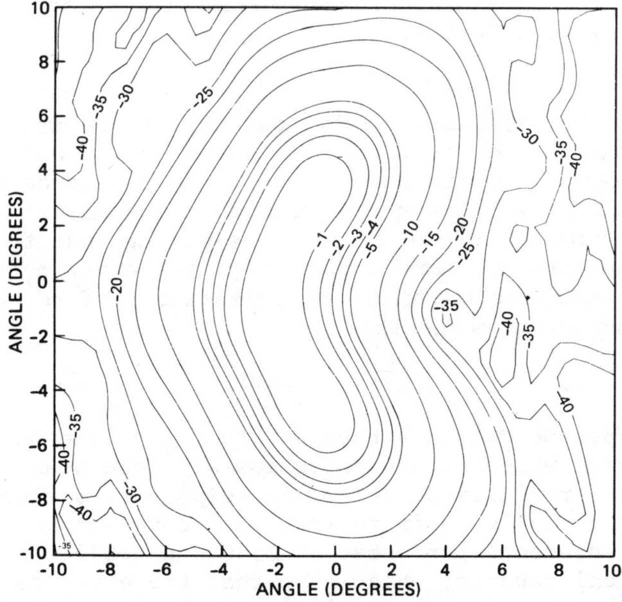

Fig. 13b Measured radiation pattern (LHCP) of experimental model antenna at 3.88 GHz.

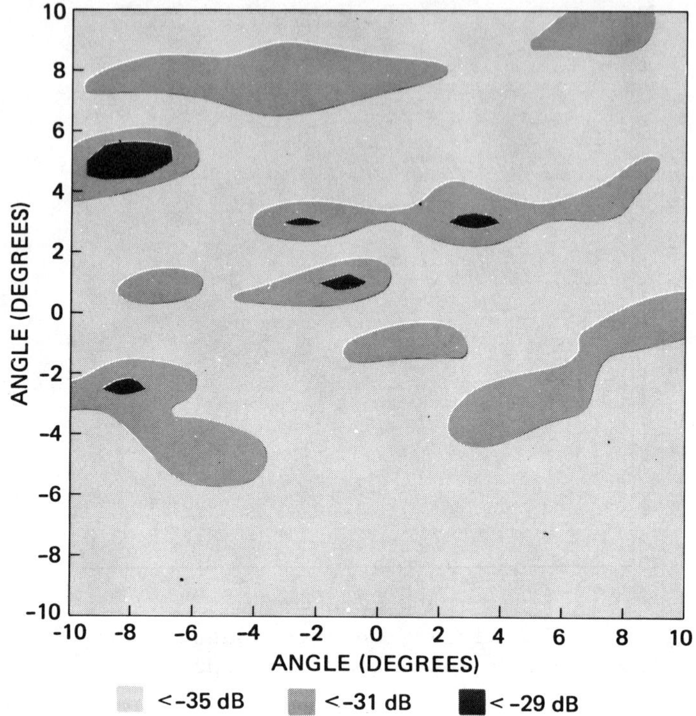

Fig. 13c Measured cross-polarization pattern (RHCP) at 3.88 GHz.

13c. Inspection of this figure shows that there are four small regions where the cross-pol level reached -31 dB and one point near the upper right-hand corner where the amplitude was -29 dB. Otherwise, the cross-pol amplitude is considerably less than -31 dB throughout the area. Considering that these data are the first test results obtained for the experimental model antenna, we consider the cross-polarization performance of the antenna to be quite acceptable.

IV. Summary

The experimental data have substantiated the basis of the antenna design which incorporates low-sidelobe beam synthesis with a dual circularly polarized cup-dipole feed element to achieve orthogonally polarized regional coverage beams. Experimental measurements performed at 3.80 and 4.075 GHz confirm our analytical studies, which show that the array feed system is not frequency-sensitive and is suitable for operation over bandwidths as great as 12%. Previous work in phased array antennas demonstrated that mutual coupling has a significant effect in circularly polarized arrays.[4-6] Anticipating similar effects for an array feed in a reflector, the cup-dipole element was developed specifically because of its inherent low axial ratio and because it can be adjusted to compensate for mutual coupling in the array environment. From our experience with the experimental model antenna, we are convinced that near-field testing of the array feed is absolutely essential to the successful development of dual-polarization array feeds.

Acknowledgment

The authors are pleased to acknowledge the value of many technical discussions with their co-worker, Chao-Chun Chen.

References

[1]Duncan, J. W., "Multiple Beam Antennas for Communication Satellites," EASCON '75 Record, Sept. 1975, Publ. 75 CHO 998-5 EASCON, Institute of Electrical and Electronics Engineers.

[2]Duncan, J. W., "Low Sidelobe Antenna Systems," Systems Docket 73-191, 1973, TRW; patent application filed with U.S. Patent Office.

[3]Chu, T. S. and Turin, R. H., "Depolarization Properties of Offset Reflectors," IEEE Transactions on Antennas and Propagation, Vol. AP-21, May 1973, pp. 339-345.

[4]Chen, C. C., "Wideband Wide-Angle Impedance Matching and Polarization Characteristics of Circular Waveguide Phased Arrays," IEEE Transactions on Antennas and Propagation, Vol. AP-22, May 1974, pp. 414-418.

[5]Tsandoulas, G. N. and Knittel, G. H., "The Analysis and Design of Dual-Polarization Square-Waveguide Phased Arrays," IEEE Transactions on Antennas and Propagation, Vol. AP-21, Nov. 1973, pp. 796-808.

[6]Lee, S. W., "Aperture Matching for an Infinite Circular Polarized Array of Rectangular Waveguides," IEEE Transactions on Antennas and Propagation, Vol. AP-19, May 1971, pp. 332-342.

A BEAM-SHAPING MULTIFEED OFFSET REFLECTOR ANTENNA

C. C. Han,* A. E. Smoll,† H. W. Bilenko,‡
C. A. Chuang,§ and C. A. Klein§

Ford Aerospace and Communications Corp., Palo Alto, Calif.

Abstract

An efficient procedure for the design of an array-fed offset reflector spacecraft antenna has been developed and employed to generate an arbitrarily shaped beam. This design procedure uses several analysis programs and an automatic optimization computer program to find the set of feed array element excitations which best meets all of the objectives. The objective function specifies the minimum coverage gain, the maximum coverage gain variation, and the maximum radiation level permitted in the outer isolated regions. The design procedure and the optimization process are described. An example is presented, along with predicted and corroborating measurements.

I. Introduction

Shaped multibeam spacecraft antennas that allow frequency reuse via both sidelobe and cross-polarization isolation have drawn much attention in recent years. This category of antennas will be one of the most important components in future communication satellite systems to satisfy increased traffic demands and to provide greater routine flexibility. Possible multibeam antenna designs are reflectors, lenses, and phased arrays. The approach adopted here is an offset reflector antenna system with multielement array feed, which has a lower cost than lens or phased array systems.

In the array-fed reflector, each feed element separately illuminates the reflector to generate a component beam in the far field.

Presented as Paper 76-249 at the AIAA/CASI 6th Communications Satellite Systems Conference, April 5-8, 1976, Montreal, Canada.

*Supervisor, Antenna Technology Section, Western Development Laboratories Division (WDL).

†Principal Engineer, WDL.

‡R&D Engineer, WDL.

§Engineering Specialist, WDL.

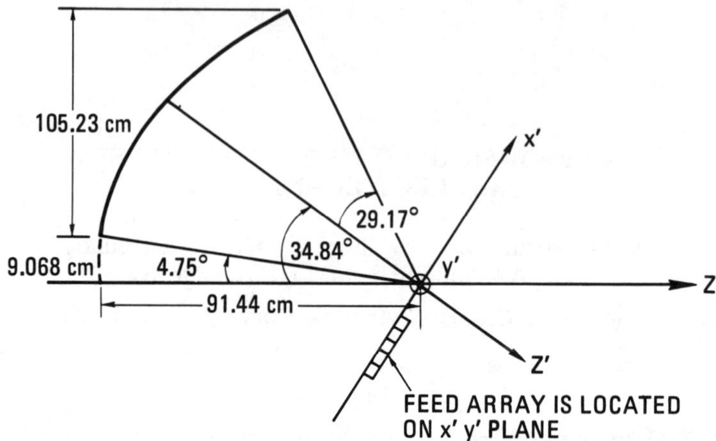

Fig. 1 Physical parameters of model antenna configuration.

By properly exciting feed elements and thus summing individual component beams, the desired shaped beam is achieved to serve a specific ground coverage area. Depending on the feed-array element, this system can be operated for any linear or circular polarization.

Circular polarization (CP) is attractive for an offset reflector when polarization diversity is used, since an offset parabolic reflector does not generate a cross-polarized signal when the feed has perfect CP pattern.[1] Thus for CP beams, good polarization isolation and axial ratio can be obtained by properly designing the element and array configuration.

Although the basic concept of this system is straightforward, implementation is not simple, especially when both left-hand (LHCP)

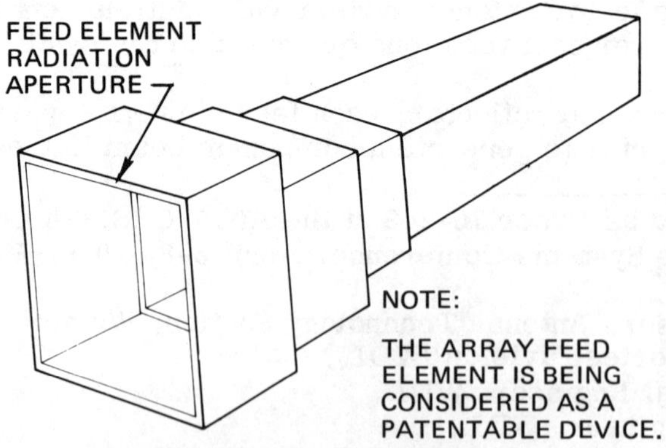

Fig. 2 Array feed element.

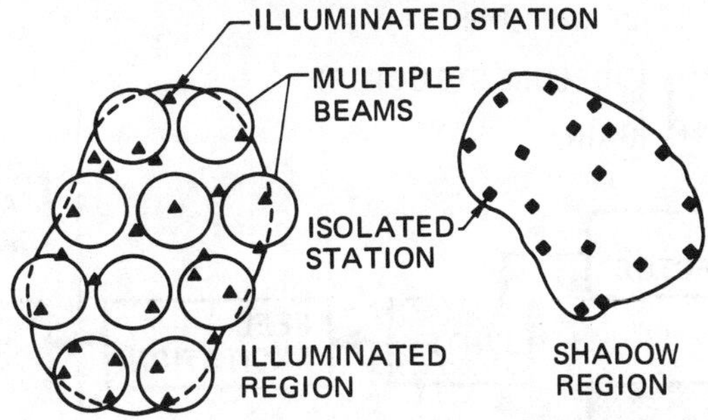

Fig. 3 Setup of multibeam synthesis problem.

and right-hand (RHCP) circular polarizations are used. As an illustration, consider an antenna system design that satisfies specific beam requirements of the INTELSAT V spacecraft. The frequency ranges from 3.704 to 4.073 GHz. There are two hemispherical beams of RHCP, East and West; and two narrower zone beams of LHCP, one in the East and the other in the West. The axial ratios of these beams within the coverage area should be no less than 0.75 dB. A gain variation over a 0.4° chord centered at any location within the coverage area must not be greater than 1.2 dB. The net isolation between beams, which is due to sidelobe and/or cross-polarization isolation, including saturated channel output powers, must be at least 27 dB.

II. A Model Configuration

The proposed design of a spacecraft antenna is an offset parabolic reflector antenna fed by a cluster of square waveguide elements. The physical parameters of the antenna are shown in Fig. 1. The reflector has a projected aperture diameter of 105.23 cm and an F/D ratio of 0.4, where F is the focal length and D is the aperture diameter of the symmetrical reflector from which the offset reflector is derived. A reflector of this size and F/D ratio is sufficient to produce appropriately shaped beams with a sharp edge rolloff, low sidelobes, and good isolation between coverage regions, while also satisfying the spacecraft physical constraints.

The feed consists of a cluster of square waveguide elements arranged on a planar, closely packed array configuration located at the optimum reflector scan plane. Each feed element consists of a coax-to-waveguide transition, a septum polarizer, and a step transformer with dual input ports that provide for both senses of circular polarization. The septum polarizer (Fig. 2) consists of a specially tapered, longitudinal, metallic-vane phase shifter in a square guide. It exhibits an extremely low axial ratio over the maximum required band. The two input ports for opposite senses of circular polariza-

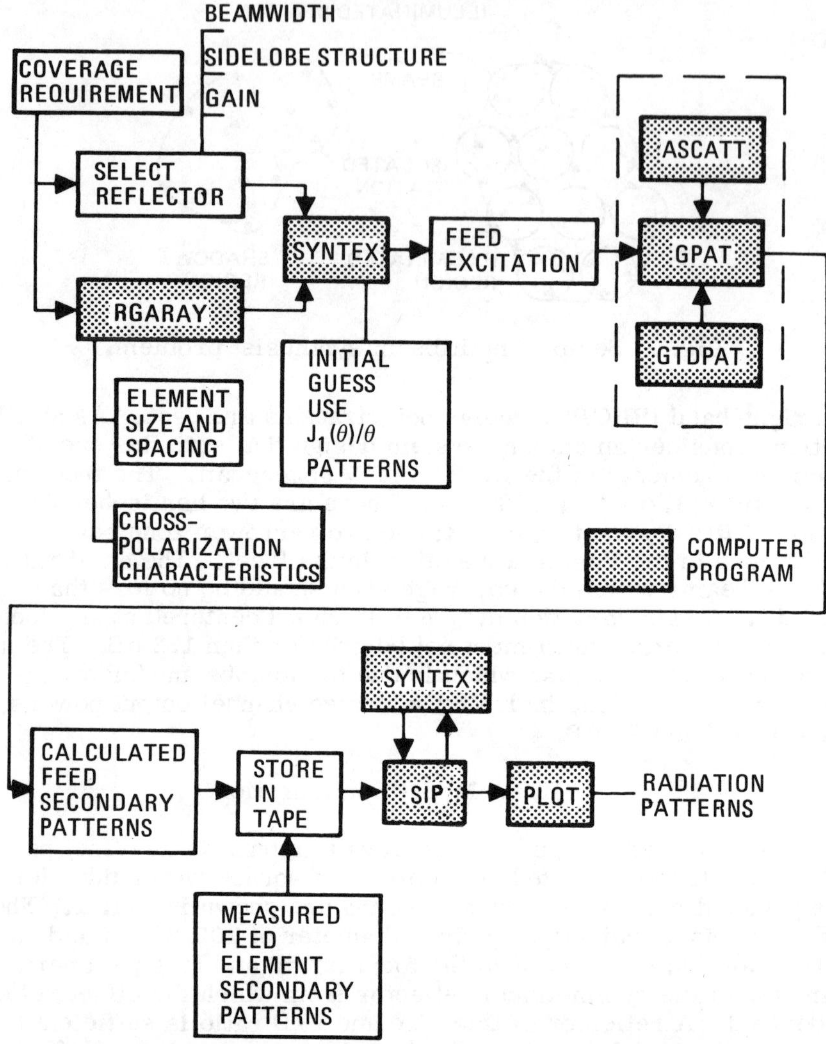

Fig. 4 Shaped-beam reflector antenna design flow chart.

tion emerge from one end of the polarizer as two rectangular guides, thus allowing the feed elements to be closely packed.

III. Design Methodology and Beam Synthesis Techniques

To design an offset reflector with multiple feeds requires an effective and practical technique for determining the feed excitation voltages (both amplitude and phase) required to achieve the desired radiation pattern. A new technique has been developed which involves a number of iterations using beam synthesis programs and reflector

Table 1 Applicable antenna computer programs

Name and function	Method
GPAT: Analyzes characteristics of large general-shape reflector antennas. Input: feed position and pattern, reflector geometry, blockage support (optional), feed polarization. Output: efficiency breakdown, secondary pattern, plots, directivity.	Aperture fields method, using geometric optics collapsed to equivalent line source. Aperture axis may be tilted to stationary phase direction.
ACATT: Computes scatter pattern from general-shape reflector antenna. Input: feed position and pattern (including all polarization components), reflector geometry (may be tilted). Output: scatter patterns (for all polarizations).	Physical optics using Ludwig's integration method. Feed pattern reduced to spherical harmonics, including near-field effects.
GTDPAT: Computes the wide-angle sidelobes of a large arbitrary-shape reflector antenna. Input: feed position, polarization and pattern, reflector geometry. Output: secondary pattern.	The geometrical theory of diffraction and equivalent (ring) current method.
RGARAY: Impedance matching and active-element pattern computation of a circularly polarized, infinite array of rectangular waveguides. Input: array configuration, excitation, dimension of matching irises and dielectric. Output: transmission and reflection coefficients, axial ratio of transmitted field, and active-element pattern.	Matching the array by using dielectric loading and irises at the open ends of the waveguides. Mode matching techniques and periodic array theory are used.
SIP: Superimposes any number of arbitrarily directed antenna beams originating from a common phase center. Input: individual radiation distributions and axis directions. Excitation inputs to each beam. Output: composite radiation distribution referenced to mainframe.	Higher-order interpolation to find the individual patterns referenced in the mainframe. Superimposition may be performed on either complex voltage or power basis.
SYNTEX: Finds the set of excitation inputs to a number of arbitrarily directed antenna beams, to achieve a specified radiation coverage in terms of illuminated and isolated areas. Input: ripple (allowable variation of intensity) over the illuminated area; isolation between illuminated and isolated areas; set of directivity of coefficients for each beam at each grid point. Output: set of excitations for the beam array, which achieves smallest maximum (minimax) weighted error over entire coverage area.	Generates set of initial beam excitations; corrects excitation to each beam in turn using a minimax weighted error criterion.
PLOTTING PROGRAMS: These several programs can produce 1) pattern amplitudes and/or phase vs angle along a specified plane, or 2) two-dimensional contour map of the total radiation distribution. Input: file selection and grid. The preceding antenna programs produce a standard output file that is acceptable as input to these plotting programs.	HW software produces graphical output on a Tektronix 4020 Graphics terminal and/or a Versatec Matrix 2000 printer.

Table 1 Applicable antenna computer programs

Fig. 5 Antenna assembly on test range.

analysis programs. The design and optimization procedure is as follows:

1) The reflector size is selected based on sidelobe isolation requirements, shaping of the coverage radiation pattern, and mechanical constraints.

2) The configuration of the feed array, including feed element size and array type, is determined by the cross-polarization characteristics of the array using the waveguide array program (RGARAY).

3) The desired area is covered with an array of beams, each produced by an individual feed element (Fig. 3). The size of each

Fig. 6 Feed array with complete feeding network.

beam is determined by the size of the reflector. The beam position and the angular separation between beams are determined by the feed element separation and the beam deviation factor of the offset reflector. For the initial run, the pattern of the form $J_1(\theta)/\theta$ is assumed, where θ is the spatial angle measured from the peak of each beam, and J_1 is the Bessel function of the first order and kind.

4) The computer program SYNTEX (synthesize excitations) is run in order to derive a set of voltage excitations for the beams covering the desired area. SYNTEX attempts to satisfy gain slope

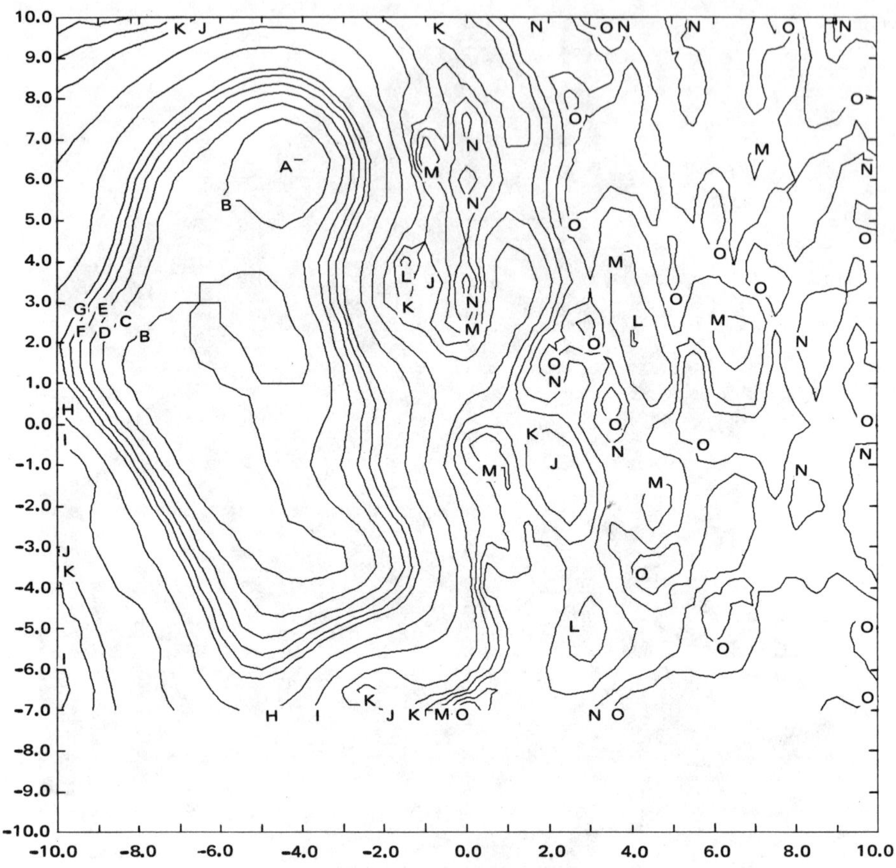

COMPOSITE NO BLOCKAGE 9.17 GHz POL = RHCP 2/15/77 REFERENCE PLOT

CONTOUR DATA*			
SYMBOL	LEVEL	SYMBOL	LEVEL
A	0.000	I	-15.000
B	-1.000	J	-21.000
C	-2.000	K	-24.000
D	-3.000	L	-27.000
E	-4.000	M	-30.000
F	-5.000	N	-35.000
G	-6.000	O	-40.000
H	-10.000		

*1.700 WAS ADDED TO INPUT DATA

Fig. 7 Measured radiation contour patterns of model antenna: principal polarization (RHCP).

and variation constraints in the shadow region. In general, the number of stations having gain and isolation constraints exceeds the number of beams by a factor of 3 or more, requiring the use of a mini-

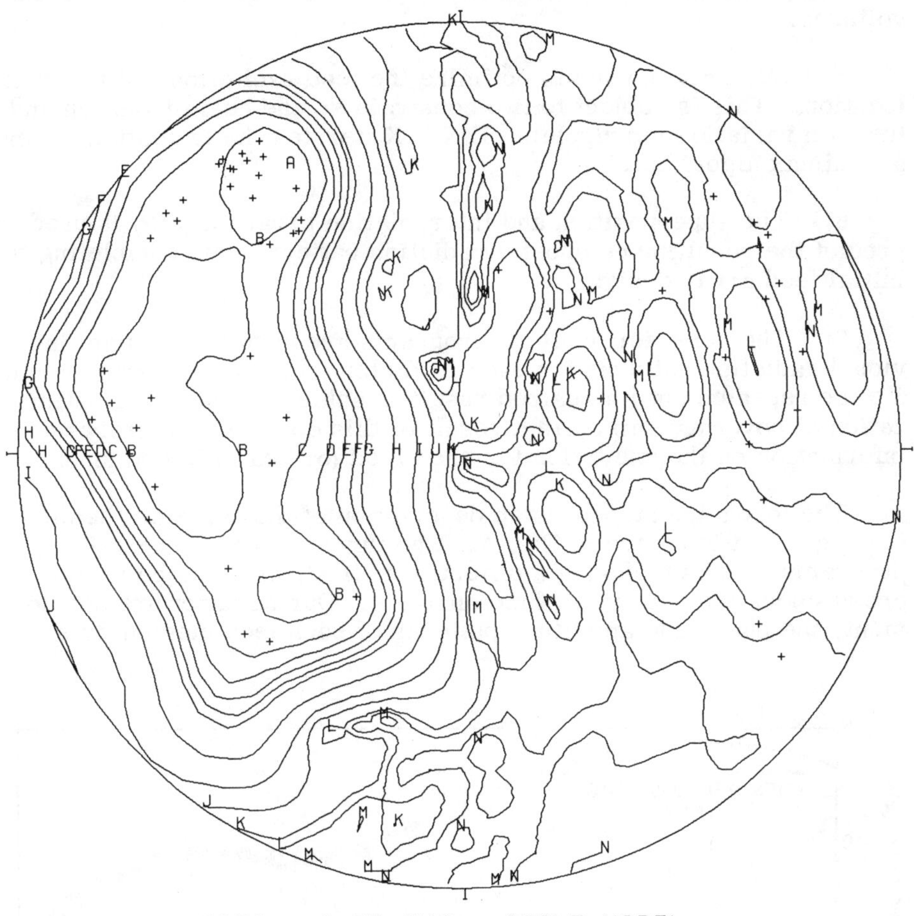

FREQ = 9.27 GHZ, SCALE MODEL
PLOT NO. 3.0 01/13/77 20.692

CONTOUR DATA	
SYMBOL	LEVEL
A	0.000
B	-1.000
C	-2.000
D	-3.000
E	-4.000
F	-5.000
G	-6.000
H	-10.000
I	-15.000
J	-20.000
K	-27.000
L	-30.000
M	-35.000
N	-40.000

MAX. -28.002 WAS ADDED
THETA TO INPUT DATA
10.000

Fig. 8 Calculated radiation contour pattern of model antenna: principal polarization (RHCP).

mum-error technique rather than an exact-solution method. The algorithm employed by SYNTEX is a minimax-error analysis. Actual received voltage is compared with desired voltage at each station, and the maximum voltage error over the entire set of stations is reduced progressively by successive iterations over the set of beam voltages.

5) SYNTEX is used to optimize the feed excitation and the beam location. This is subject to two constraints: the beam isolation and the gain variation (or ripple). SYNTEX iterates the excitations using a minimax algorithm.

6) The superposition and interpolation program (SIP) is used to predict the resulting composite radiation pattern, thus completing the initial feed array design.

7) The reflector analysis programs are used to calculate individual radiation patterns for each feed element, allowing steps 5 and 6 to be repeated until a desired result is achieved, including optimization of coverage gain. As a result of this optimization process, information on the excitation for each feed (or beam) is obtained.

The foregoing step-by-step description of the beam synthesis technique is illustrated in Fig. 4. The three reflector analysis programs, each employing a different method, also can be used to cross-check the accuracy of the analysis. During hardware development, the phase and amplitude patterns of each feed element in an

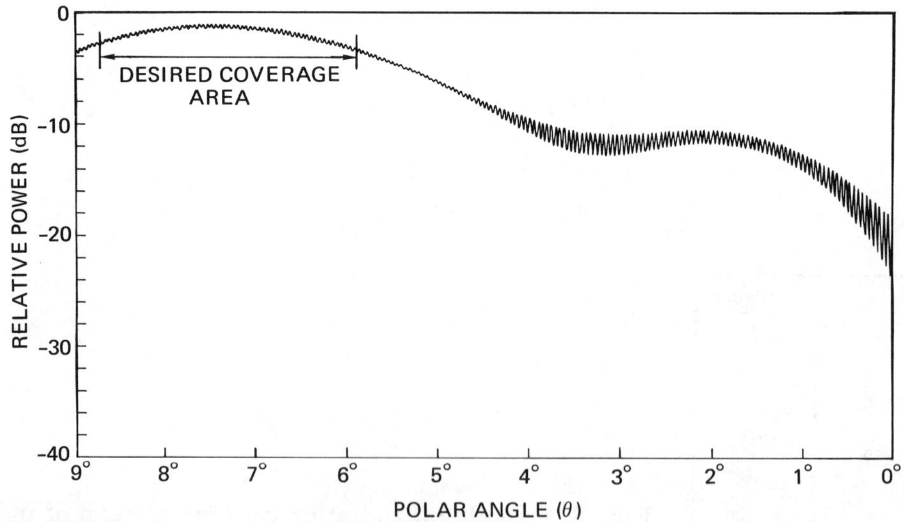

Fig. 9 Typical measured axial ratio pattern ($\phi = 130°$).

array environment will be measured and supplied to the reflector analysis program to obtain the secondary beam (component beam). At that time, the optimization program will take all these secondary patterns and go through the optimization process to generate the proper excitation of each feed horn for a coverage beam. This stage has been described in the design flow chart of Fig. 4. The functions and methodology of the programs just described are summarized in Table 1.

IV. Antenna Model Test Results

A model antenna was designed. Figure 5 shows the antenna assembly on the test range, and Fig. 6 shows the feed array with a complete feeding network. The physical parameters of the model antenna were shown in Fig. 1. To obtain low circular cross polarization, it is necessary to tune the aperture of each feed element in the array cluster. A technique using metallic tuning mechanisms in the aperture was developed[2] to reduce mutual coupling and to equalize E- and H-plane patterns of the active element patterns. A feed distribution network consisting of stripline power dividers, attenuators, and line stretchers was used to provide feed elements with proper amplitude and phase.

The test was conducted on a range of $3D^2/\lambda$ in length at the testing frequency. The range reflection level was found to be better than 42 dB. A measured two-dimensional radiation contour for principal polarization is shown in Fig. 7. The calculated -4 dB contour is included for comparison. As shown, the shape of the measured principal polarization is very close to that of the calculated radiation pattern. The sidelobe level over the shadow region is below -30 dB. Figure 8 shows the calculated pattern, and Fig. 9 shows a typical axial ratio pattern.

V. Conclusion

An efficient design procedure has been developed for array-fed reflector antennas to synthesize an arbitrarily shaped pattern with isolation ripple and gain requirements. The computer optimization system uses assumed patterns only for an initial guess; later beam excitations are based on actual computed patterns. In the final hardware phase, measured individual patterns can be used. Comparison of calculated and measured results demonstrates the validity of this design approach.

Acknowledgment

The authors wish to acknowledge the consultation services of S. W. Lee of the University of Illinois during the course of developing this design technique.

References

[1]Chu, T. S. and Turrin, R. H., "Depolarization Properties of Offset Reflector Antennas," IEEE Transactions on Antennas and Propagation, Vol. AP-21, May 1973, pp. 339-345.

[2]Schennum, G. H., Han, C. C., and Gould, H. J., "Use of Equalizers to Reduce the Cross-Polarization Response in a Waveguide Array," to be presented at the 1977 Antenna Applications Symposium, April 27-29, 1977, Monticello, Illinois.

DEVELOPMENT OF MULTIPLE-BEAM LENS ANTENNAS

W. G. Scott,* H. S. Luh,+ T. M. Smith,≠ and R. H. Grace§

Aeronutronic Ford Corporation, Palo Alto, Calif.

Abstract

Two different microwave lens antennas are described with feed arrays: 1) a TEM lens intended for simultaneous frequency reuse of multiple pencil beams at C band, transmit, and receive; and 2) a waveguide lens intended for variable shaped-beam operation for narrow-band applications. In each case, an experimental pattern test model has been built and subjected to radiation pattern tests. The TEM lens test patterns verify that -30-dB sidelobe levels can be attained for beam positions over an 18° conical field of view for wide frequency band. The waveguide lens measured patterns show that, for an 18° conical field of view, beam shape can be varied from a narrow pencil beam in any direction to an earth-coverage, 18° wide, flat-topped conical beam with low axial ratio circular polarization. For both lenses, it was found that mutual coupling and lens/feed array scattering phenomena become significant, particularly if pattern shapes at low amplitude levels are important.

Introduction

This paper describes the design and electrical development testing of two different microwave constrained lens antennas intended for different "single-aperture multiple-beam" appli-

Presented as Paper 76-250 at the AIAA/CASI 6th Communications Satellite Systems Conference, April 5-8, 1976, Montreal, Canada.

*Multi-beam Antenna Section Supervisor, Western Development Laboratories Division.

+Senior Engineering Specialist, Western Development Laboratories Division.

≠R&D Engineer, Western Development Laboratories Division.

§Formerly, Western Development Laboratories Division; currently with Avantek Corporation.

cations onboard future synchronous orbit communication satellites. The first antenna is called a TEM lens for the transverse electromagnetic (mode) transmission lines that interconnect a pickup array of small radiators on the lens primary, or inner, surface (S_1) with a similar array of radiators on the secondary, or outer, surface (S_2) of the lens. An array of 127 small feed elements driven from a multiple-port beam-forming network (MPBFN) illuminates S_1 (see Fig. 1). This lens is characterized by broad frequency bandwidth (30% to 60%) and simultaneous multiple pencil ($\sim 3.5^o$) beams designed for low sidelobes (-30 dB) for spatial frequency-reuse satellite applications.

The second lens is for narrower-band applications of typically 5%. It consists of a waveguide lens formed of thin-walled square hollow conducting waveguides nested into a square grid wherein each straight pipe is a transmission line interconnecting its two open-end radiators, which form the S_1 and S_2 lens surfaces. The lens, illustrated in Fig. 2, is fed by an array of 61 waveguide horns driven from a single port by a command-controlled variable power dividing beam-forming network (VBFN). This lens can be commanded to produce a large variety of beam shapes, ranging from a single narrow (2°) pencil beam pointed anywhere over the Earth, to intermediate "beams" of multiple noncontiguous lobes, to broader pencil beams, up to an entire Earth coverage flat-topped beam, 18° in width. The variable shaped-beam control is afforded by the incorporation of voltage-controlled solid-state microwave variable power dividers (VPD's) into the VBFN.

Both lens antennas have an Earth coverage 18° field of view (FOV). Both may be termed space-fed arrays, and both lenses are constrained in that ray paths through each lens are constrained to follow the geometric paths of the transmission

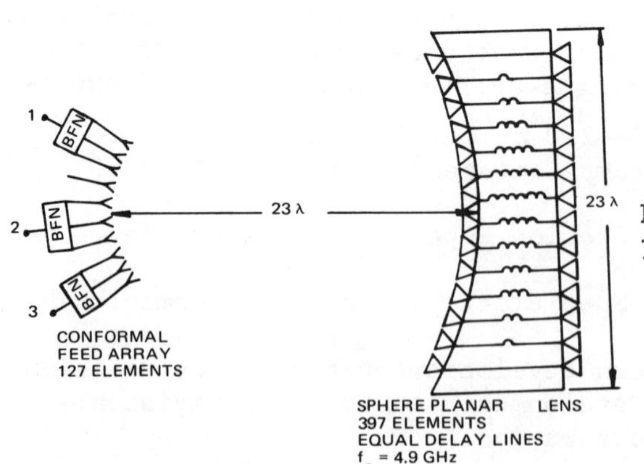

Fig. 1 Multiple-beam TEM lens diagram.

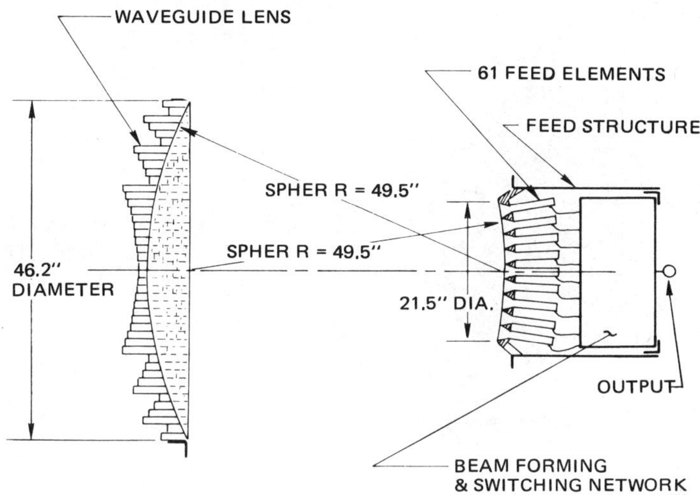

Fig. 2 Waveguide lens and variable shaped-beam feed array.

lines interconnecting S_1 and S_2 rather than obeying Snell's law of refraction at the S_1 and S_2 interfaces, as do the more familiar solid dielectric optical lenses. The following sections of the paper describe briefly the results of pattern testing of an electrical feasibility model of each of these antennas.

TEM Lens Multiple-Beam Antenna

The TEM line lens has a spherical S_1 surface and a planar S_2 surface, with all interconnection lines of equal length. Such geometry provides excellent small-angle scan focusing necessary to maintain low sidelobes for all beam positions over the lens FOV. The common realization of the TEM line is round coaxial cable, and a lens consisting of literally thousands of such cables interconnecting small dipole or horn radiators on S_1 and S_2 is the obvious form of this antenna. However, for satellite use, such a lens tends to be heavy, is difficult to assemble, and would have production and connector reliability problems. The use of printed-circuit techniques, such as open microstrip lines, permits a close approximation to the exact TEM configuration while offering the advantages of low weight (thin PC cards), high reliability (no connectors), and ease of quantity precision production (photoetching). We have described previously the preliminary design of this lens.[1] Briefly, each triad (consisting of one S_1 radiator, one snaked microstrip line, and the corresponding S_2 radiator) is photoetched into the 1-mil copper surfacing the two sides

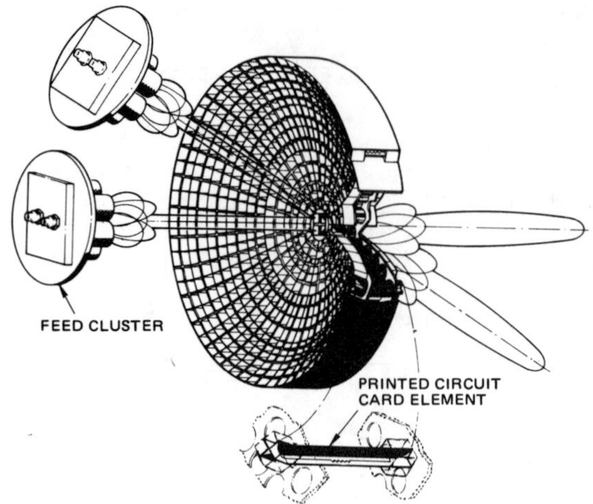

Fig. 3 Printed circuit card, TEM lens, and feed clusters.

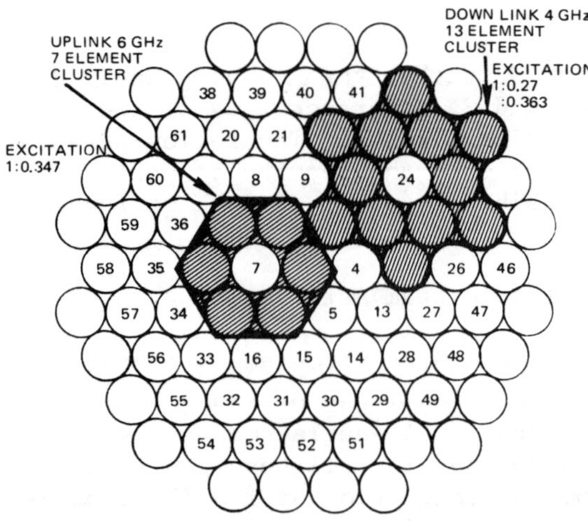

Fig. 4 Feed cluster geometry for TEM lens.

of a thin (0.020 in. thick) flat plastic card. Each radiator is a specially developed tapered slot, etched on one surface, and flared to the ends of the narrow dimension of the rectangular flat card. Each card is mated with a second orthogonal card into a cross configuration. A collection of 396 such card pairs was mounted through cross slots in a pair of circular coaxial metallic sheets (the S_2 sheet being planar and the S_1 sheet spherical), so that each radiator protruded "outward" beyond its respective sheet, which acts as a ground plane for the radiators. In effect, the S_1 and S_2 sheets are slotted racks serving to hold the 396 card pairs in a circular array matrix. This collection of cards and conducting sheets is the TEM lens, as illustrated in Fig. 3.

The feed array was designed to consist of 127 small radiators in a nearly triangular lattice on a spherical focal scan surface. To produce each low-sidelobe pencil beam, a symmetrical cluster of feed elements was driven as an array, using 7 elements at 6 GHz and 13 elements at 4 GHz as shown in Fig. 4. A computerized synthesis technique was developed to calculate optimized phase and amplitude excitation coefficients of the 7- and 13-element clusters.

A vector computer model of this lens was written by A. E. Smoll of Western Development Laboratories, permitting radiation patterns to be calculated using measured patterns of lens and feed elemental radiators as inputs. This program, called Conlens was used to calculate radiation patterns for different feed clusters of the feed array, each cluster producing a differently pointed pencil beam. The set of cluster dividers constitutes the M-port BFN. (An M x N microwave switch matrix could be introduced to allow beam position switching of N transponders to the M-beam ports of the BFN.) The theoretical design of such a lens for a 12-times frequency-reuse, single-aperture antenna for synchronous satellite application (covering the transmit and receive bands) was performed using the computer model, which calculates both pattern and interbeam isolation performance. The results of this work are described elsewhere.[3]

TEM Lens Test Model

An electrical test model of this antenna was constructed for experimental pattern testing at C band, as illustrated in Fig. 3. This model had a diameter of 5 ft, as well as a 5 ft local length, and used 396 card pairs in the TEM lens. It includes a 19-element linearly polarized feed cluster on a mechanical mount that permits positioning the cluster to any position within the Earth coverage FOV. Either a 7-way (6-GHz band) or 13-way (4-GHz band) power divider could be connected to the 19-element feed cluster for tests, with unconnected elements terminated in matched loads. The lens and positionable feed cluster were mounted within an absorber-lined weatherproof metal cylinder, which was mounted on a three-axis rotator on an outdoor antenna range for far-field ($2D^2/\lambda$) pattern tests. The aperture was covered with a flat transparent nylon skin, 0.002 in. thick, for moisture protection.

TEM Lens Pattern Test Results

Although the lens is designed to be unpolarized (by the use of independent crossed card pairs for each element), the experimental feed array was polarized linearly for simplicity

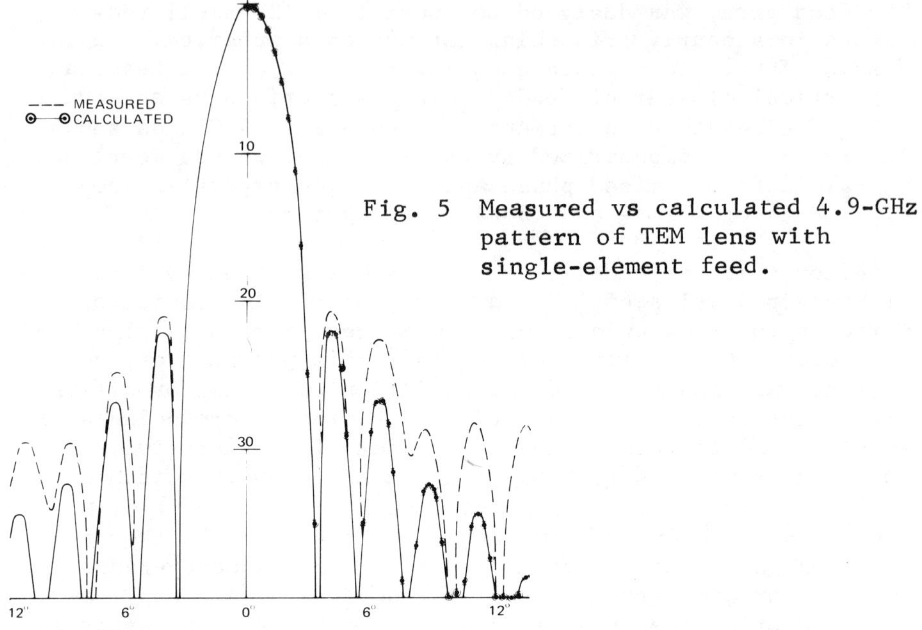

Fig. 5 Measured vs calculated 4.9-GHz pattern of TEM lens with single-element feed.

and economy. Patterns were measured at the design center frequency of 4.9 GHz and at various frequencies from 3.7 to 6.4 GHz in order to explore the potential of this concept.

Figure 5 shows an H-plane (relative to feed cluster polarization) secondary pattern measured at 4.9 GHz, with a single on-axis feed element energized to produce a basic singlet pattern for this case; the agreement is satisfactory. For low-sidelobe feed illumination, the center feed element and the six elements in a ring around it (triangular grid) were energized from a variable-power divider circuit that was adjusted to provide -8-dB and +12° excitation of each element in the ring relative to 0-dB and 0° excitation of the central element. This excitation was predicted to be optimum by the synthesis program mentioned earlier. Figure 6 shows the resulting measured pattern and the corresponding calculated pattern. The assymetry of the measured -31.5 (left) and -35.5-dB (right) sidelobes reflects fabrication tolerances and some minor range reflections at these low signal levels. However, the achievement of the desired low sidelobes was gratifying. Experimental adjustments of the feed cluster excitations showed that the synthesized values could not be improved upon. Figure 7 shows an edge-of-FOV beam position achieved by moving the feed cluster on a circular arc to the desired scan angle. The on-Earth sidelobes stayed below

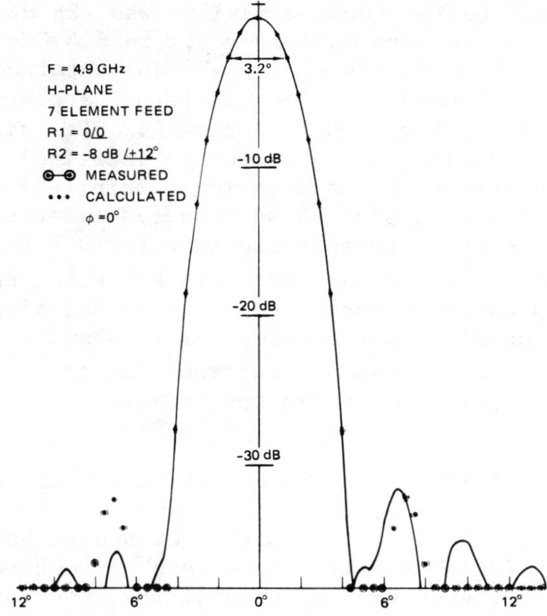

Fig. 6 Measured vs calculated 4.9-GHz pattern of TEM lens with seven-element feed cluster (on-axis).

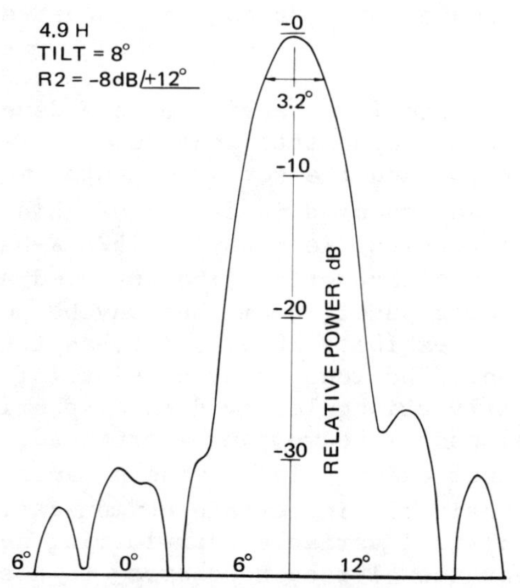

Fig. 7 Measured 4.9-GHz pattern of TEM lens with seven-element feed cluster (8.4° scan).

-30 dB. The off-Earth sidelobe maximum was -26 dB. Other pattern measurements were made from 3.7 to 6.4 GHz with similar low sidelobe levels (below -30 dB) obtained from 4.2 to 6.4 GHz. However, in some pattern cut planes, sidelobes were considerably higher. Below 4.2 GHz, sidelobes were higher and filled in. Subsequent tests indicated that higher-mode generation and scattering within lens elements were the probable causes of these measured effects, not accounted for in the corresponding calculated data. Currently, electrically smaller card elements are being designed via active impedance measurement sub array techniques to rectify this anomalous performance for two-band (3.7 to 4.2 and 5.9 to 6.4 GHz) operation, representing transmit and receive C band operation in the one antenna.

Variable-Beam Waveguide Lens Antenna

In this antenna, the lens is a matrix of square hollow metallic waveguides forming a metallic "egg-crate" structure. S_1 is spherical, with the sphere centered at the lens focal point (F/D=1) as in the TEM lens. S_2 is elliptical, with three ring-steps (setbacks) to prevent excessive lens edge thickness and consequent unacceptable narrow frequency bandwidth. The usable bandwidth of the stepped lens is about 5 to 10% because of the dispersive nature of the hollow waveguides. Dispersion causes defocusing of the lens at frequencies sufficiently far from the design frequency f_o, resulting in pattern shape deterioration.

The lens described here is a version of one developed by Dion and Ricardi of the Massachusetts Institute of Technology Lincoln Laboratory,[4] and the scalar computer model developed by Dion was modified and used in design of this lens. The design is for a three-step lens having 1528 X-band waveguide cells arranged in a square grid. The 46-in.-diam lens produces a 2° half-power pencil beam that may be positioned over a $\pm 9^o$ cone FOV. It exhibits 20-dB sidelobes for single-feed element excitation. The feed array consists of 61 horns arranged in a nearly triangular grid on a spherical focal surface. Thus 61 pencil beams can be produced, one from each small feed horn excitation. To provide a variable shaped-beam capability desirable in certain communication satellite applications, a 61-to-1 variable beam-forming network (VBFN) is used to connect the 61-feed horn array to a single antenna port. The VBFN is formed of a tree of waveguide lines interconnecting 60 variable (two-way) power dividers (VPD's), which are specially developed current-controlled ferrite devices.[5] Control of the current to each VPD permits variation of power split, with no change in impedance match or insertion phase delay. In this fashion, any one of the 61 pencil beams may be excited alone, or any sub set from 1 to 61 at a time may be chosen with equal or unequal power excitation. If all 61 are excited equally, an 18° flat-topped conical beam illuminates the Earth FOV. Thus the antenna can beam its energy to one 2° spot on the Earth or split it between two, three, or more spots, contiguous or noncontiguous, as desired by the satellite controller.

Waveguide Lens Pattern Test Model

An electrical feasibility model of this antenna was constructed with the 1528-cell lens and a 61-horn circularly polarized feed array. This feed array was excited from a 61-to-1 VBFN formed a semirigid coaxial lines, power dividers, variable attenuators, and phase shifters, for experimental testing. (Waveguide components of minimum loss would be used in a flight design.) The test model is illustrated in Fig. 8. The feed array is positioned on a spherical surface (see Fig. 9) for simplifying shaped-beam excitations (no phase corrections needed for off-axis beams, as required for a flat array). The lens was mounted between two absorber-covered vertical posts on a platform, as shown. The circularly polarized feed array and BFN were mounted from a cantilevered focusing mechanism, to a vertical post on the same absorber covered platform. The platform is mounted on a three-axis positioner on a $> 2D^2/\lambda$ outdoor pattern range.

Fig. 8 Multiple-beam waveguide lens antenna.

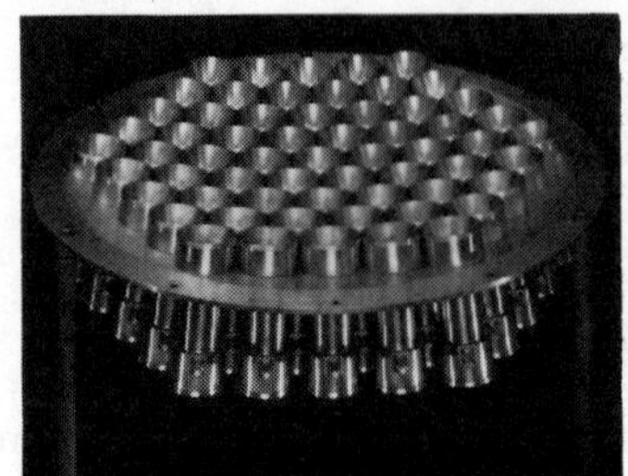

Fig. 9 61-element feed spherical array.

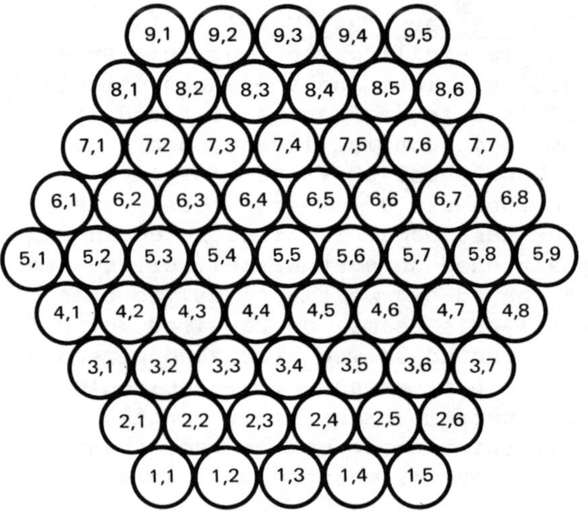

Fig. 10 MBA beam designation.

Waveguide Lens Pattern Test Results

Figure 10 shows the element numbering designation scheme. Element (5,5) is the on-axis beam. Figure 11 shows the measured and calculated circularly polarized pattern, with element (5,5) energized and all other elements terminated in matched loads. The abreement is satisfactory. Figure 12 shows five adjacent single beam measured patterns. Scan effects are very small.

Figure 13 shows a spinning linear-measured pattern of an off-axis beam (5,7). At an angle equal but opposite to the (5,7) beam, a large "conjugate" lobe is evident which has a very high axial ratio. Very similar patterns were measured over a 4% band centered at f_o. This lobe is not predicted by the computer program, which neglects the scattering effects of all feeds. It was established by experiments that the "conjugate" lobe, which always is present at an equal but opposite

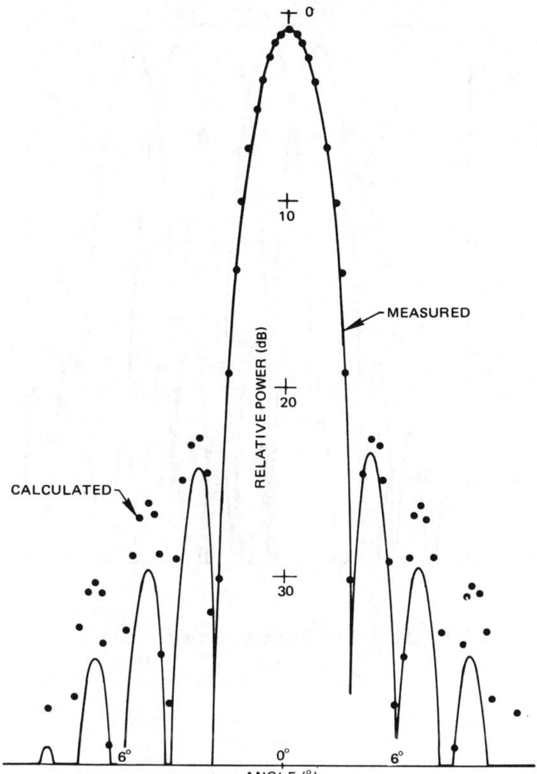

Fig. 11 Measured vs calculated beam pattern $(5,5)f_o$, RHCP.

angle to each excited feed, was the result of scatter from the spherical inner surface of the lens. This scattered energy approximately focuses at the conjugate horn and then rescatters through the lens, producing a beam as a weakly energized conjugate horn might do. Further experiments indicated that both "antenna mode" and "structural" scattering occurred from the conjugate feed. These two modes of scattering from the feed horn occur with approximately equal magnitudes but opposite senses of circular polarization, resulting in the high ellipticity of the conjugate lobe. By design of improved feed scattering parameters, the image lobe magnitude (on a spin linear pattern) was reduced to -27 dB. Better matching of the lens surfaces would, of course, further reduce this lobe.

Figure 14 shows a measured vs calculated Earth coverage beam formed by equal excitation of all 61 feed elements. Agree-

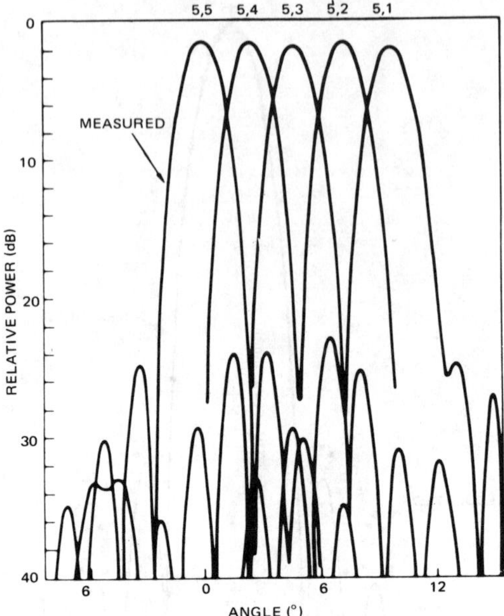

Fig. 12 Adjacent scanned beam patterns f_o, RHCP.

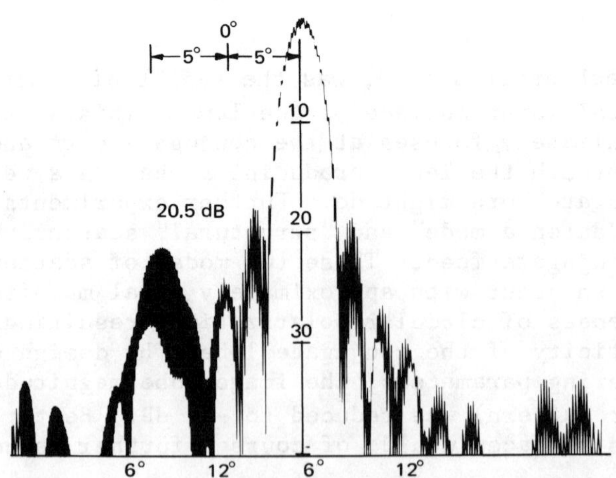

Fig. 13 Spinning linear measured patterns f_o, RHCP.

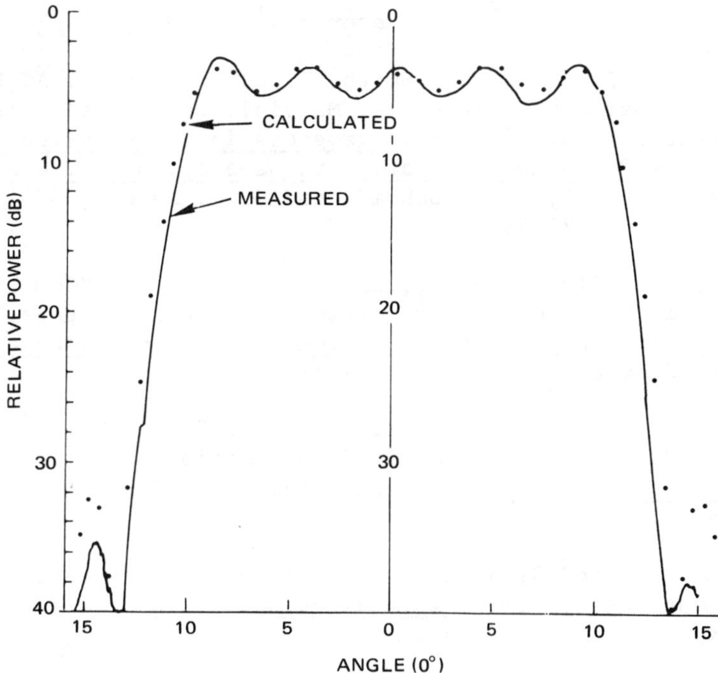

Fig. 14 Earth coverage beam, 61 feeds equally excited.

ment is also satisfactory. Similar patterns were measured over a 4% band. Measured axial ratios were under 1.5 dB over the full 18° FOV. Amplitude ripple of 3 dB over the beam top can be reduced by careful changes from uniform excitation of the 61 elements. Many other beam shapes have been created by various beam excitations of the VBFN.

Conclusion

Pattern test results have been presented from experimental models of two different types of multiple-beam lens antennas, ultimately intended for application on synchronous communication satellites. The TEM lens antenna can be used where wide bandwidth and very low sidelobe performance (i.e., simultaneous multiple-beam frequency reuse) are required. The waveguide lens is simpler mechanically but suitable only where narrow-frequency-band operation is acceptable. For either lens, beam shaping can be achieved through the use of specially designed microwave networks. For the more critical applications, careful attention must be paid to scattering properties of lens and feed arrays.

References

[1] Luh, H. S., Scott, W. G., Smith, T. M., and Smoll, A. E., "A Constrained Lens Antenna for Multiple Beam Satellites," AIAA Paper 74-465, 1974; also Progress in Astronautics and Aeronautics-Communication Satellite Developments: Technology, Vol. 42, edited by W. G. Schmidt and G. E. LaVean, AIAA, New York, 1976, pp. 25-33.

[2] Han, C. C., Bilenko, H. W., and Wickert, A. N., "Computer-Aided Array Feed Design for Multiple Beam Lens Antenna," 1975 International Symposium of the Antenna and Propagation Society, June 2-4, 1975, pp. 374-377.

[3] Scott, W. G., Luh, H. S., and Matthews, E. W., "Design Trade-offs for Multibeam Antennas in Communication Satellites," International Conference on Communications, June 14, 1976, Philadelphia, Pa.

[4] Dion, A. R., and Ricardi, L. J., "A Variable Coverage Satellite Antenna System," Proceedings of the IEEE, Vol. 59 February 1971, pp. 252-262.

[5] Matthews, E. W., "Variable Power Dividers in Satellite Systems," 1976 IEEE & MTT-S Microwave Symposium, June 1976.

COMMUNICATIONS ANTENNA DESIGN FOR THE JAPANESE COMMUNICATION SATELLITE

M. Kudo*
National Space Development Agency, Tokyo, Japan

Y. Nemoto†
Mitsubishi Electric Corporation, Kamakura, Japan

A. N. Wickert,‡ C. C. Han,§ and D. Ford¶
Ford Aerospace & Communications Corporation,
Western Development Laboratories Division,
Palo Alto, California

Abstract

A wide-band, shaped-beam, horn-reflector antenna has been developed for the Japanese Medium-Capacity Communication Satellite for Experimental Purposes. The mechanically despun antenna will provide high-gain communication links within the Japanese Islands at the 4/6- and 20/30-GHz communication bands. The antenna design extends existing horn-reflector technology through 1) the development of a single-feed system capable of operating over four separate frequency bands, and 2) the realization of a doubly curved, shaped-reflector surface providing radiation pattern characteristics that closely approximate the outline of the main Japanese Islands at the 20/30-GHz bands. Tests performed on both the feed and the complete antenna system demonstrate that the design results in a compact, high-performance antenna that is in close agreement with theory.

Presented as Paper 76-252 at the AIAA/CASI 6th Communications Satellite Systems Conference, April 5-8, 1976, Montreal, Canada.

*Senior Engineer; presently Staff Engineer of Yokosuka Electrical Communication Laboratory, Nippon Telegraph and Telephone Public Corporation.

†Staff Engineer.

‡Manager, Antenna Engineering Department.

§Supervisor, Antenna Technology Section.

¶Engineering Specialist, Antenna Technology Section.

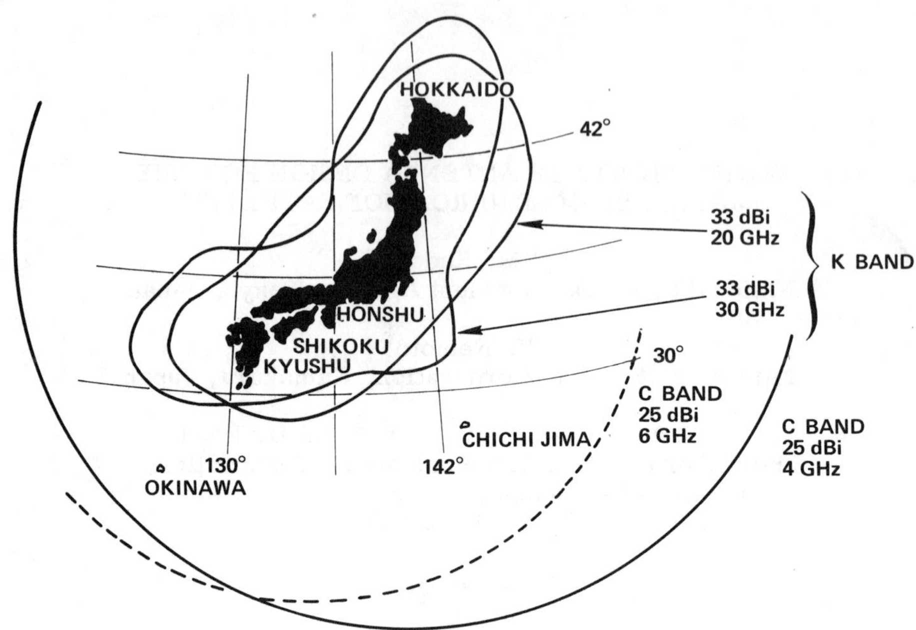

Fig. 1 CS antenna coverage areas.

I. Introduction

For a communication satellite antenna to illuminate an irregularly shaped target area on the Earth efficiently, it is necessary to employ some method of beam shaping. If the target area is also small in angular extent as viewed by the antenna, the antenna's directivity, and hence the effective aperture size, must be large in order to provide maximum gain over the desired coverage region. An efficient method for achieving both high gain and the desired beam-shaping performance is to employ a shaped reflector and a single feed system. Such a configuration minimizes both the rf losses associated with the feed system and the size and weight of the antenna system.

The shaped-beam, horn-reflector antenna, developed for the Japanese Medium-Capacity Communication Satellite for Experimental Purposes (CS) to meet the preceding objectives, will provide high-performance communication links within Japan at the 4/6- and 20/30-GHz communication bands. The horn-reflector antenna has been employed successfully in several ground terminal applications over the last several decades to generate a narrow, low-noise pencil beam.[1,2]

In 1969 the Western Development Laboratories (WDL) Division of Ford Aerospace & Communications Corporation applied the horn-

Table 1. CS antenna requirements

Parameter	Requirements	
	C band	K band
Frequency, GHz		
Transmit	3.7-4.2	17.7-21.2
Receive	5.925-6.425	27.5-31.0
Polarization		
Transmit	Right-hand circular	Same
Receive	Left-hand circular	Same
Coverage area	All Japanese territory	Main Japanese Islands
Minimum gain over coverage area, dBi	$\geq$ 25	$\geq$ 33

reflector antenna concept to an earth-coverage mechanically despun antenna for the United Kingdom's Skynet satellite.[3] This antenna consisted of a flat-plate reflector fed by a lens-compensated horn, and provided a 19° earth-coverage beam at X-band frequencies. The antenna implementation described in this paper extends the previous horn-reflector antenna technology by 1) utilizing a single feed system that operates at four widely separated bands, 2) providing two-dimensional beam shaping through a specially contoured reflecting surface, and 3) employing a method for mechanically despinning the antenna without the need for a wide-band rotary joint.

The initial electrical design concept for this antenna was developed by Nippon Telegraph and Telephone Public Corporation and Mitsubishi Electric Corporation (MELCO),[4,5] and the baseline electrical design was performed by MELCO. The final antenna design and its manufacture were undertaken by WDL. This paper presents an overview of the electrical design and the measured performance of the antennas and feed system.

II. Antenna Design

As viewed by the antenna, the main Japanese Islands represent an irregularly shaped area (Fig. 1). To optimize the antenna gain over this coverage area, a shaped-beam horn-reflector antenna was selected, since this design provides the necessary two-dimensional beam shaping. The final reflector contour was selected to satisfy the basic performance requirements given in Table 1. An aperture size of 1 m was chosen to provide a minimum of 25-dBi gain at C-band frequencies, whereas the reflector shape was designed to provide a minimum gain of 33 dBi over the main Japanese Islands at K-band frequencies.

The antenna (Fig. 2) consists of three distinct parts: a feed system for exciting or extracting signals, a conical feed horn to guide the spherical waves from the feed system, and a shaped reflector to produce the desired two-dimensional wavefront. Both the feed and the reflector, however, are interrelated electrically and both affect overall antenna performance. The feed provides the mechanism for launching or extracting signals from the horn/reflector combination with the correct sense of polarization. Because the feed acts as a quadruplexer in separating four distinct frequency bands, it must provide good interchannel isolation and minimize the excitation of higher-order waveguide modes in the conical horn structure.

To separate signals in the four operating bands, the feed system uses both frequency and polarization discrimination techniques. The high-frequency (K-band) feed network consists of an orthomode transducer (OMT) followed by a wide-band polarizer. The OMT separates receive and transmit signals on the basis of polarization by producing vertical polarization at the receive band and horizontal polarization at the transmit band. The polarizer then converts the linear polarization into circularly polarized waves. This implemen-

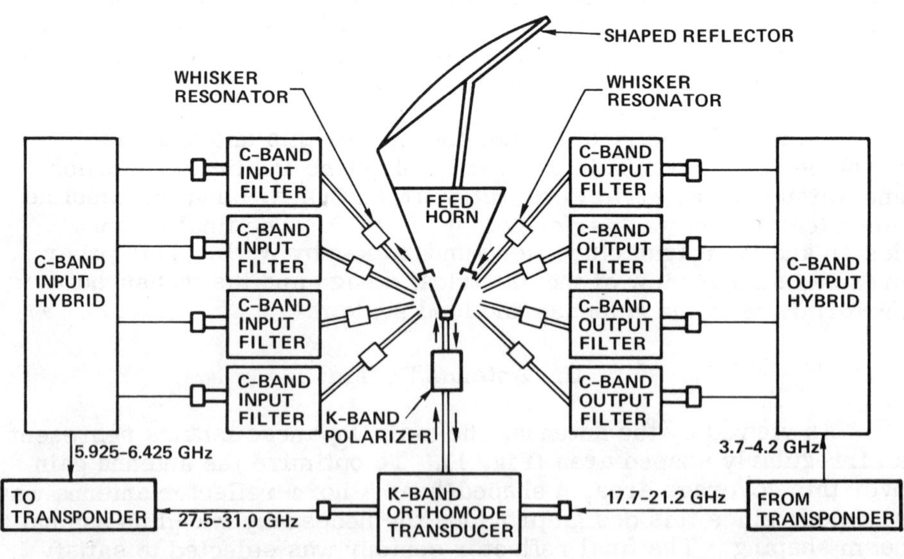

Fig. 2 Communications antenna schematic diagram.

tation results in opposite senses of circular polarization for receive and transmit signals. Because the polarizer must provide low axial ratio performance over both the receive and transmit bands, a differential type of polarizer was developed. This device consists of two differential phasing sections with different dispersion characteristics to achieve a nearly constant 90° differential phase shift over both frequency bands. The polarizer was realized in circular waveguide with closely spaced corrugations to provide phase retardation of the applied signal. The size of the circular waveguide was selected to suppress the TM_{11} mode, which can be excited by the corrugated structure. Since the waveguide size is small as compared to C-band wavelengths, there is no cross-coupling between the C-band transmit ports and the K-band receive ports.

C-band energy is coupled to the conical horn through two sets of four symmetrical, longitudinal slots that are excited in phase progression by a hybrid network to obtain circular polarization. To prevent coupling of K-band transmit energy into the C-band receive and transmit slots, a low-pass waffle-iron filter is placed in each coupling arm adjacent to the slot. In addition, two sets of dipole-like resonators are placed between the waffle-iron filter and the coupling slot. In the C-band transmit network, these resonators are configured to prevent coupling of C-band receive signals and to suppress excitation of higher-order modes at the slot discontinuity. A similar set of resonators is used in the C-band receive network in order to suppress the excitation of higher-order modes at K-band frequencies. With the exception of the hybrid networks, all C-band components were realized in waveguide to minimize feed loss. The

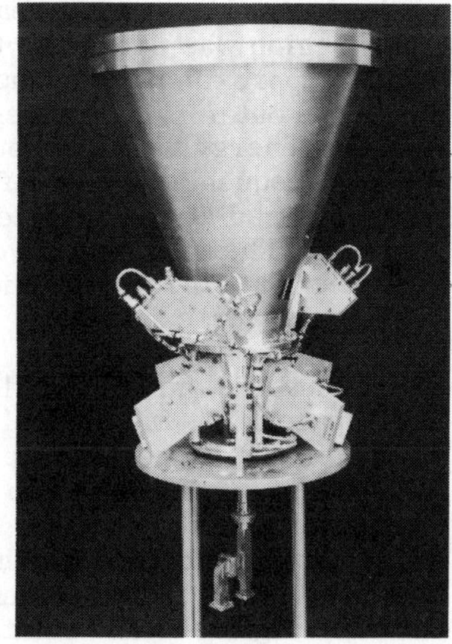

Fig. 3 Feed assembly with inner horn, protoflight model.

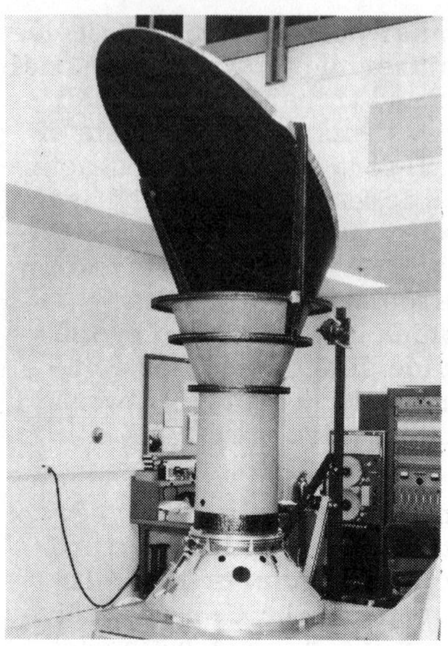

Fig. 4 Protoflight communications antenna model.

hybrid networks were constructed from air-suspended stripline and joined to the coupling arms through coaxial-to-waveguide adapters. The major components in the C- and K-band feed network are shown in Fig. 3, and the complete antenna system is shown in Fig. 4.

After a circularly polarized signal is excited in the feed, it propagates up the conical horn structure to illuminate a shaped-surface offset reflector. The reflector contour basically consists of two regions: an inner part that generates a triangularly shaped spherical wavefront and an outer part that directs energy to the edge of the desired coverage area.[5] The wavefront shape produced by the inner region was chosen as triangular, since the main Japanese Islands approximate a triangularly shaped region on the Earth. The inner and outer reflector regions were adjusted so that most of the radiated energy was distributed over the main Japanese Islands. At C-band frequencies, the reflector shaping is less dominant and the antenna's radiation characteristics approximate those of a parabolic antenna. The radiation coverage at C band is thus broader and more symmetrical, thereby providing coverage to both the main and the remote Japanese Islands.

To utilize the horn-reflector antenna design with a spin-stabilized spacecraft, the conical horn was separated at a relatively large diameter and connected to a despun drive motor assembly (DMA). A 0.030-in. (0.76-mm) gap between the two horns was chosen as the best compromise between electrical and mechanical performance. An rf choke is included at this gap to minimize energy leakage and reflected power. The antenna components below the rotary gap are attached rigidly to and rotate with the spacecraft, whereas those above the gap are driven by the DMA at an equal, but opposite, radial velocity. In this manner, the antenna beam is pointed continuously at the same location on the Earth.

In practice, neither the feed system nor the reflector will generate a perfect circularly polarized wave. Consequently, the resultant polarization of the antenna system is a vector combination of the fields produced by the feed and the reflector. Since the reflector is despun with respect to the feed system, the relative phase between the two circularly polarized components is changed to produce a different polarization response of the antenna system. Thus, the polarization transmission loss between the horn-reflector antenna and a perfect circularly polarized antenna will vary continuously over a period corresponding to 1 rev of the spacecraft.

This situation is complicated further by the presence of higher-order modes, which are excited in the feed system and radiated by the antenna. These modes, which are present mainly at the higher frequencies (K-band receive), are derived from two major sources: the K-band feed system, which may excite TM_{01} and TE_{21} energy,

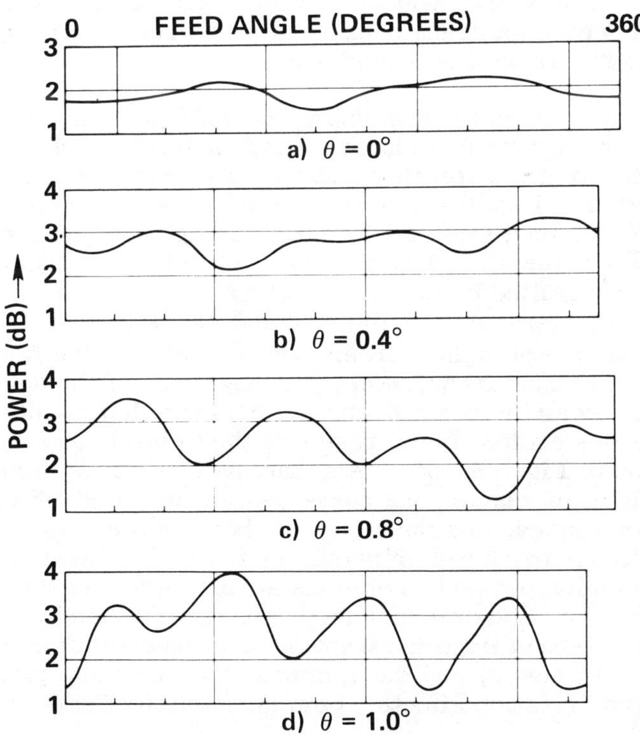

Fig. 5 Spin modulation waveforms (29.25 GHz, ϕ = 105°).

and the C-band coupling slots which, at K-band frequencies, may excite $TE_{2n+1,1}$ modes. Although the CS antenna has been configured to minimize higher-order mode excitation, low-level higher-order mode radiation does occur. Since the relative phase of this energy is also changed by rotation of the reflector with respect to the feed, the net polarization response of the antenna will change in a complex fashion as a function of spacecraft rotation. The resulting fluctuation in receive power, as seen by the output of a perfect circularly polarized receive antenna, is due to the varying polarization purity of the antenna. This effect, termed "spin modulation," generally increases in magnitude as one moves further away from the beam center.

III. Measured Performance

The performance of the CS antenna was measured on a specially designed far-field range facility. Because the beam-shaping property of the antenna results in a non-constant phase distribution over the aperture plane, the distance between the transmit and test antennas was made larger than the normal $2D^2/\lambda$ far-field criteria to minimize the effect of phase error on pattern shape and antenna gain.[6] A distance of $5D^2/\lambda$ at 31 GHz was selected as the best compromise between antenna near-field measurement error and a practical antenna range implementation.

As stated previously, despinning the reflector with respect to the feed system results in a change to the radiated copolarized power as a function of satellite rotation. To illustrate this spin modulation effect, Fig. 5 shows the measured power variation at 29.25 GHz as the feed system is rotated through 360° with respect to a stationary reflector. The magnitude of the spin modulation waveform is controlled by two main factors: 1) the amount of cross-polarized energy generated by the feed and the reflector, and 2) the magnitude and type of higher-order modes excited in the feed system. At the receive K-band frequencies, the magnitude of the spin modulation waveform near the beam maximum is controlled mainly by the $TE_{2n+1,1}$ modes excited in the region of the C-band coupling slots. A comparison of Figs. 5a-5d shows that, as one moves further off-axis from the beam center, the peak-to-peak magnitude of the spin modulation increases, and the waveform becomes less symmetrical. The first effect is produced primarily by the higher level of cross-polarized radiation present at off-axis angles, whereas the second effect results from higher-order mode energy radiated by the antenna. At the K-band transmit frequencies, the excitation of higher-order modes is weaker, and the spin modulation results primarily from cross-polarization effects. Spin modulation effects at C band are negligible.

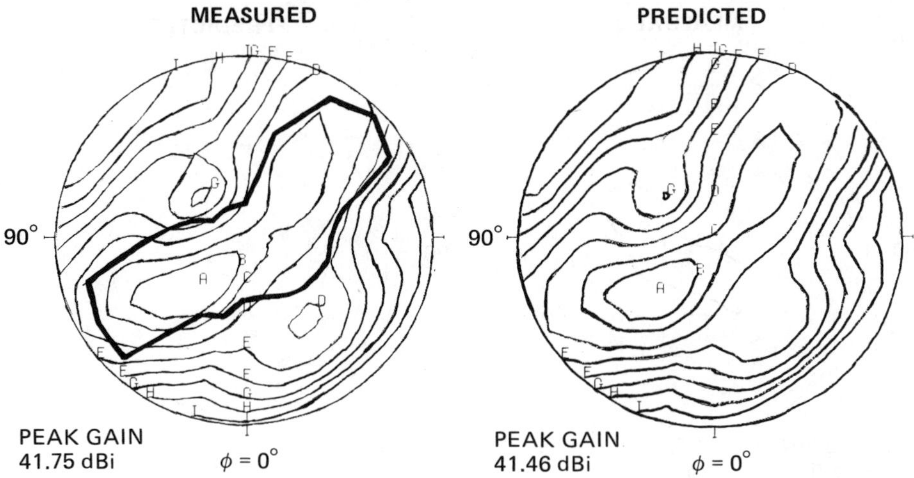

a) 29.25 GHz left-hand circular polarization (max theta 1.500)

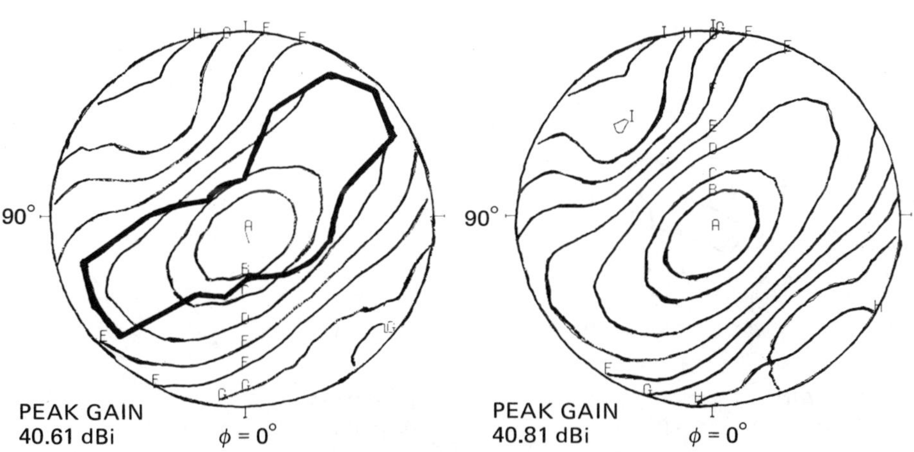

b) 19.45 GHz right-hand circular polarization (max theta 1.500)

CONTOUR DATA			
SYMBOL	LEVEL	SYMBOL	LEVEL
A	0.000	F	-8.000
B	-1.000	G	-10.000
C	-2.000	H	-12.000
D	-4.000	I	-14.000
E	-6.000		

Fig. 6 Measured and predicted radiation characteristics.

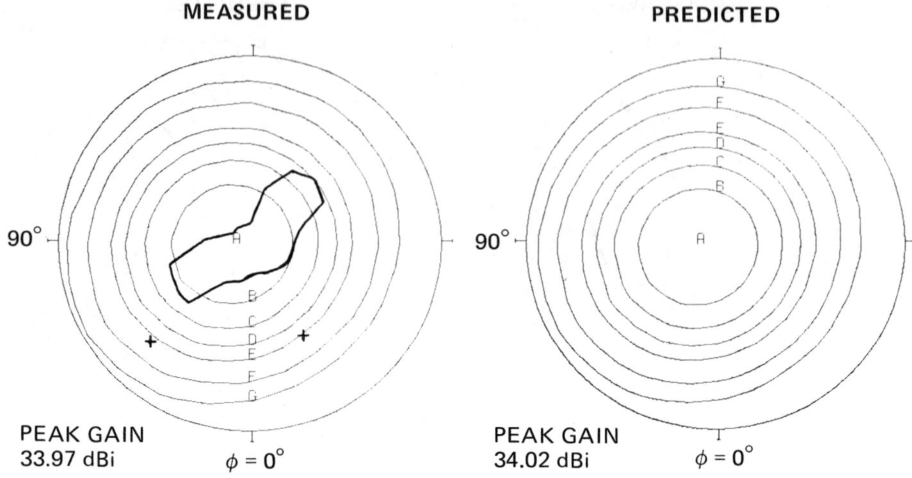

c) 6.175 GHz left-hand circular polarization (max theta 3.000)

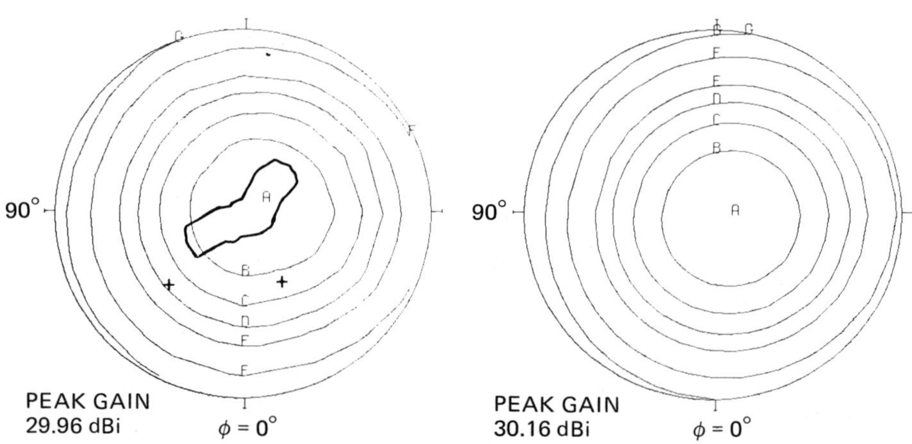

d) 3.95 GHz right-hand circular polarization (max theta 4.000)

CONTOUR DATA			
SYMBOL	LEVEL	SYMBOL	LEVEL
A	0.000	E	-4.000
B	-1.000	F	-6.000
C	-2.000	G	-8.000
D	-3.000		

Fig. 6 Measured and predicted radiation characteristics (continued).

Figure 6 shows measured and calculated radiation characteristics of the shaped-beam antenna expressed in equal-power polar contour plots. The angular position of a point in space is given by two spherical angles. The angle θ is measured radially from the center of the contour plot, whereas the angle ϕ is measured circumferentially. For each contour plot, the angular radius of the circle is given by "max. theta." In each plot the power at a particular point in space corresponds to the time-average power level of the spin modulation waveform. Also shown in these plots is the outline of the required coverage area.

The measured contours shown in Figs. 6a and 6b show that the antenna provides a beam shape which, at K-band frequencies, closely matches the coverage area. The antenna gain shown in each figure was measured at the beam peak and includes all losses in the horn-reflector antenna. As can be seen, the net antenna gain is in excess of the 33-dBi requirement at all points within the coverage area. At C-band frequencies the radiation patterns are broader and more symmetrical, thereby providing coverage of both the main and remote Japanese Islands. At all points within Japanese territory, the antenna provides gain in excess of the 25-dBi requirement.

The theoretical performance of the antenna was analyzed by both the induced current and the aperture field methods. Over the

Table 2 Protoflight model feed system performance summary

Electrical parameter	Frequency, GHz	Worst case value	Nominal value
Input VSWR	3.7-4.2	1.35:1	1.2:1
	5.925-6.425	1.22:1	1.15:1
	17.7-21.2	1.32:1	1.2:1
	27.5-31.0	1.12:1	1.1:1
Insertion loss, dB	3.7-4.2	0.9	0.7
	5.925-6.425	0.95	0.85
	17.7-21.2	0.4	0.18
	27.5-31.0	0.2	0.1
Axial ratio, dB	3.7-4.2	0.75	0.5
	5.925-6.425	0.65	0.4
	17.7-21.2	2.7	1.5
	27.5-31.0	2.0	1.2
Isolation, dB	3.7-4.2 into 6-GHz port	41.5	55
	17.7-21.2 into 6-GHz port	57	>60
	17.7-21.2 into 30-GHz port	40	50
	5.925-6.425 into 4-GHz port	38	55
	27.5-31.0 into 20-GHz port	30	38

antenna coverage region the agreement between these two techniques is excellent. The predicted antenna gain shown in these plots includes the measured feed loss. A comparison of the measured and calculated patterns and gains confirms that the antenna performance can be predicted accurately.

The measured performance of the feed system is summarized in Table 2. For each test parameter, both the worst case and nominal performance values are shown over the frequency band of interest. The test data show that the feed design provides high transmit-to-receive isolation, a low axial ratio, and a low input voltage standing-wave ratio over both the receive and transmit bands. The K-band axial ratio performance also demonstrates that the differential type of polarizer design is feasible for wide-band performance requirements.

IV. Conclusion

Performance measurements of the CS horn-reflector antenna show that the antenna design provides both high-gain and shaped-beam performance that are in close agreement with theory. It also demonstrates that a wide-band feed can be realized which has excellent performance characteristics and little interaction between the four operating frequency bands.

References

[1]Crawford, A. B., Hogg, D. C., and Hunt, L. E., "A Horn-Reflector Antenna for Space Communication," Bell System Technical Journal, Vol. 40, July 1961, pp. 1095-1116.

[2]Hines, J. N., Li, Tingye, and Turrin, R. H., "The Electrical Characteristics of the Conical Horn-Reflector Antenna," Bell System Technical Journal, Vol. 42, Pt. 2, July 1963, pp. 1187-1211.

[3]Gregorwich, W. S., "A Mechanically Despun Antenna for the Skynet (IDCSP/A) Communications Satellite," Communication Satellites for the 70's: Technology," edited by N. E. Feldman and C. M. Kelly, in Progress in Astronautics and Aeronautics, Vol. 25, MIT, 1971, pp. 241-254.

[4]Shinji, M. et al, "A Shaped Beam Horn-Reflector Antenna for Domestic Communications Satellite System," Transactions of the Institute of Electronics Communication Engineers, Vol. 57-B, June 1974, p. 355.

[5]Katagi, T. and Takeichi, Y., "Shaped Beam Horn-Reflector Antennas," IEEE Transactions, Vol. AP-23, Nov. 1975, pp. 757-763.

[6]DiFonzo, D. and English, W., "Far-Field Criteria for Reflectors with Phased Array Feeds," International IEEE/AP-S Symposium, 1974 Georgia Institute of Technology, Atlanta, Ga.

Chapter V – Spacecraft RF Subsystems

TECHNOLOGY ADVANCES IN REALIZATION OF FILTER NETWORKS

S. Kallianteris* and M. V. O'Donovan+

Com Dev Ltd., Dorval, Quebec, Canada

Abstract

New types of low-loss filters employing $TE_{11}n$ and $TE_{10}n$ dual mode cavities are described. Results are presented which demonstrate a reduction in passband loss of 60% relative to conventional TE_{111} mode filters. Techniques for the suppression of unwanted spurious responses in the stopband regions and procedures for the optimization of cavity sizes also are considered. Experimental results emphasizing the advantages accruing from the use of such filters as output multiplexing elements in satellite transponders at frequencies above 10 GHz are presented.

1. Introduction

The development of dual TE_{111} mode Chebychev and elliptic filters at 4 GHz[1] and their realization in graphite fiber material[2] probably form a benchmark in terms of the weight and performance that can be achieved at 4 GHz. Filters of this type employ square or cylindrical resonators, each approximately half-guide wavelength long at center frequency, operating in TE_{101} and TE_{111} modes, respectively. The unloaded Q of such filters is approximately equivalent to that of rectangular waveguide designs. Unloaded Q is a quality that determines, and is inversely proportional to, the midband insertion loss of a specific filter.

The types of filter just described exhibit unloaded Q's of 10,000 at 4 GHz and 5000 at 12 GHz. This means that a filter at 12 GHz, with the same percentage bandwidth and other performance characteristics as one at 4 GHz, will have twice the midband insertion loss. In communications satellites, the

Presented as Paper 76-292 at the AIAA/CASI 6th Communications Satellite Systems Conference, April 5-8, 1976, Montreal, Canada.

*Member, Technical Staff.
+President.

passband loss is not an important parameter in the multiplexing networks used prior to the final amplification stage, provided that the inband gain slope requirements are not exceeded. It is, however, a critical characteristic of the output multiplexing networks, which interconnect the final amplifiers with the antenna subsystem. RF power is probably the most expensive satellite resource, and loss in the output circuits reduces the spacecraft EIRP, which decreases the communications capacity of the overall system.

The skirt selectivity provided by the elliptic function filter at frequencies close to the passband is sharper than that provided by the Chebychev function. For non-contiguous output multiplexers of the type described herein, either a five-pole Chebychev or four-pole elliptic filter could be used. The elliptic filter is chosen as optimum for this application since it provides minimum passband loss. It should be kept in mind, however, that the generally held belief that the elliptic function is universally superior to the Chebychev function is incorrect. It has been shown[12] that the traffic requirements of the communication channel determine which filter type is optimum.

A number of communications satellites scheduled for launch in the late seventies and early eighties stress increasing use of the 11-, 12-, and 14-GHz communications bands. The main purpose of this paper is to describe the development and realization of optimum amplitude filters, which exhibit lower insertion losses than those currently available. Emphasis is placed on their applications in the output circuits of communications satellites with downlink frequencies above 10 GHz. Experimental results are presented for filters designed for use in the frequency band 11.7-12.2 GHz. The performance of a typical 12-GHz output multiplexing network using these filters is analyzed and compared with that of existing networks used in satellite transponders operating in the 3.7-4.2-GHz band.

2. Circular Waveguide Filters Utilizing TE_{011} Mode Resonators

This type of filter is the most obvious low-loss filter candidate because of the extremely high unloaded Q's that can be realized (≃ 15,000 at 12 GHz). Analytic and experimental work on this design type exposed a number of significant electrical and mechanical problem areas.

2.1 Electrical Problem Areas

The larger cavity diameter required to support the TE_{011} mode results in unwanted spurious mode resonances, including a degenerate TM_{111} mode very close to the passband. These resonances can be reduced in amplitude by incorporating in the design one or more of the following strategems: 1) alternate right-angle coupling between adjacent cavities[3]; 2) the addition of dissipative material to the end-wall tuning plungers[4]; 3) the use of mixed mode resonators[5]; and 4) the insertion of a conventional wideband filter in series with the TE_{011} filter. The reduction in unloaded Q resulting from use of the preceding techniques was found to be related directly to their efficiency as mode suppressors.

Another area of difficulty lay in the realization of filters of this type with passbandwidths of 0.5 to 1.0%, the range within which most satellite communications requirements are found. An inherent bandwidth limitation of 0.2% to 0.3%, because of resonance breakdown at the input and output coupling elements, was found to be extremely difficult to overcome, particularly for filters with few resonant cavities (four or less). A study of the literature shows that experimental results reported are for filters with passbandwidths of less than 0.2%.

2.2 Mechanical Problem Areas

The necessity of large end-wall tuning plungers and the use of intercavity coupling through the cylindrical walls result in a cumbersome mechanical design, difficult to manufacture and to qualify for space applications. The addition of engineering fixes, of the type just described, to minimize spurious mode resonances adds further to the mechanical complexity. Since this filter is realized in a single-mode configuration, its size and weight, even at 12 GHz, are substantial. Taking into account all of the preceding factors, it was concluded that the suitability of the TE_{011} filter for multiplexer applications in communications satellites was questionable.

3. Dual Mode Filters Employing Cylindrical or Square Cavities Operating in $TE_{11}n$ and $TE_{10}n$ Modes, Where n is >1

Conventional dual mode filters employ cavities of cylindrical or square cross section operating in the TE_{111} and TE_{101} modes. At 12 GHz, unloaded Q's of 5000 can be realized with such filters. Each cavity is approximately one-half guide wavelength long at the center frequency of the filter. An increase in the third index of the mode subscript implies an increase in the length of each cavity by one-half wavelength, so that a TE_{112} filter will have each cavity equal to approximately one guide wavelength. Conceptually, such filters will exhibit higher unloaded Q's and hence lower passband

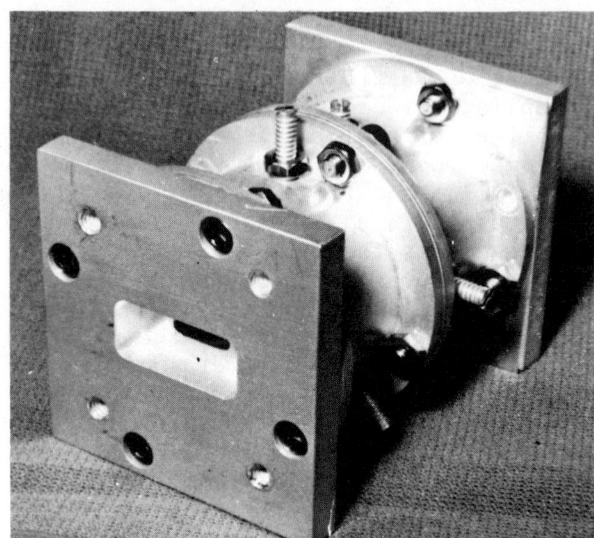

Fig. 1 Four-pole TE_{112} elliptic filter at 12 GHz.

losses. The losses of such filters can be minimized by optimizing the volume of each cavity so that a specific ratio of the cross sectional area to length is achieved. This optimization process will result in an increase in the height or diameter of each cavity and cause a number of additional modes to be supported at frequencies close to the passband. One way of eliminating such spurious modes is to cascade cavities of slightly different cross sectional area. The onset frequency of unwanted modes will not, therefore, coincide and will be attenuated. The spurious mode attenuation level for any specific filter design will, of course, be dependent on the number of filter sections and the relative dimensions of each cavity.

3.1 Dual Mode TE_{112} Filters

Synthesis techniques were developed and a number of experimental models built to test the suitability of this type of filter for multiplexer applications. The photograph of Fig. 1 shows a four-pole elliptic filter design to exhibit performance characteristics typical of a satellite output multiplexer at 12 GHz. An unloaded Q of 7500 was achieved with this unit. One of the requirements of such a filter is that no spurious responses should be present across the 11.70-12.20-GHz communications band. To satisfy this requirement,

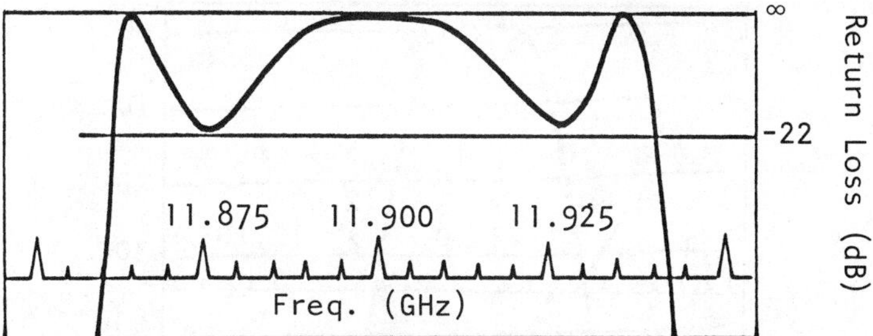

Fig. 2 Passband return loss TE_{103} filter.

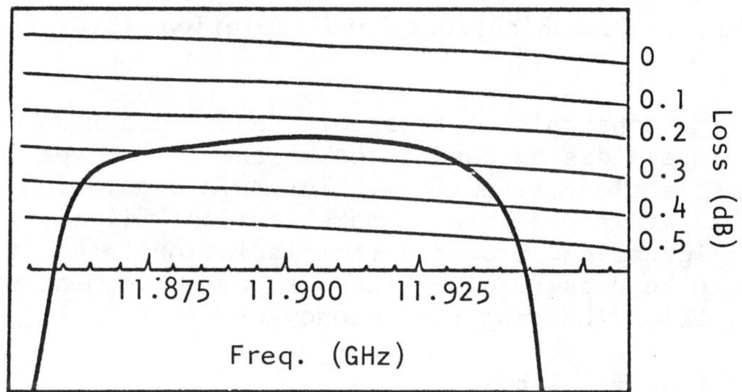

Fig. 3 Passband insertion loss TE_{103} filter.

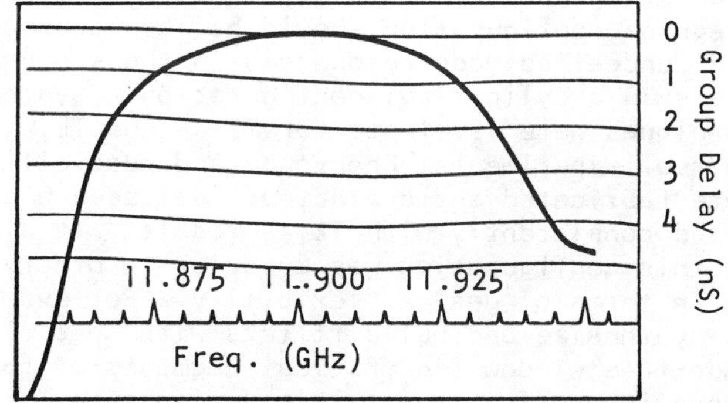

Fig. 4 Passband relative group delay TE_{103} filter.

it was necessary to use slightly different diameters in each of the dual mode cavities. A second requirement is that the isolation characteristic should not be transparent in the 14.0-14.5-GHz satellite receive band. This requirement was

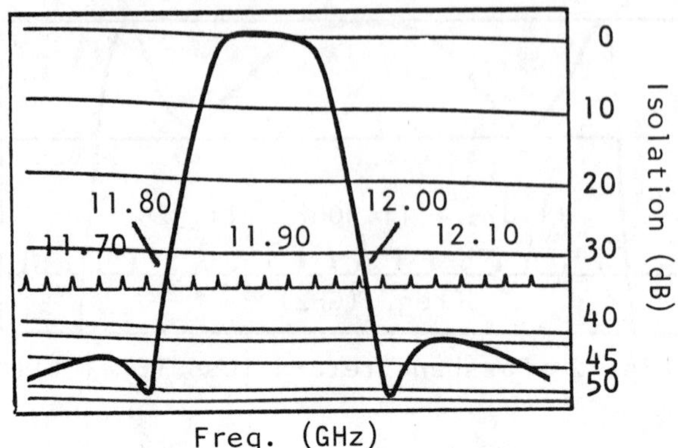

Fig. 5 Communications-band isolation TE_{103} Filter

the limiting constraint because of the crowded mode chart inherent in all design candidates of the $TE_{11}n$ type in which the volume is close to optimum. The main conclusion arrived at was that it is possible to make low-loss filters using a $TE_{11}n$ configuration, provided that isolation integrity does not have to be preserved outside a spectrum segment of 5% above and below the center frequency of the filter.

3.2 Dual Mode $TE_{10}n$ Filters

A close scrutiny of the mode lattice charts for square and circular waveguides indicates that $TE_{10}n$ filters in a square waveguide configuration should be less susceptible to the onset of undesired mode resonances in the stopband than $TE_{11}n$ filters in a cylindrical configuration. Synthesis and design techniques were developed for TE_{102} and TE_{103} filters, and a number of experimental Chebychev and quasielliptic designs were fabricated and evaluated. Unloaded Q's of 9000 were realized consistently with TE_{103} models. It also was found that this configuration was superior to other design candidates in terms of design flexibility. For example, the ability to synthesize and build filters with specifically defined mode-free windows in critical segments of the out-of-band response is one feature of this design type.

A total of six elliptic function filters were designed to comply with the specific satellite output multiplexer requirements. The measured passband characteristics of one of these filters are shown in Figs. 2-4. These results agree with those predicted from theoretical considerations. It can be seen from Fig. 3 that the center frequency insertion loss is 0.265 dB, which represents a power loss of less than 6%. A TE_{111} dual mode filter designed to comply with the same performance characteristics will have a center band loss of 0.5 dB (10.8% of transmitted power).

The isolation characteristic across the 11.70-12.20-GHz communications band is shown in Fig. 5. No mode resonances are present, and the response agrees with theory except for a slight asymmetry, which is particularly noticeable in the measured notch level. This asymmetry is present in all dual mode filters and stems from the frequency selectivity of the intercavity coupling irises.

The isolation characteristic between 10 and 11.7 GHz is shown in Fig. 6. All unwanted mode resonances are more than

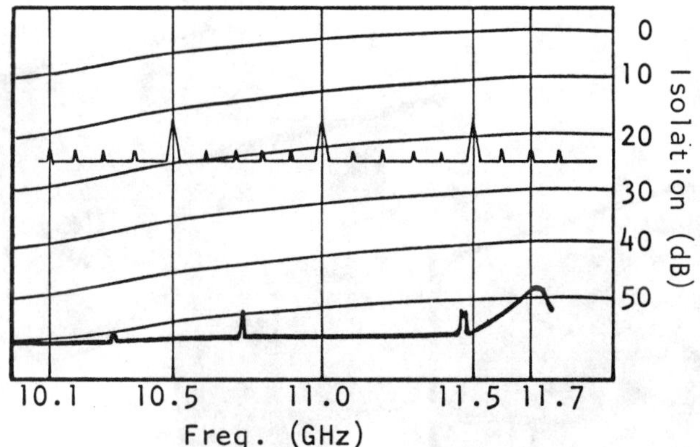

Fig. 6 Isolation 10-11.7-GHz TE_{103} filter.

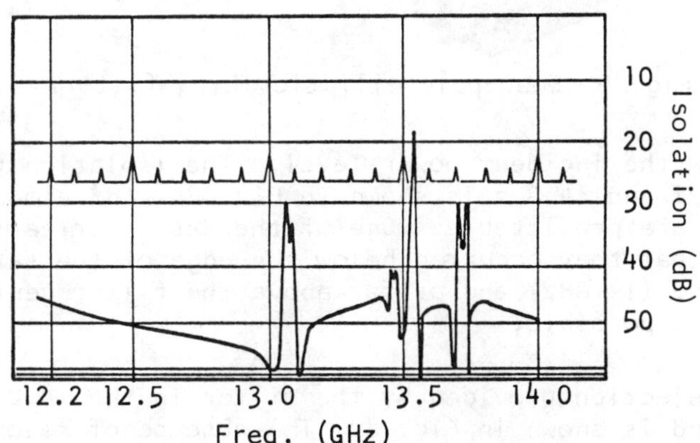

Fig. 7 Isolation 12.2-14.0-GHz TE_{103} filter.

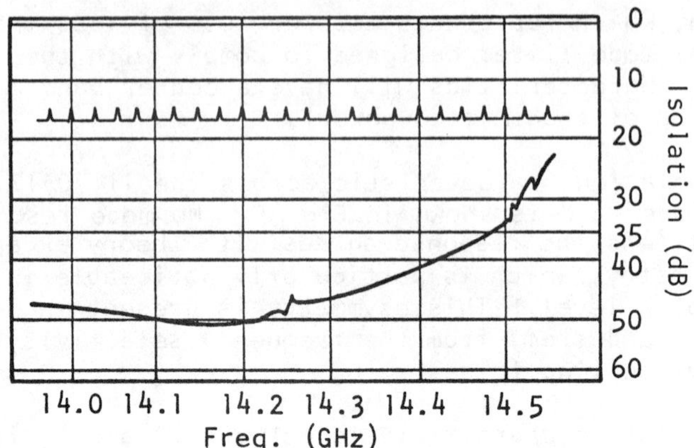

Fig. 8 Rx-band isolation 14.0-14.5 GHz TE_{103} filter.

Fig. 9 Four-pole elliptic TE_{103} filter.

50 dB below the incident power level. The isolation response between 12.2 and 14 GHz is shown in Fig. 7. Three mode resonances are predictable. One of the design criteria was to insure that they occurred below the edge of the satellite receive band (14 GHz) and as far above the filter center frequency as possible.

The rejection provided by the filter in the 14.0-14.5 receive band is shown in Fig. 8. The absence of resonances in this region is an essential requirement, since the output TWT's in a satellite emit broadband noise, which must be sup-

Fig. 10 12-GHz TE_{111} and TE_{103} filters.

pressed in the receive band by the output circuits to prevent any breakthrough that would degrade the satellite receiver noise temperature. As shown in Fig. 8, a mode-free window occurs right across the receive band. A photograph of the filter, whose characteristics were just described, is shown in Fig. 9.

Five-section Chebychev and six-section quasielliptic filters also were designed in the TE_{103} configuration. The five-section Chebychev filter was realized with the end cavities excited with dual modes and the central cavity with a single mode. It was found that the undesired orthogonal mode in the central cavity introduced a spurious resonance in the stopband region approximately 100 MHz above the center frequency. It was eliminated completely by reducing slightly the height of the central cavity. The relative sizes of a conventional TE_{111} dual mode filter and an equivalent TE_{103} filter are shown in Fig. 10. Both of these filters were designed to operate at 12 GHz.

4. Multiplexing Design Considerations

The primary functions of the filter elements in the output multiplexer of a satellite transponder are to introduce minimum loss in the passband of the transmitted signals and to provide sufficient rejection in the stopband to insure that the other channels sharing the same manifold are multiplexed with minimum degradation to their transmission characteristics. In-band group delay is not normally a serious constraint in this type of application, since its contribution is small and can be equalized in the ground receiver. Future satellite systems may require the multiplexing of contiguous channels in the output circuits, in which case the in-band delay and amplitude characteristics will be significant design parameters.

Table 1 Frequency plan for 12-GHz system (channel passband 72 MHz)

Channel	Frequency, GHz
1	11.74
2	11.82
3	11.90
4	11.98
5	12.06
6	12.14

A typical downlink frequency plan for a satellite operating in the 11.7-12.2-GHz band is shown in Table 1. Channels 1, 3, and 5 are multiplexed into one antenna port and 2, 4, and 6 into a second antenna port. At least 40 dB of isolation must be provided by each filter at the band edge of the closest channel multiplexed on the same manifold to insure minimum in-band degradation at the manifold-filter interface.

The passband amplitude and group delay characteristics of the filters required for multiplexing the channels shown in Table 1 were derived[6] and a set of specifications developed. These are shown in Table 2. It is worth noting that the requirements for minimum usable bandwidth and edge of band isolation at the closest multiplexed channel cannot both be

Table 2 Filter specification based on alternate channel multiplexing

Filter function	Elliptic
No. of cavities	4
Passbandwidth	80 MHz
Passband return loss	-22 dB
Center frequency	Designated channel fo
Notch level	-42 dB-min
Edge of band group delay	4 nsec maximum at fo ±36 MHz

met with a four-section Chebychev function filter. If the single mode TE_{011} filter described earlier in this paper had been used, it would be necessary to use a five-section filter, with the resultant increase in mass.

In discussing the primary functions of output multiplexing filters, no mention was made of the fact that, to minimize multipath distortion, significant rejection of adjacent channel frequencies has to be provided within the transponders.[11] Virtually all of this rejection usually is provided in the input multiplexer. It generally is not appreciated that, if elliptic filters are used in the output multiplexers, the sharpness of rolloff outside of the passband for a specific number of sections is a function of the stopband ripple level. If the out-of-band isolation level is adjusted to provide the minimum rejection necessary for multiplexing alternate channels, a significant amount of adjacent channel rejection will result. An accurate knowledge of the tradeoffs available in this area is an important element in the optimization of the in-band and out-of-band characteristics of the input multiplexer.

Elliptic filters in a dual mode realization do not behave exactly according to theory. It is necessary, therefore, to compile the required data from a large number of experimental models. A total of eight four-section elliptic filters with different slope factors were built at 4 and 12 GHz. The measured data were reduced and a unified tradeoff design chart compiled. This chart is shown in Fig. 11. Its use is explained in the Appendix.

Table 3 Output multiplexer loss budget

	4 GHz maximum loss, dB	12 GHz maximum loss, dB
Multiplexing filter[a]	0.25	0.27
Manifold	0.05	0.05
Low-pass filter	0.10	0.20
Isolator	0.10	0.10
Total	0.50	0.62

[a]Based on passbandwidth of 0.66% and Qo of 10,000 at 4 GHz and 9000 at 12 GHz.

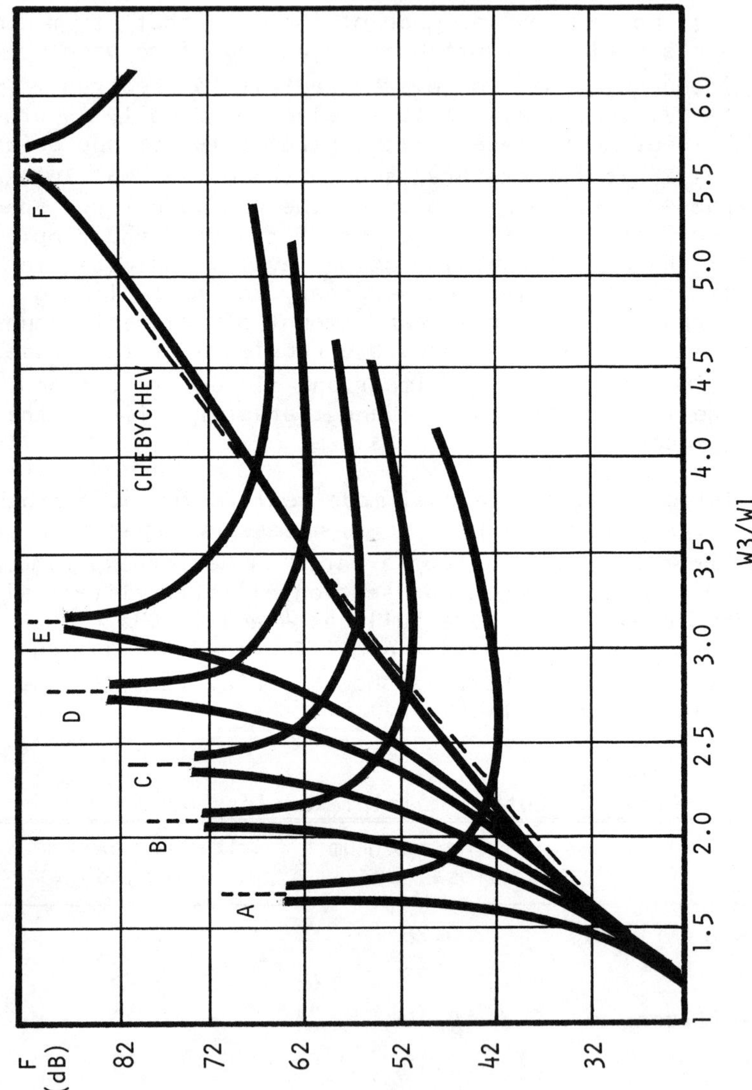

Fig. 11 Tradeoff design chart four-pole elliptic filters.

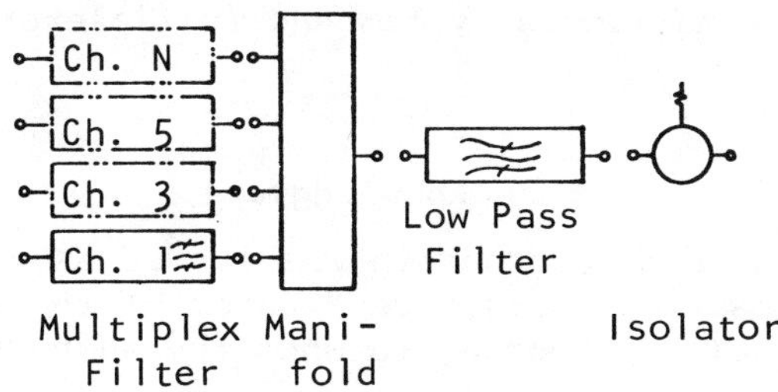

Fig. 12 Output multiplexer schematic.

5. Output Multiplexer Comparison at 4 and 12 GHz

A typical communications satellite output multiplexer schematic is shown in Fig. 12. In this configuration, up to six TWT outputs are combined using bandpass filters and a waveguide manifold. A low-pass filter to suppress harmonic outputs and an isolator to provide a good output impedance and to absorb reflections from the antenna subsystem are inserted between the manifold output and the antenna port. In 4-GHz systems, unloaded Q's of 10,000 have been realized using either Chebychev rectangular waveguide filters[7] or dual TE_{111} mode filters.[8,9] As shown in this paper, unloaded Q's of 9000 can be achieved at 12 GHz. Losses in the low-pass filters are typically 0.1 dB at 4 GHz. An output isolator loss of 0.1 dB can be achieved at both 4 and 12 GHz.[10] A loss budget for equivalent multiplexers at 4 and 12 GHz is shown in Table 3. It should be noted that in a real-life system there will be additional feeder losses and variations because of differences in the percentage bandwidths of the multiplexing filters.

6. Conclusions

The results presented in this paper are the outcome of an ongoing research and development program aimed at developing new types of filters and multiplexing networks for communications and broadcast satellite systems. Current work in progress is aimed at refining design techniques, developing complete multiplexer subsystems, and evaluating the power-handling capability of different filter candidates.

TE_{112} and TE_{103} filters described in this paper exhibit significantly lower loss than existing designs with TE_{101} or TE_{111} resonators, and possess the size and weight advantages inherent in the dual mode realization. The tradeoff design chart compiled from experimental data provides hitherto unavailable information to system and multiplexer design engineers.

Acknowledgment

The work described herein was funded as part of the Industrial Research Assistance Program of the National Research Council of Canada, to whom the authors are indebted.

Appendix: Use of Tradeoff Design Chart for Four-Pole Elliptic Filters

The design chart of Fig. 11 was constructed using experimental data and a previously developed normalization technique:

F = arithmetic sum (dB) of passband return loss and stopband isolation

W3 = frequency at which isolation is required minus the passband center frequency (MHz)

W1 = [return loss passbandwidths (MHz)]/2

As an example, a four-pole elliptic filter with a passband return loss of -22 dB and a center frequency of 12 GHz must provide a minimum of 42-dB isolation in its stopband. If the return loss passbandwidth is 80 MHz, what are the edge of band frequencies of the closest channels to be multiplexed?

$F = 22 + 42 = 64$ dB

$W3/W1 = 2.85$ (response E of design chart)

$W1 = 40$ MHz $\therefore$ $W3 = 2.85 \times 40 = 114$ MHz

Edge of band for closest multiplexed channels are 12,114 and 11,886 MHz.

As can be seen from the preceding example, if any three of the four parameters, passbandwidth, return loss, stopbandwidth, and stopband isolation, are known, then the remaining parameter can be found easily. In fact, if only two are specified, the tradeoffs between the remaining two can be ascertained without difficulty. The theoretical background and derivation of the premises on which this type of normalized design chart are based can be found in Ref. 6.

References

[1]Atia, E. A. and Williams, A. E., "New Types of Waveguide Bandpass Filters for Satellite Transponders," Comsat Technical Review, Vol. 1, Fall 1971, pp. 21-41.

[2]Kudsia, C. M., Kallianteris, S., and Epstein, N., "Dual Mode Filters for Terrestrial and Satellite Communications," Proceedings of the 18th Midwest Symposium on Circuits and Systems, Aug. 1975, pp. 354-359.

[3]Matthaei, G. L., Yound, L., and Jones, E. M. T., Microwave Filters. Impedance-Matching Networks and Coupling Structures, McGraw-Hill, New York, 1964, pp. 921-934.

[4]Matthaei, G. L. and Weller, D. B., "Circular TE_{011} Trapped-Mode Bandpass Filters," IEEE Transactions on Microwave Theory and Techniques, Vol. MTT-31, Sept. 1965, pp. 581-589.

[5]Taggart, D. A. and Wanselow, R. D., "Mixed Mode Filters," IEEE Transactions on Microwave Theory and Techniques, Vol. MTT-22, Oct. 1974, pp. 898-902.

[6]Kudsia, C. M. and O'Donovan, M. V., Microwave Filters for Communications Systems, Artech House, Jan. 1974, pp. 35-45.

[7]Kudsia, C. M. and O'Donovan, M. V., "A Lightweight Graphite Fiber Epoxy Composite Waveguide Multiplexer for Satellite Applications," Proceedings of 4th European Microwave Conference, Sept. 1974.

[8]Atia, A. E., "Computer Aided Design of Waveguide Multiplexers," Transactions on Microwave Theory and Techniques, Vol. MTT-22, March 1974, pp. 332-336.

[9]Pfitzenmaier, G., "A Waveguide Multiplexer with Dual Mode Filters for Satellite Use," Proceedings of 5th European Microwave Conference, Sept. 1975.

[10]O'Donovan, M. V., Kallianteris, S., Singer, S., and Woods, G., "Design of High Power 12 GHz Microwave Components for Use in Broadcast Communications Satellite Systems," Proceedings of 5th European Microwave Conference, Sept. 1975.

[11]O'Donovan, M. V., Kudsia, C. M., and Keyes, L. A., "Design of a Lightweight Microwave Repeater for a 24 Channel Domestic Satellite System," RCA Review, Vol. 34, Sept. 1973, pp. 506-

538; also AIAA Selected Reprint Series, Satellite Communications Systems, Vol. XVIII, edited by I. Kadar, New York, Jan. 1976, pp. 165-181.

[12]Kudsia, C. M. and Swamy, M. N. S., "Prototype Characteristics for Non Equi-ripple Filters for Communications Applications," Proceedings of the IEEE Communications and Power Conference, Oct. 1976.

DESIGN OF THE CS COMMUNICATIONS SUBSYSTEM

H. C. Hyams *

Ford Aerospace & Communications, Palo Alto, Calif.

With the increasingly high density telephony, TV and other traffic in Japan, it is necessary to exploit multichannel satellite systems to cope with the demand. A satellite employing a transponder of six channels in the 30 GHz/20 GHz band and 2 channels in the 6/4 GHz band with channel bandwidths of 200 MHz has been developed for this requirement. The transponder makes use of unique circular directional filters, specially developed TWTA's and a field effect transistor for low noise figure.

The CS satellite now being constructed by Ford Aerospace (as a subcontractor to Mitsibishi Electronics Corporation) for the Japanese Space Agency presents a unique challenge in the requirement to utilize for space communications a frequency band where the only known previous utilization was for propagation measurements. The communications subsystem employs 200-MHz channels: six using the 30-GHz band for uplink and the 20-GHz band for downlink; and the other two using the conventional 6-GHz up and 4-GHz down. A simplified block diagram is shown in Fig. 1.

With the increasing high density telephony, TV and other traffic in Japan, it is necessary to exploit multi-channel satellite systems to cope with the demand. Therefore, the bandwidth rich 30-GHz/20-GHz band is of great interest compared with the 14-GHz/11-GHz and 6-GHz/4-GHz bands each of which is constrained by the CCIR to 500-MHz bandwidth. Although it is planned to experiment with many types of signals, the passage of 100-megabit bi-phase signals was used as a design criterion.

Presented as Paper 76-293 at the AIAA/CASI 6th Communications Satellite Systems Conference, April 5-8, 1976, Montreal, Canada.

*Manager, Advanced Communication Systems Engineering Department.

Table 1 Design requirements of Ford supplied channels per NASDA-SPC-371B

Parameter	K-band channels	C-band channels
Bandwidth (3dB down)	Assigned Frequency ± 100 MHz	Assigned Frequency ± 100 MHz
Amplitude Characteristics (dB)		
± 50 MHz	Within ± 0.7	Within ± 0.7
± 70 MHz	≥ -2	≥ -2
± 100 MHz	-3 nom.	-3 nom.
± 140 MHz	≤ -15	---
± 200 MHz	---	-20
Noise Figure	< 13 dB	< 9 dB
Input Level (dBm)	-71 to -55	-70 to -60
Input Impedance (VSWR)	≤ 1.3:1	≤ 1.5:1
Level Control Range	Less than 1 dB output power variation for input level variation of 16 dB with AGC	Less than 1 dB output power variation for input level variation of 10 dB with AGC
Transmit Power Output	> 34 dBm	> 34.5 dBm
Output Impedance (VSWR)	≤ 1.5:1	≤ 1.5:1
Input/Output Linearity	Output level for 1 dB compression shall be 32 dBm (Design goal)	Output level for 1 dB compression shall be 32.5 dBm (Design goal)
Transmit Spurious L Level	< -40 dB	< -40 dB
Receive Spurious Level	< -40 dBm	< -40 dBm
Interchannel Interference D/U Ratio	> 26 dB	> 26 dB
AM/PM Conversion	< 4.5°/dB	< 4.5°/dB
Group Delay		
± 25 MHz	± 1 ns	± 1 ns
± 50 MHz	± 0.5 to 3 ns	± 0.5 to 4 ns
± 70 MHz	+ 2 to + 5.5 ns	+ 3 to + 8 ns
± 100 MHz	+ 4 to 10 ns	+ 6 to + 15 ns
Long Term Stability	± 1 x 10^{-5}/yr	± 1 x 10^{-5}/yr
Short Term Stability (Including effect of temp. & power supply variation)	± 6 x 10^{-6}	± 6 x 10^{-6}
Beacon Frequency	19.45 GHz	
Beacon Power	30 MW (Min)	

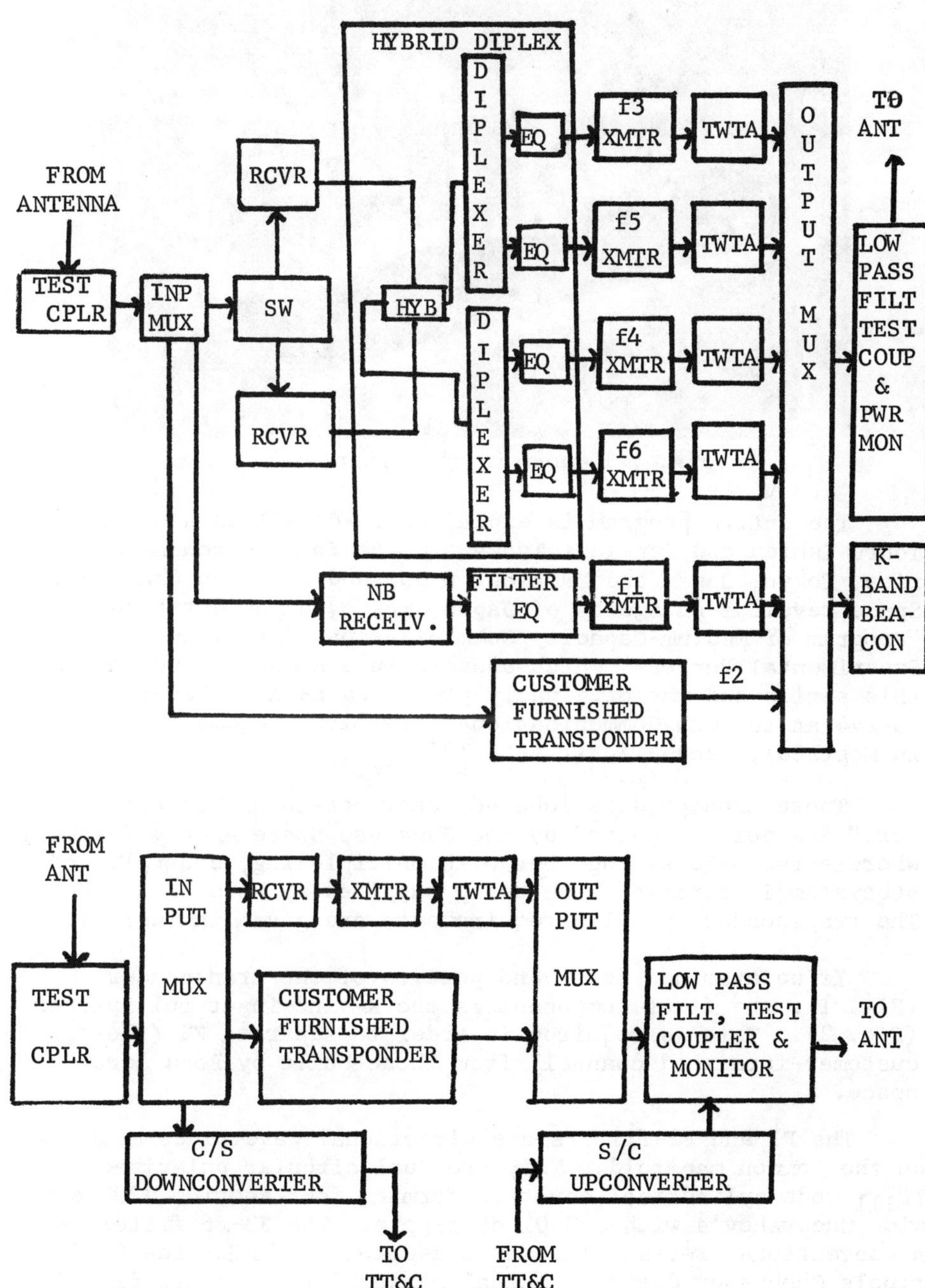

Fig. 1 K- and C-band transponder simplified block diagram.

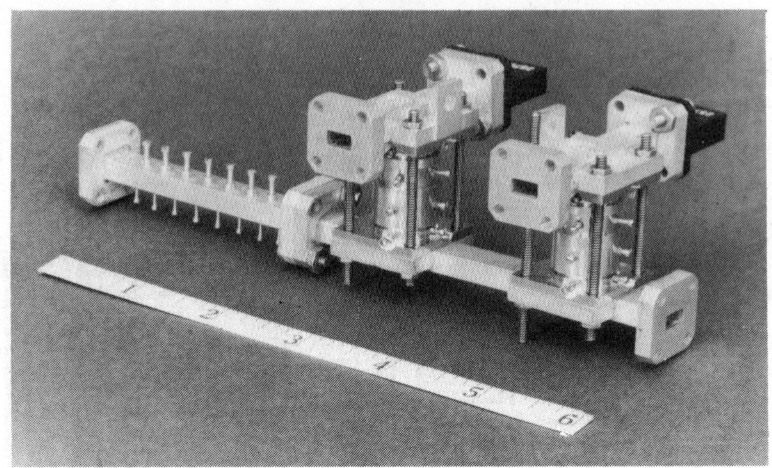

Fig. 2 K-band input multiplexer.

The entire program is explained in detail in a paper by Tohru Ishida and Ken-Ichi Tsukamoto, Radio Research Laboratories, Tokyo, Japan and Masaichi Hirai and Hironubu Okamoto, Space Development Agency of Japan, Tokyo, Japan entitled "Program of Medium-Capacity Communications Satellite for Experimental Purpose" which appears in a companion volume of this series and was originally presented as AIAA Paper No. 76-244 at the 6th Communications Satellite Systems Conference in Montreal, Canada April 5-8, 1976.

Those transponders labeled "customer-furnished transponders" are being supplied by the Japanese Space Agency (NASDA), whereas the others, together with multiplexing, control, and subsystem integration, are being supplied by Ford Aerospace. The transponder detailed requirements are given in Table 1.

If we examine the K-band portion of the transponder (Fig. 1), the first component is the K-band input multiplexer (Fig. 2). This is required in order to separate F2 (the customer-furnished channel) from those built by Ford Aerospace.

The F1 and F2 filters are directional waveguides mounted on the common manifold. They are dual circular polarized TE_{111} mode cylindrical cavities forming four-section, 200-MHz-wide Chebyshev's with a 0.01-dB ripple. The F3-F6 filter is a conventional rectangular seven-section 1726-MHz-wide 0.01 dB ripple Chebyshev design. Actual measured results are given in Table 2.

A block diagram of a redundant half of the wide-band receiver is shown in Fig. 3 and a photograph in Fig. 4. The

Table 2 K-band input multiplexer measured results

CH	Insertion loss, dB	Flatness, dB	Differential group delay, nsec
F1	1.1	1.0	2.2
F2	1.0	0.5	1.0
F3	0.6	0.2	0.3
F4	0.5	0.1	0.1
F5	0.4	0.0	0.0
F6	0.4	0.1	0.2

Table 3 K-band wide-band receiver performance (Unit A)

Parameter	F3	F4	F5	F6
Gain, dB	25.5	25.4	24.8	24.7
Frequency Response, dB	0.6	0.4	0.1	0.3
Group delay variation, nsec	0.4	0.3	0.1	0.2
Noise Figure, dB	10.5	10.5	10.0	10.0
Third order intermods				
(Po = -30 dBm), dBm	< -70 dBm			
Spurious signals	None observed			

redundant receivers are separated by the input switch. From here, the signals go to a low noise mixer where the 28.45 - 30.05 GHz signals are down converted to a 3.15 - 4.75 GHz intermediate frequency. The local oscillator for this is generated from a 210.83 MHz crystal oscillator and multiplied (with careful filtering) up to 25.3 GHz. The received signals after down conversion are amplified by 30 dB in a wide-band pre-amplifier. Measured results are given in Table 3. A unique feature of this receiver is the use of a GaAs FET for the first amplifier stage. A photograph of the pre-amp is shown in Fig. 5. Referring back to Fig. 1, the redundant receivers are rejoined by a hybrid. A diplexer separates the channels into four separate transmitters.

The K-band transmitters are packaged in pairs (Fig. 6), as are the redundant receivers. This is done to conserve

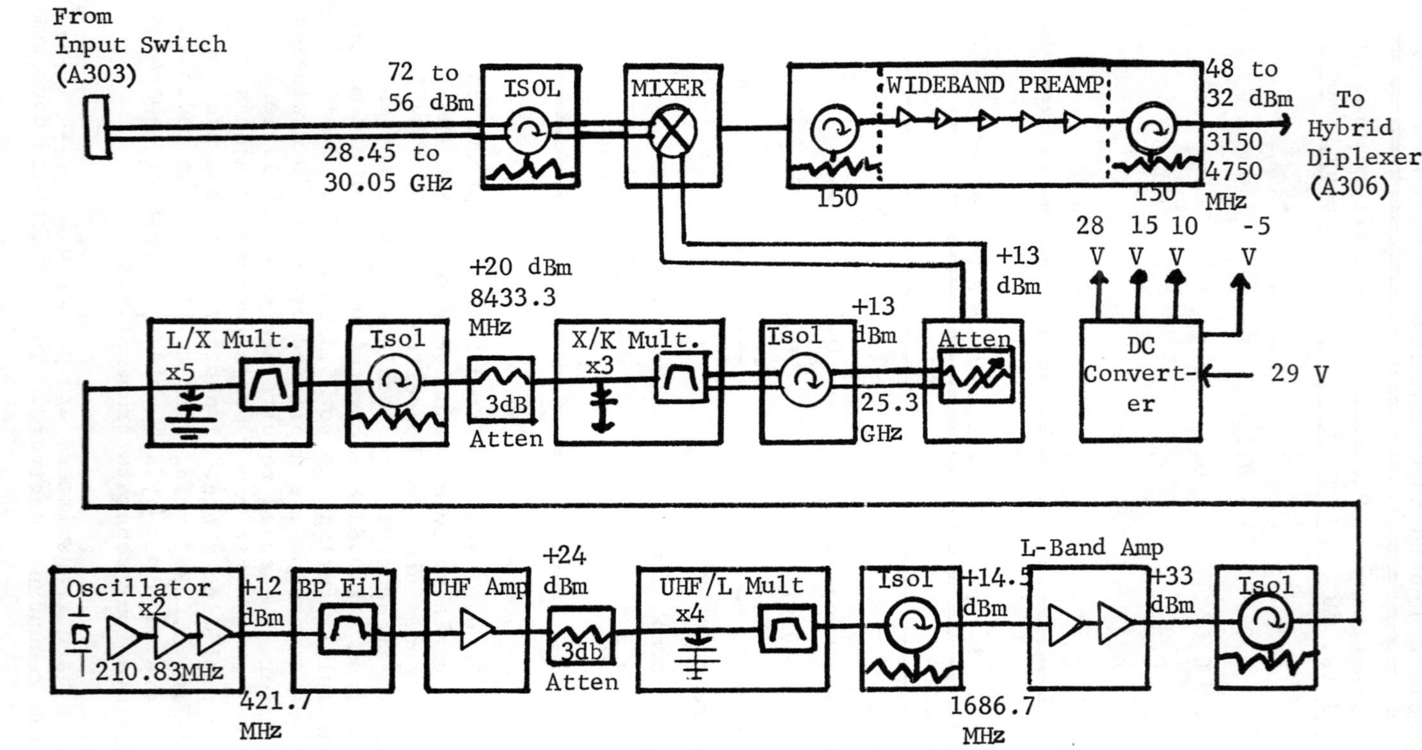

Fig. 3 K-band wide-band receivers block diagram.

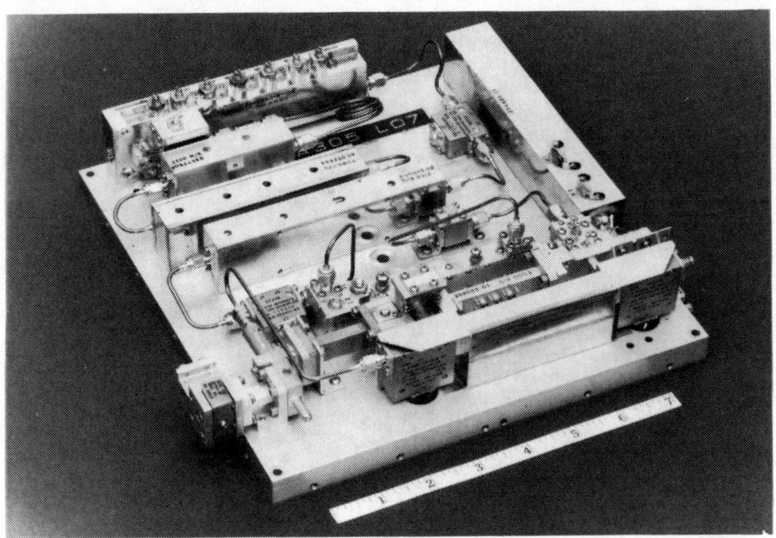

Fig. 4 Wide-band K-band receiver.

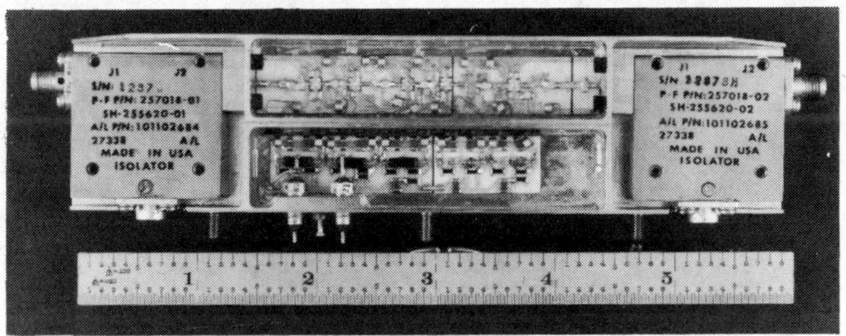

Fig. 5 Wide-band K-band pre-amp.

footprint space. Fig. 7 shows a single transmitter and Fig. 8 its block diagram. The signals enter a transmitter via a four-stage IF amplifier.. Following this, signals pass through another IF amplifier which has automatic gain control (AGC) accomplished by coupling off some of the signal and rectifying it in an AGC amplifier and feeding the d.c. back to control a pin diode attenuator. The signal is then heterodyned up to the 20 GHz band by a mixer and local oscillator similar to those in the receiver. Transmitter performance summary is given in Table 4.

The 20-GHz TWTA's are procured under a subcontract from Hughes Electron Dynamics Division. A photograph is shown in Fig. 9. The TWTA's weigh 2.13 Kg each. TWTA characteristics are shown in Table 5.

Fig. 6 K-band transmitters in pair.

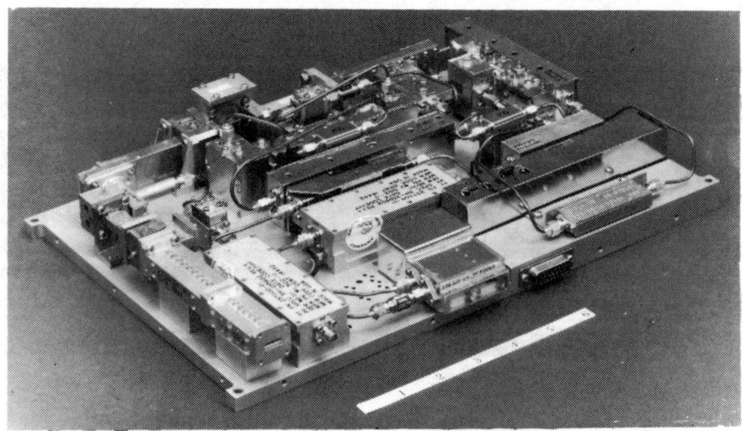

Fig. 7 Single transmitter.

Each TWTA plus a beacon feed into the K-band output multiplexer (Figs. 10 and 11). This consists of seven directional waveguide filters coupled into a common output manifold. As with the input multiplexer, these are dual circular polarized TE_{111} mode cylindrical cavities. The filters are comprised of four section 182 MHz 0.01 dB ripple Chebyshev's. Design goals and actual performance are shown in Table 6.

The C-band channels follow a similar path. The C-band input multiplexer (Figs. 12 and 13) separates the incoming signals into their three paths receiver G1, G2, and the C to

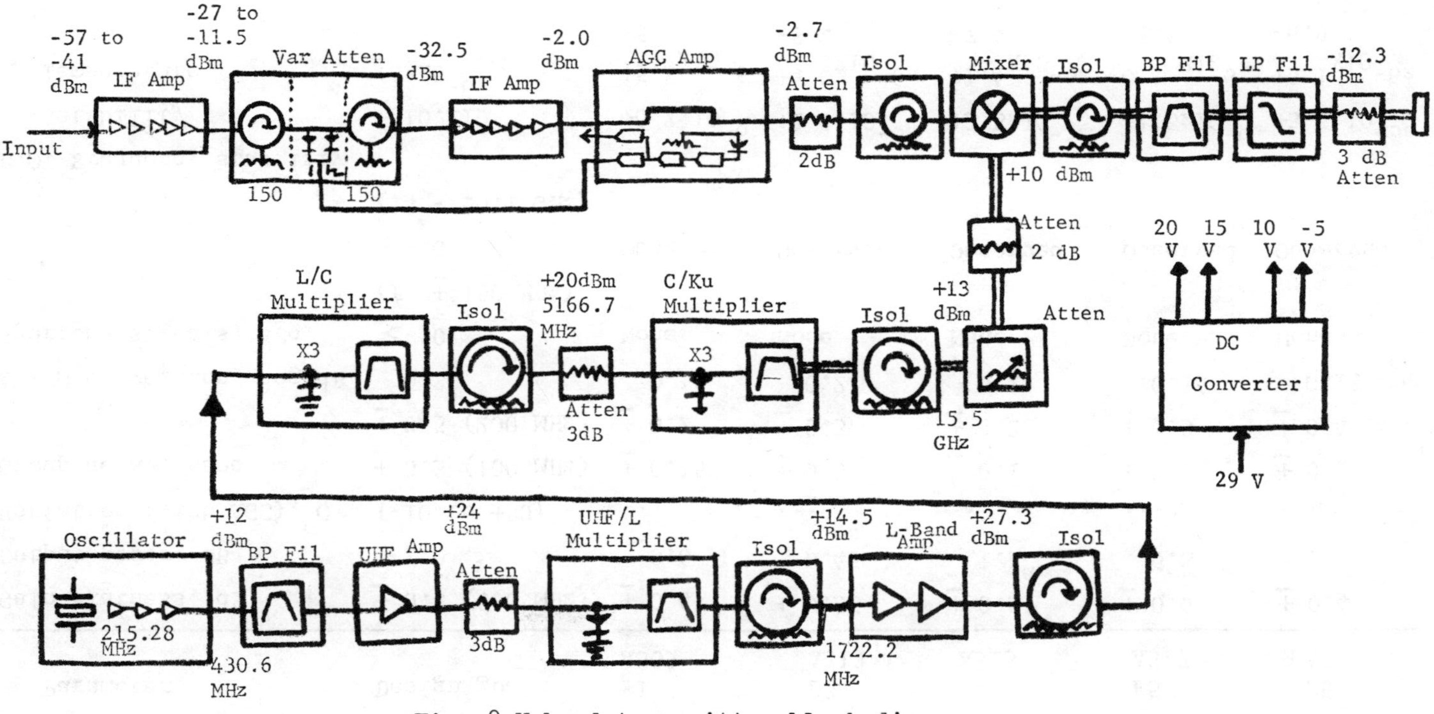

Fig. 8 K-band transmitter block diagram.

Table 4 K-band transmitter performance summary

Parameter	Design goal	F1 A309	F3 A313	F4 A315	F5 A317	F6 A319
Gain Flatness, dB	± 0.6 (200 MHz)	± 0.47	± 0.35	± 0.6	± 0.6	± 0.4
Output power, dB	1.0	0.6	0.45	1.1[a]	0.8	0.6
Variation (with AGC), ^{o}C	(-10 to +50)	...	...	...	...	...
Group delay, nsec	± 0.3 (100 MHz)	± 0.15	± 0.3	± 0.1	± 0.1	± 0.2
	± 0.5 (200 MHz)	± 0.2	± 0.3	± 0.2	± 0.3	± 0.4
AM/PM conversion, deg/dB	0.3	0.12	0.2	0.2	0.16	0.15
Spurious signals, dBc	< -50 (f_o ± 150 MHz)	None	None	None	None	None
	< -30 (14 - 30.1 GHz)	Observed	Observed	Observed	Observed	Observed
L.O. Frequency stability						
Setability	± $10.x10^{-5}$	$+6.2x10^{-6}$	$+2.9x10^{-5}$[a]	$+3.8x10^{-5}$[a]	$-1.3x10^{-6}$	$+3.5x10^{-5}$[a]
Temperature	$\pm 6x10^{-6}$	+5.8 -5.7 $X10^{-6}$	+4.0 -8.2 $X10^{-6}$[a]	+1.4 -7.4 $X10^{-6}$[a]	0 -3.6 $X10^{-6}$	+5.1 -6.9 $X10^{-6}$[a]
dc power, W	7.5	6.6	6.1	6.4	6.4	6.25

[a] Exceeds design goal limit.

Table 5 TWTA performance data

Parameter	Specification	Channel and unit				
		f1[a] (FM1)	f3[b] (EM3)	f4 (EM2)	f5[c] (FM3)	f6 (FM5)
Output power (SAT), dBm	35	36.5	35.6	36.2	36.1	36.6
Input Drive, dBm	-16 ± 3	-17.0	-19.0	-18.5	-11.0	-14.9
Saturated gain	...	53.5	54.6	54.7	47.1	51.5
Frequency response variation (200 MHz), dB	0.2	0.1	0.2	0.15	0.2	0.18
Input VSWR						
Hot	2.5:1	1.30	1.37	1.44	2.22	1.40
Cold	1.5:1	1.25	1.26	1.37	1.25	1.38
Output VSWR						
Hot	2.5:1	1.70	1.99	1.67	1.41	1.17
Cold	1.5:1	1.08	1.44	1.22	1.27	1.17
AM/PM conversion, deg/dB	4.5	...	...	...	...	3.0
Harmonic output (total), dBc	-12	-24.9	...	...	-40.0	-19.9
dc power, W	21.7	20.7	22.2	21.0	19.5	20.7

[a] Measurements taken at f3, except rf input and output power at f1.

[b] Measurements taken at f4, except rf input power, output power, and frequency response at f3.

[c] Gain was low, since this was a -01 model TWTA used to replace a defective -02 unit.

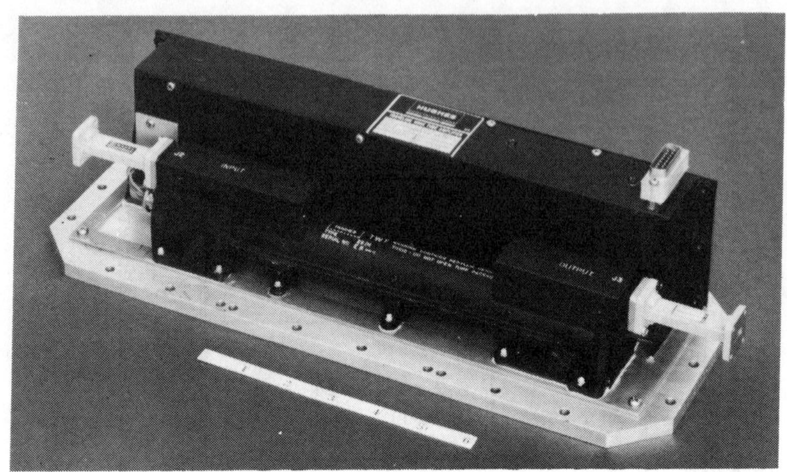

Fig. 9 20-GHz TWTA.

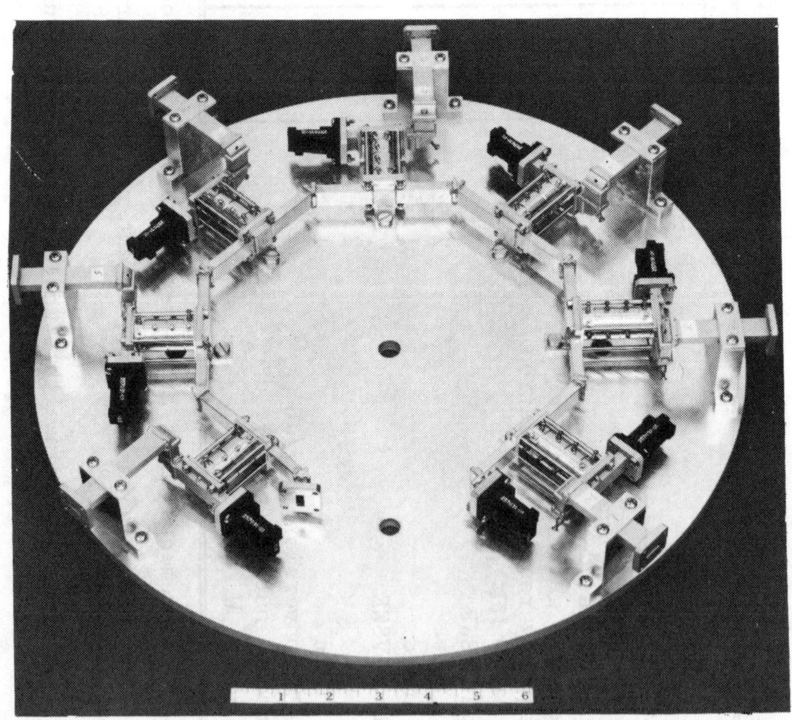

Fig. 10 TWTA with beacon feed into K-band output multiplexer.

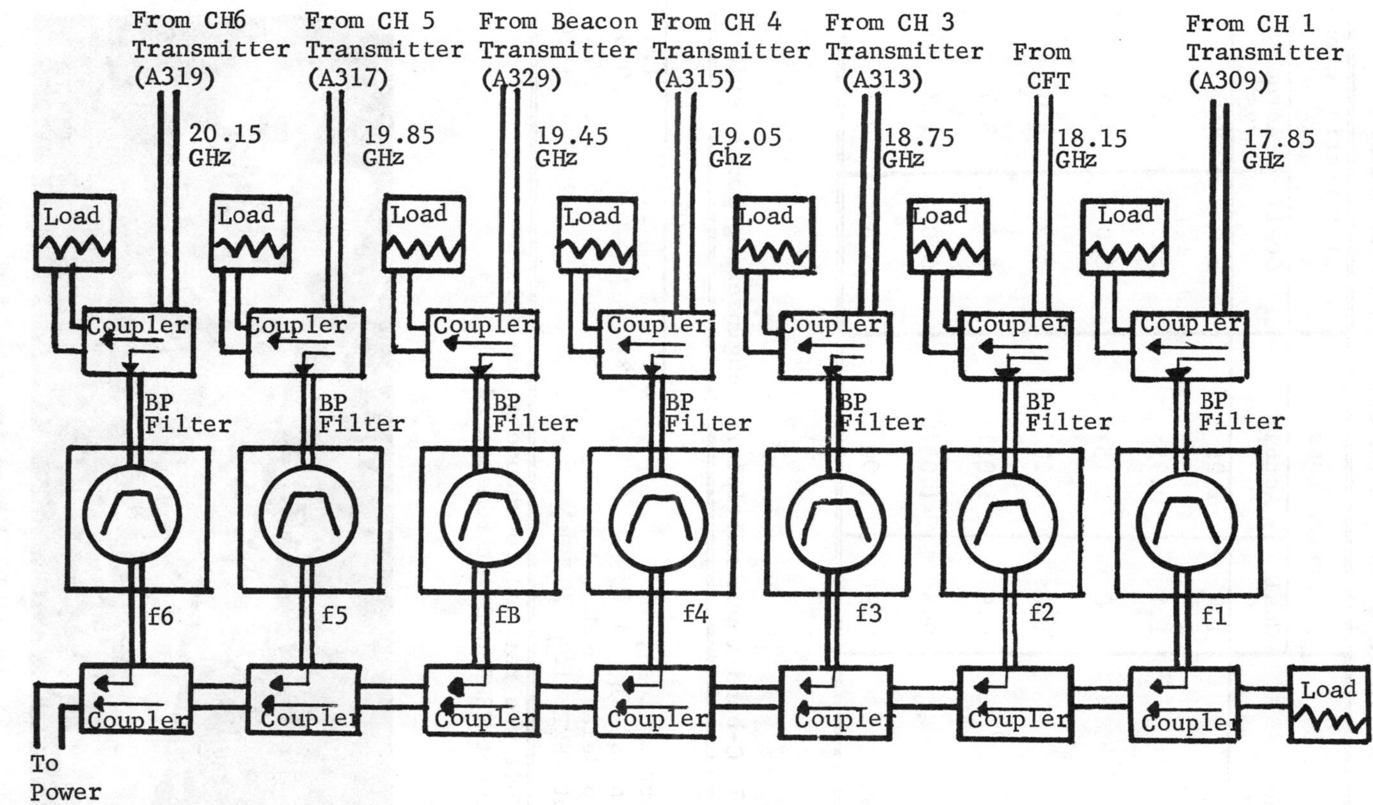

Fig. 11 K-band output multiplexer block diagram.

Table 6 Output multiplexer performance

Channel	Insertion Loss		Differential group delay, nsec	
	Goal	Actual	Goal	Actual (25°C)
f_1	1.2	1.05	3.0	2.0
f_2	1.1	0.95	3.0	2.0
f_3	1.0	1.15	3.0	3.9
f_4	1.0	1.10	3.0	2.0
f_5	0.9	1.05	3.0	3.0
f_6	0.8	1.05	3.0	1.8
Beacon	0.9	1.00	N/A	N/A

Table 7 C-band input multiplexer engineering model performance

	G1	G2	G3
Insertion loss, dB	1.0	1.15	1.5
Flatness, dB	0.3	0.4	0.5
Group delay, nsec	5.5	4.5	N/Z
Input VSWR (fo± 70 MHz)	1.20	1.20	1.20

a Unit was retuned to reduce group delay. Input VSWR not yet remeasured.

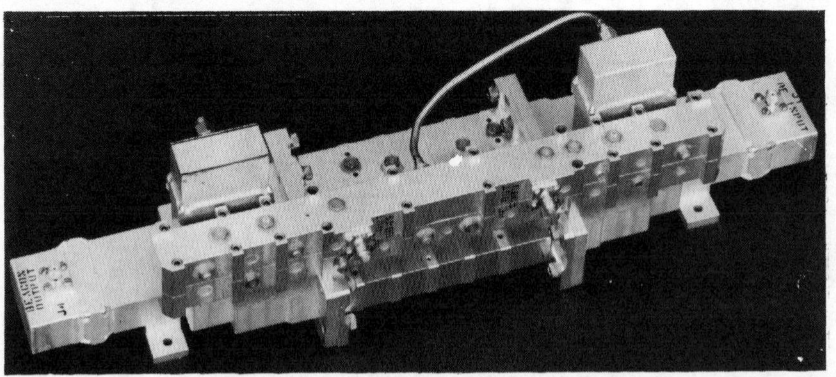

Fig. 12 C-band input multiplexer.

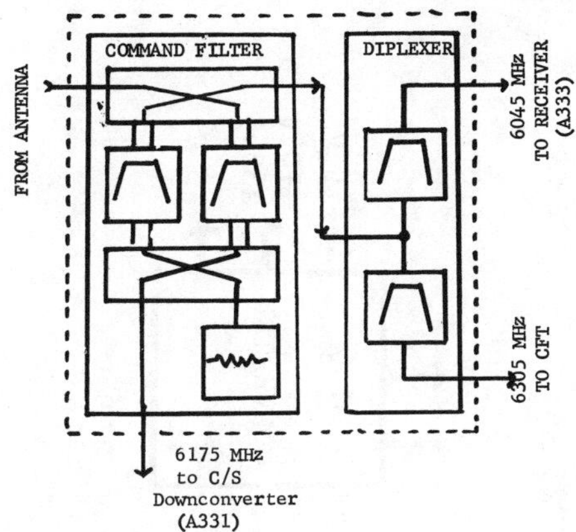

Fig. 13 Input multiplexer (A332).

Table 8 Receiver performance summary

Parameter	Design Goal	Measured
Gain, dB	19 Minimum	21.9
Gain variation (-10° to $+55^{\circ}$C), dB	1 Maximum	0.2
Gain ripple (200 MHz), dB	1 Maximum	1.0 (-10°C)
Group delay variation, nsec		
100 MHz	0.5	0.2
200 MHz	2.0	1.8
Noise figure, dB	7 maximum	6.2
L.O. power, dBm	+10 to +12	+10.7
L. O. Frequency, MHz	2225	2225.000
Setability	$\pm 1 \times 10^{-5}$	$+1.5 \times 10^{-6}$
Stability	$\pm 6 \times 10^{-6}$	$+5.7 \times 10^{-6}$ (-10°) -2.2×10^{-6} ($+50^{\circ}$C)
AM/PM conversion, deg/dB	0.2	0.03
Spurious signals	< -107 dBm at input at G2 and G2 image	OK to < -92 dBm (measured limit)
	< -73 dBm at output (no input signal)	None
	< -50 dBc at output (with signal in)	None
dc power (includes transmitter, W		4.12

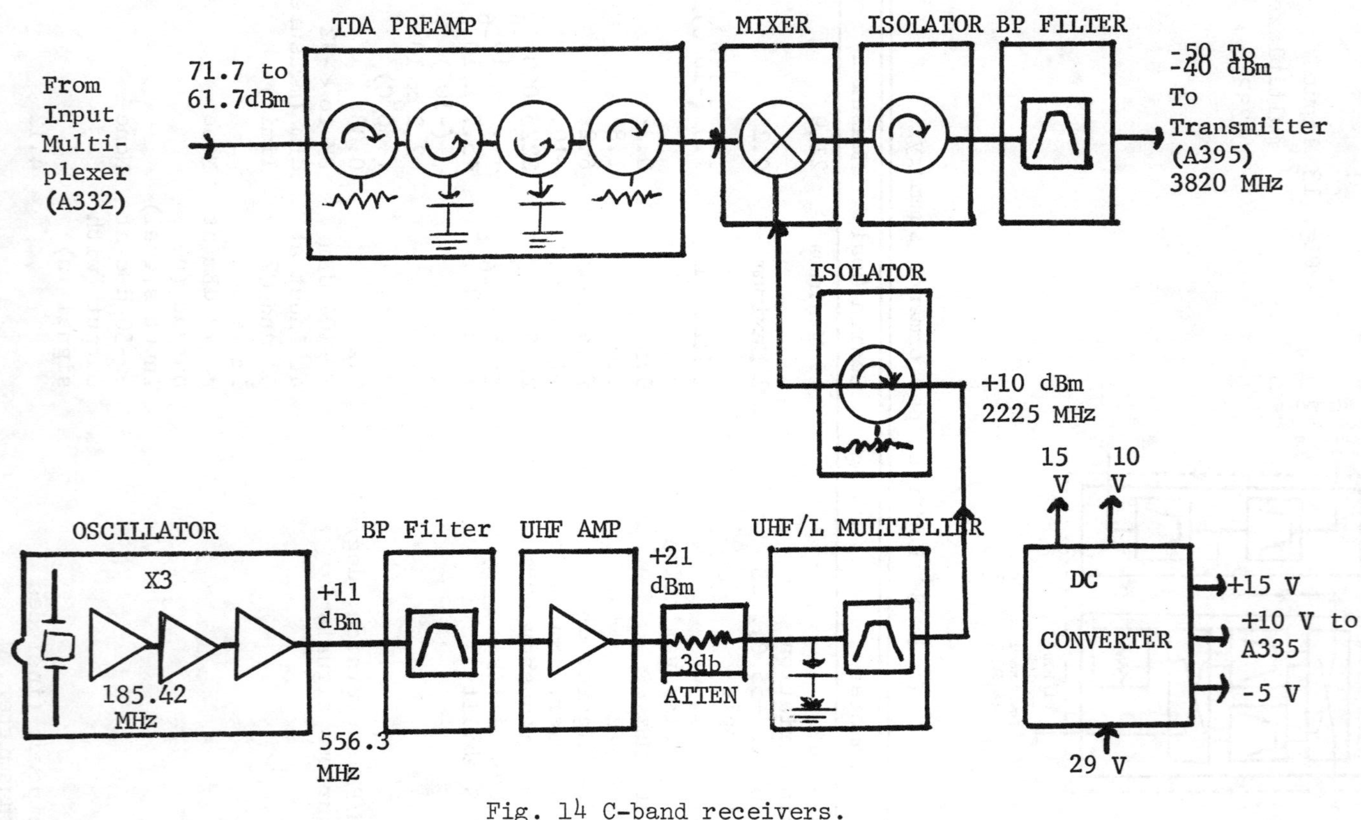

Fig. 14 C-band receivers.

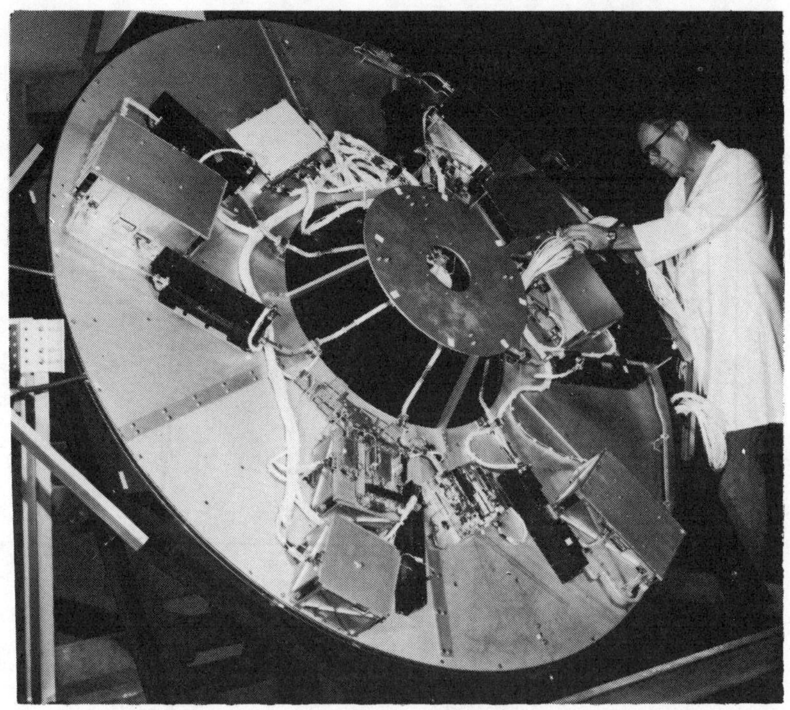

Fig. 15 Transponder mounted for thermal vacuum testing.

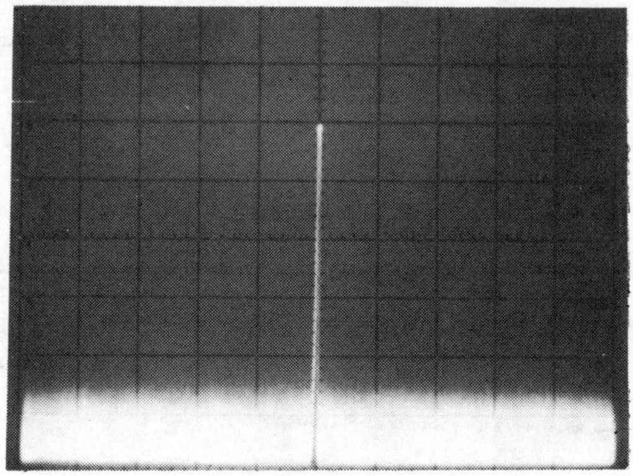

Fig. 16 Spectrum analyzer.

S-band command downconverter. The multiplexer consists of a dual magic-T hybrid-type command frequency filter and linear phase 10-section interdigital filters in the channel diplexer. Results are shown in Table 7.

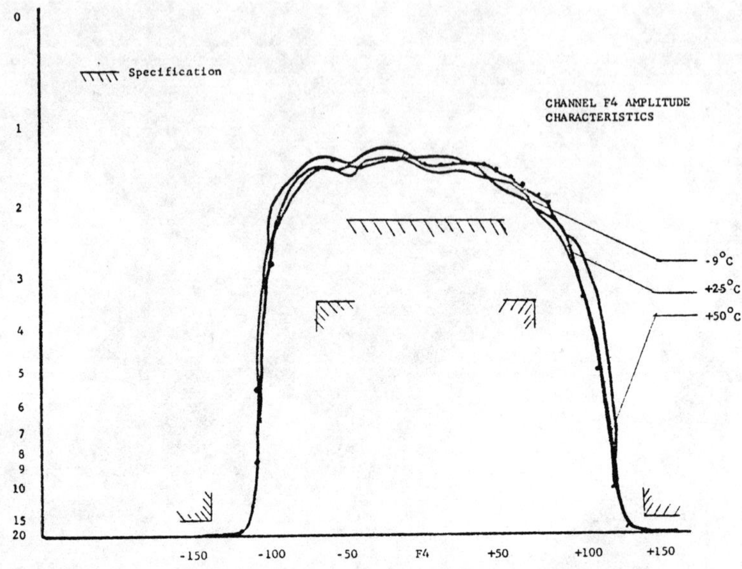

Fig. 17 Channel F4 AGC loops.

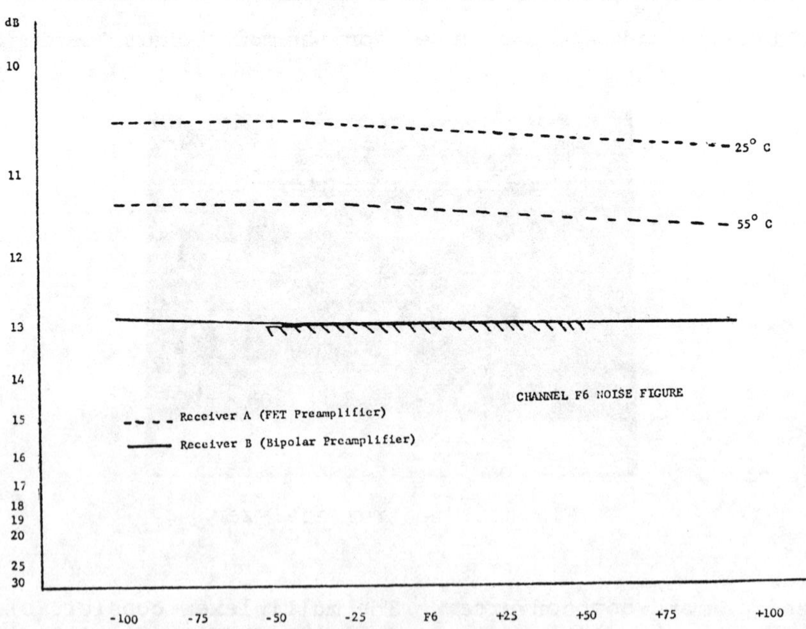

Fig. 18 Channel 6 noise figure.

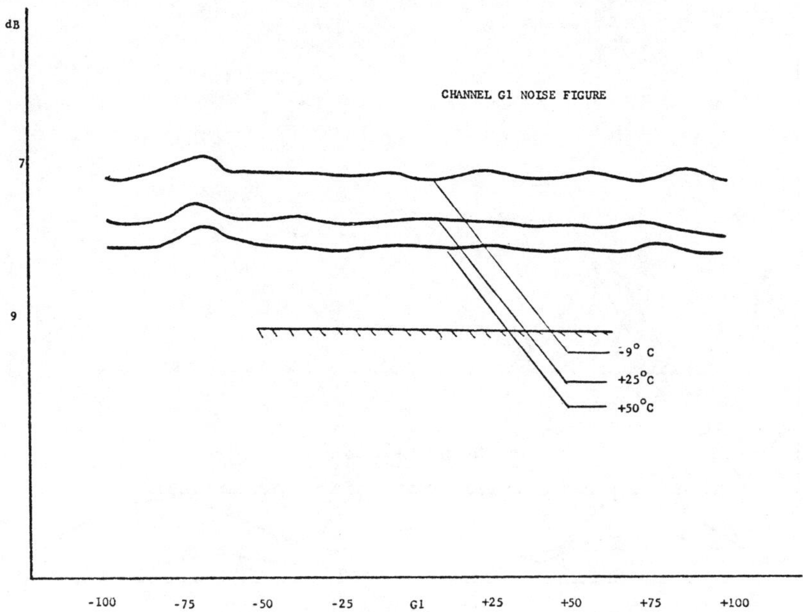

Fig. 19 C-band channel G1 employing TDA.

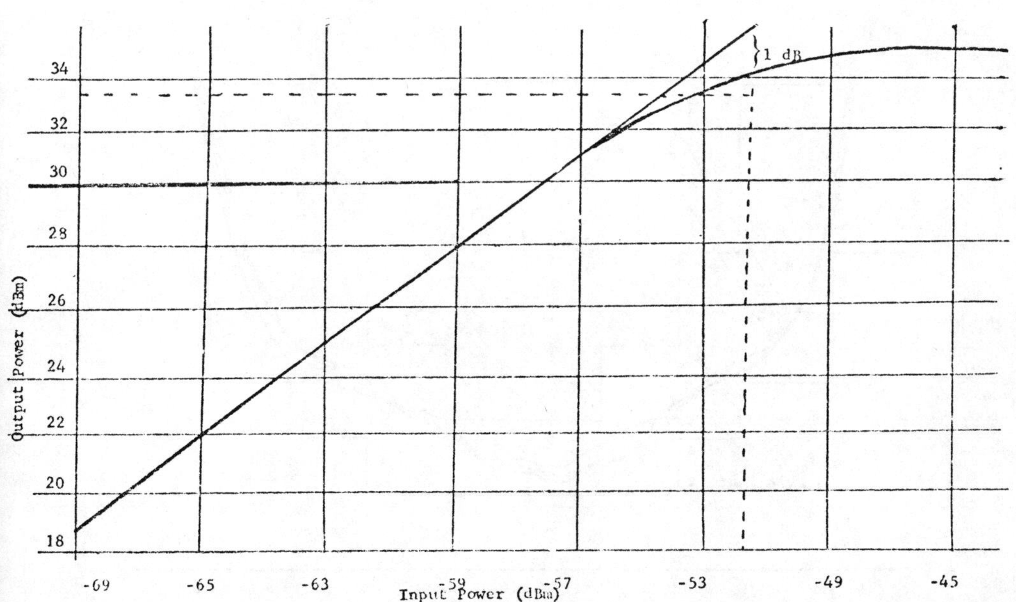

Fig. 20 Input/output linearity & impedance curves.

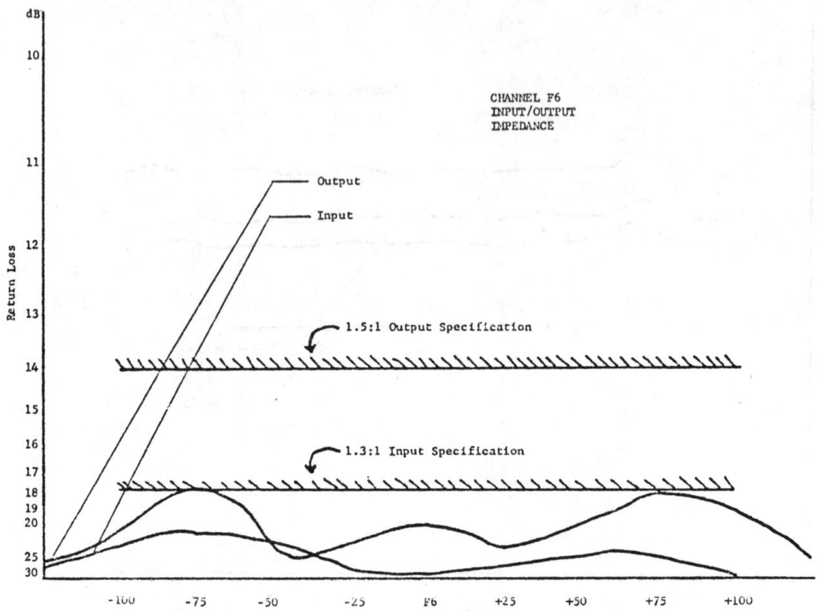

Fig. 21 Input/output linearity & impedance curves.

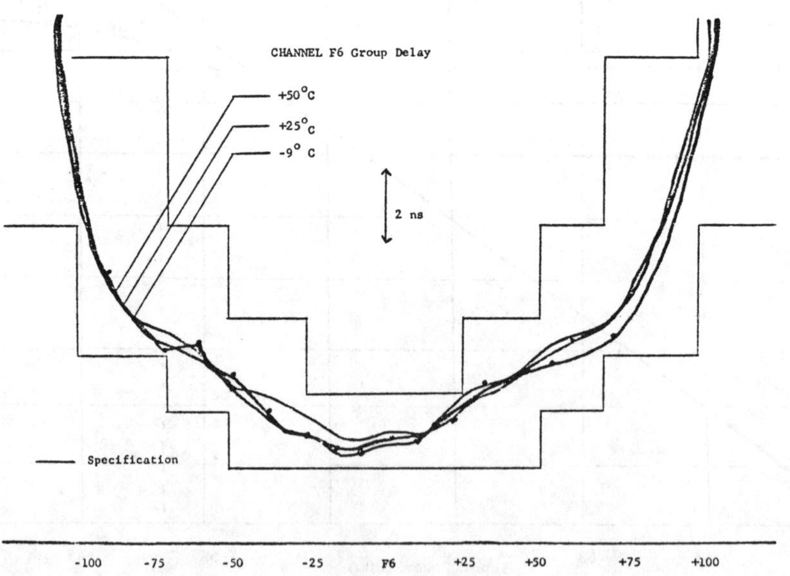

Fig. 22 Channel 6 group delay.

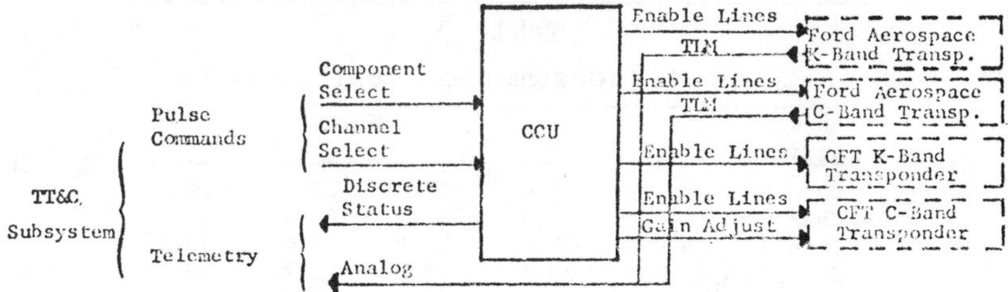

Fig. 23 Communication control unit functional block diagram.

The C-band receiver (G1) is shown in Fig. 14. The TDA preamp (two-stage) provides gain and sets noise figure before downconversion to 4 GHz. Performance details are shown in Table 8. A photograph of the complete transponder mounted on a test fixture for thermal vacuum testing is shown in Fig. 15.

Tests of the telemetry, tracking, and command capability were made with excellent results. No false locks, spurious signals, or phase noise changes occurred. The threashold measured -132.5 dBm against a specification of -127 dBm, swept acquisition and power output spectrum were normal, and no degradation of S-band noise figure by C-band were found. Interchannel interference measurements were made by simulation of a 100-mb/sec biphase signal. No interference was found with typical spectrum analyzer photographs, as shown in Fig. 16.

Bandwith and amplitude were tested on a channel-by-channel basis with AGC loops opened. As seen from the results of channel F4 (Fig. 17), amplitude characteristics were well within specification, and little change occurs with temperature. Input level and level control range were well within specification. During the engineering model phase of the program, the use of an FET first-stage IF preamplifier vs a bipolar stage had not been approved, and so one redundant amplifier was made each way. Fortunately, approval was received subsequently to fly the FET. The noise figure in each case is shown for channel F6 (Fig. 18), whereas the results for C-band channel G1 (employing a TDA) are shown in Fig. 19. Typical input/output linearity and impedance curves are shown in Figs. 20 and 21.

Group delay measurements were made for each channel. Optimization was accomplished by means of variable group delay equalizers in each channel. Results for F6 are shown in Fig. 22. Note that temperature has little effect. The communication control unit (CCU) provides the interface between the T&C subsystem and the components of the transpon-

Table 9

Dc power consumption test results

Equipment	Watts -9°C	+25°C	+37°C	+50°C
Chan F1	27.4	28.0	28.6	29.5
Chan F2	40.3	42.3	39.8	40.6
Chan F3	27.0	27.8	27.0	28.8
Chan F4	25.2	25.2	25.2	25.2
Chan F5	23.4	25.0	23.8	26.1
Chan F6	31.7	32.2	32.3	32.5
Beacon Transmitter	8.0	7.7	7.3	7.3
Chan G1	21.8	22.4	21.2	22.7
Chan G2	32.8	32.5	31.3	32.8
Down/Up Converter	5.1	4.8	4.8	5.0

Table 10 Transponder subsystem weight summary

ITEM	ID NO.	QTY/ SYSTEM	UNIT WEIGHT LBS (KG)	TOTAL WEIGHT LBS (KG)
C-BAND TRANSPONDER				
TEST COUPLER	A370	.	0.09 (0.04)	0.09 (0.04)
INPUT MULTIPLEXER	A322	1	2.56 (1.16)	2.56 (1.16)
RECEIVER	A333	1	3.80 (1.72)	3.80 (1.72)
TRANSMITTER	A335	1	2.80 (1.27)	2.80 (1.27)
TWTA AND ISOLATOR	A336	1	3.61 (1.64)	3.61 (1.64)
OUTPUT MULTIPLEXER	A341	1	2.00 (0.91)	2.00 (0.91)
POWER MONITOR	A343	1	0.68 (0.31)	0.68 (0.31)
C/S DOWNCONVERTER	A331	1	3.45 (1.56)	3.45 (1.56)
S/C UPCONVERTER	A344	1	3.15 (1.43)	3.15 (1.43)
RF COAX CABLE		1	0.91 (0.41)	0.91 (0.41)
SUBTOTAL				23.05 (10.45)
K-BAND TRANSPONDER				
TEST COUPLER	A369	1	0.09 (0.04)	0.09 (0.04)
INPUT MULTIPLEXER	A302	1	0.88 (0.04)	0.88 (0.04)
INPUT SWITCH	A303	1	0.22 (0.10)	0.22 (0.10)
RECEIVERS	A304, 305, 321	3	4.95 (2.25)	14.85 (6.74)
HYBRID/DIPLEXER	A306	1	3.00 (1.36)	3.00 (1.36)
FILTER/EQUALIZER	A322	1	0.06 (0.27)	0.60 (0.27)
TRANSMITTERS	A313, 315, 317, 319, 321	5	6.75 (3.06)	33.75 (15.30)
TWTA'S AND ISOLATORS	A314, 316, 318, 320, 310, 372, 374, 375, 371, 373	5	4.70 (2.13)	23.50 (10.66)
OUTPUT MULTIPLEXER	A326	1	2.20 (1.00)	2.20 (1.00)
POWER MONITOR	A330	1	0.79 (0.36)	0.79 (0.36)
BEACON TRANSMITTER	A329	1	2.92 (1.32)	2.92 (1.32)
RF WAVEGUIDE AND COAX		1	4.81 (2.18)	4.81 (2.18)
SUBTOTAL				87.61 (39.74)
COMMUNICATION CONTROL UNIT		1	3.58 (1.62)	3.58 (1.62)
			TOTAL*	114.24 (51.82)

*EXCLUDING CFT EQUIPMENT

der, as shown functionally in Fig. 23. The CCU receives pulse commands from the T&C system, which drives switching and latching relays in the unit. These provide enabling voltage and/or switch control signals to selected components upon receipt of the proper command and provides telemetry status indications of components that have been commanded. Finally, the dc power consumption and weight tables are shown in Tables 9 and 10.

TRAVELING WAVE TUBE AMPLIFIERS ABOVE 10 GHz FOR SPACE APPLICATIONS

J. H. Herman*

Hughes Aircraft Company, Torrance, Calif.

Abstract

A variety of space traveling tubes and amplifiers have been developed and manufactured from 10 through 60 GHz. The tubes, which employ either a helix or coupled cavity slow-wave structure, provide output power of 1 to 100 W or more in some cases. High-efficiency, lightweight power supplies have been provided with certain amplifiers. Phase and gain performance characteristics are presented which demonstrate applicability to communications systems. Prospects for future increases in power and efficiency offer expanding horizons to the system engineer.

Introduction

In the relatively few years since the beginning of space communications, advances in the traveling wave tube amplifier (TWTA) art have resulted in space devices operating in once "unheard of" specification "boxes." Syncom in 1963 provided 2 W at L band. Today a number of devices are available from 10 through 60 GHz to power levels over 100 W.

Traveling wave tubes (TWT's) utilizing helix slow-wave structures (Fig. 1) have dominated the lower frequency regimes. To date, for the rf power required, this class of tube has been ideally suitable. They are smaller, lighter in weight, lower in operating voltages, and lower in cost than other types of traveling wave tubes.

Presented as Paper 76-294 at the AIAA/CASI 6th Communications Satellite Systems Conference, April 5-8, 1976, Montreal, Canada.

*Manager of Programs, Electron Dynamics Division.

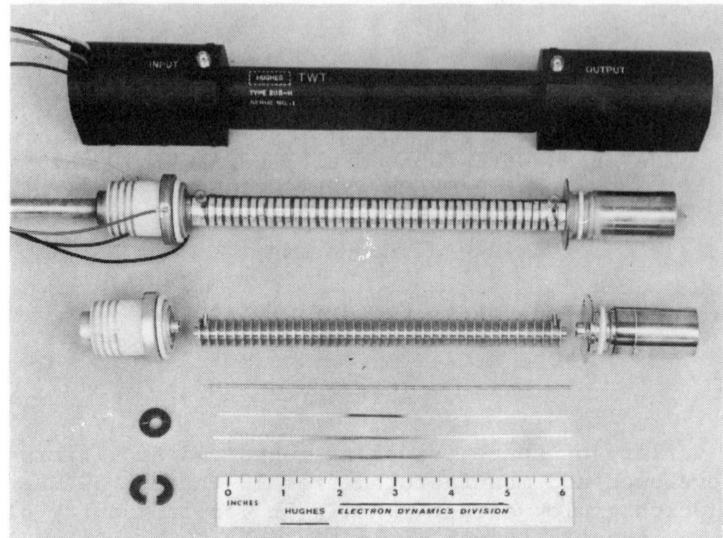

Fig. 1 Typical helix TWT construction.

As operating frequencies increase, the suitability of the helix circuit decreases, since physical dimensions are inversely proportional to frequency. The helix diameter, helix wire size, and electron beam diameter become so small that the power cannot be dissipated adequately nor the electron beam diameter controlled properly. At higher frequencies and power levels, the coupled cavity slow-wave structure becomes more appropriate (Fig. 2). This device operates at higher voltages and, therefore, larger beam and circuit sizes. Thus, with the more massive circuit, much higher power handling capability is obtained. Although this structure is inherently narrow band in comparison to the helix, it provides sufficient bandwidth for typical communications application.

The lower operating voltage of the helix tube (usually less than 5 kV) offers advantages in the electronic power conditioner (EPC). The portion of the EPC supplying the voltages to the tube can be constructed utilizing solid or semisolid encapsulant with reasonably small component size and spacings. Coupled cavity TWT's operate typically above 8 kV and approaching 20 kV in space applications. Although the former can utilize solid encapsulant in the high-voltage module, the component size and spacing are much larger than for the voltages encountered in helix TWT's. When operating at voltages greater than 10 kV or so, an oil dielectric has been the approach used in the past which, with the considerably larger component sizes, becomes a heavier and more complex assembly. Extending

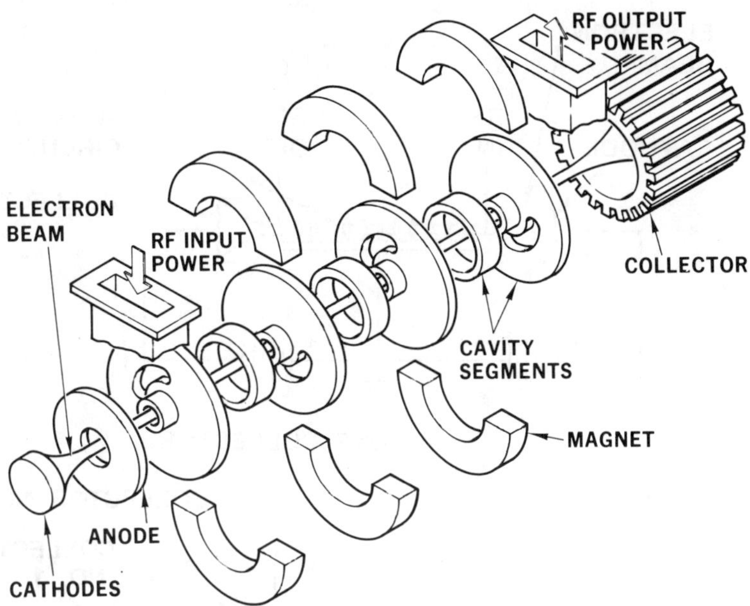

Fig. 2 Coupled cavity tube.

the operating frequency higher, operating voltages must be increased further in order to realize physical dimensions that are practical and capable of handling the required power.

Because of the usual restrictions in spacecraft, the systems designer continually is seeking higher efficiencies. As one increases frequency, however, the ohmic losses of the slow-wave circuits increase. This, plus other design compromises, causes TWT efficiencies to decrease with frequency. Therefore, although TWT efficiencies of greater than 50% are attainable in S-band applications, efficiencies generally are lower at Ku band and above.

Higher tube efficiencies are attainable with a collector design different from those seen in the previous figures. Fig. 3a schematically shows a single collector operating at a potential between the cathode and helix (ground). This collector is operating "depressed," and, the closer to the cathode the potential can be depressed without reflecting electrons back down the beam, the higher is the efficiency. By designing a second collector as in Fig. 3b, it can collect the portion of the beam which tends to be reflected while the main collector is depressed further as compared to the single-collector case. Thus, several percentage points efficiency improvement can be achieved. Additional collectors may be added for fur-

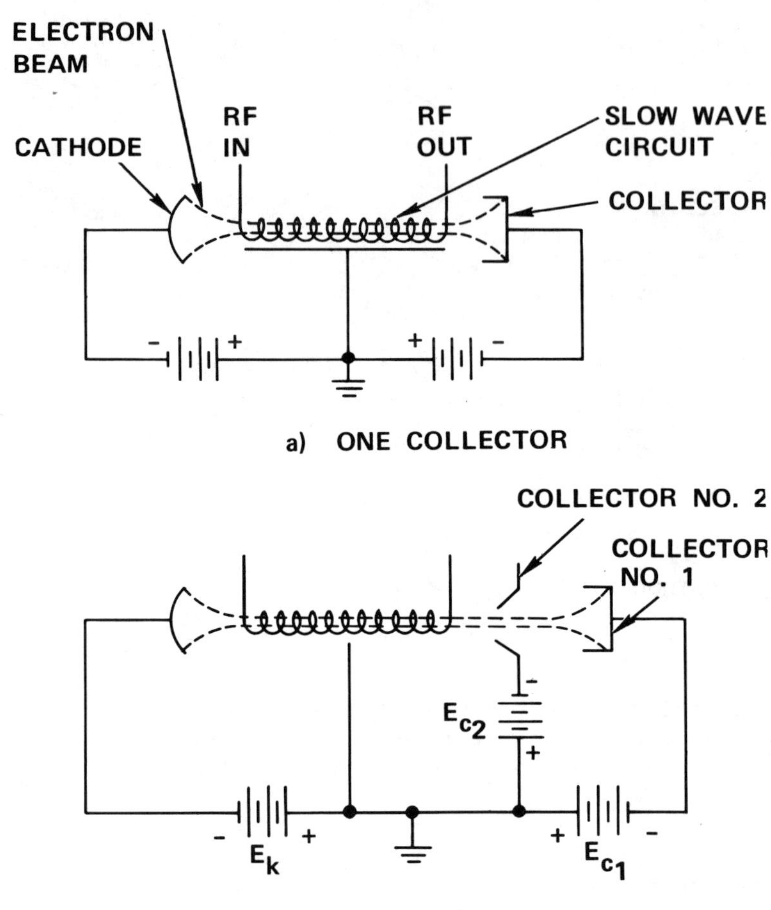

Fig.3 Schematic representations of one and two collector TWTA's.

ther efficiency improvement, but returns are limited above three collectors. Some additional complexity and, perhaps, small efficiency loss in the EPC is the penalty paid in this approach. Because of the way currents in the two collectors vary as a function of various parameters, apparent negative resistances in this collector scheme must be handled in special ways so as not to cause the TWTA to become unstable. In addition, at very-low-amplitude modulating schemes, the spacecraft designer must be prepared for large-amplitude modulations in the current on the power bus.

A major economic consideration in spacecraft systems is life. At frequencies below X band and power levels less than

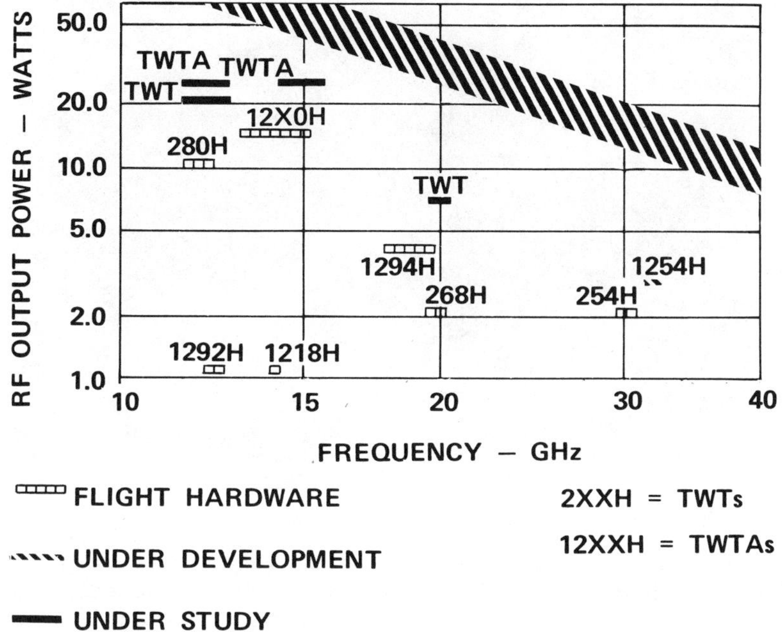

Fig. 4 Helix traveling wave tubes and amplifiers.

tens of watts, lifetimes of 10 yr have been demonstrated. With incursions into frequencies above 10 GHz occurring only relatively recently, the wealth of life data does not exist as it does at lower frequencies. However, in most TWTA's, the life-limiting factor is the wearout of the cathode of the TWT. Low-frequency tubes in space have utilized oxide-coated cathodes operating at low temperatures (approximately 700°C) and low current densities (<200 mA/cm^2). This regime typically has a theoretical wearout lifetime of 100,000 hr or more. Millions of hours of operational performance have been accumulated in support of this value. As frequency and power levels increase, the current density in the cathode must increase within the realm of available electron gun and beam focusing design. As current density increases above 200 to 300 mA/cm^2, oxide-coated cathode life is no longer acceptable, and a different cathode, the impregnated or dispenser cathode, is used. This cathode operates at higher temperatures as well as higher current densities and is a much more rugged cathode as compared to oxide cathodes with respect to recovery from abuse by contaminations. Data to support theoretical life expectancies of the impregnated cathode just now are being accumulated. Theory indicates lifetimes of greater than 5 yr.

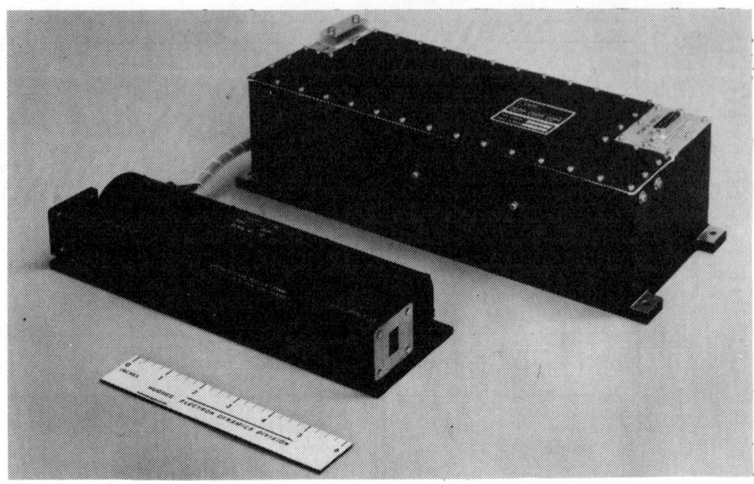

Fig. 5 1230H TWTA.

A variety of TWT's and TWTA's above 10 GHz for space application now exist. The history in this frequency range started with the ATS and Skylab programs. Currently experimental communication satellite, broadcast satellite, and other programs are using or readying hardware at higher frequencies. With the approaching saturation of the 4/6-GHz communication band, a great deal of shifting to higher frequencies will be taking place. With it, of course, will be advances in the TWTA art.

The following sections describe some of the devices available today and show their performance. Also described are expected future developments and trends.

Helix Tubes and Amplifiers

To date, a variety of devices have been developed in the frequency range of 10 through 40 GHz. Operating frequencies and powers for some of these are indicated in Fig. 4. Aside from existing hardware, the figure also shows devices under development and those under study and consideration for future development. Assuming current technology in beam control and thermal considerations, a tentative power/frequency upper limit is shown for cw communications traveling wave tubes as the dashed diagonal line on the figure. This upper limit is difficult to quantify, and currently some activity is under consideration within this "gray area." The following paragraphs describe a few of the amplifiers that provide a general characterization.

Table 1 Characteristics of the 12XOH series TWTA's

Frequency range	13 to 15 GHz
Bandwidth	1 GHz
Output power	16 W
Gain, saturated	46 dB
AM-to-PM conversion	6 deg/dB
Input voltage	23 to 33 Vdc
Input power	80 W
Weight	10 lb

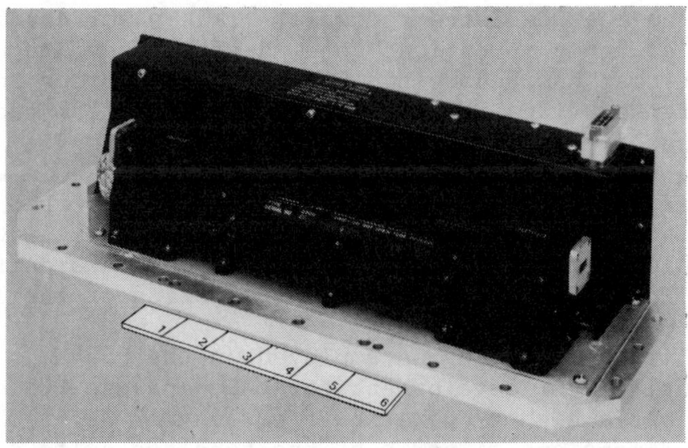

Fig. 6 1294H TWTA.

The 12XOH series of amplifiers was developed as an outgrowth of the Skylab 20-W amplifier used in the altimeter scatterometer experiment. Operating in different segments of the frequency range of 13 to 15 GHz, this series provides approximately 15 W in wide-band, long-life applications. Utilizing a power conditioner developed for the DSCS II high-level TWTA, these amplifiers operate from a widely varying unregulated power bus with overall efficiencies of about 20%. Telemetries of five operating parameters, protective circuits, and means for performing cathode activity tests are incorporated in this unit, which is shown in Fig. 5. Some of the characteristics are shown in Table 1.

Table 2 Characteristics of 1294H TWTA

Frequency range	18 to 20 GHz
Bandwidth	>1 GHz
Output power	4 W
Gain, saturated	50 dB
Input power	21 W
Weight	4 3/4 lb

At about 20 GHz, amplifiers have been provided to Aeronutronic-Ford for the Japanese Communication Satellite (JCS) program. The TWT is an outgrowth of a tube developed for the ATS program, and the power conditioner is a modification of one used on the RCA Satcom program. This power conditioner utilizes a Hughes invention: a high-efficiency, regulating converter that has become the workhorse of current designs. This design provides EPC efficiencies between 85% and 90%. A photograph of this TWTA, the 1294H, is shown in Fig. 6. As the output power in TWT's is lowered, a decrease in efficiency must be expected, since the heater power (power that is required to maintain the cathode at the proper emitting temperature), which is on the order of 2 W, becomes a significant factor. Table 2 lists a few characteristics of the 1294H.

Utilizing technologies employed on the 1294H, a device is currently under development for the Japanese ECS satellite program. At a higher operating frequency of 31.65 GHz, the physical dimensions of parts of the tube are quite small and give an insight into the reasons behind the upper limit line shown in Fig. 4. The helix dimensions of the tube in the 1254H are a 0.0025-in. i.d., with an 0.006-in. diam. wire. This compares with 0.032 and 0.007 in. respectively, on the JCS tube.

In the near future, assaults on the upper limit area line will occur. Applications for 30 to 40 W between 10 and 15 GHz and 30 W at 20 GHz are developing. This will result in more applications for impregnated cathodes generating more data for supporting life estimates. At these power levels, high efficiency is crucial, and activities will be oriented to efficiency enhancement techniques such as multiple depressed collector operation. Future technology development most likely

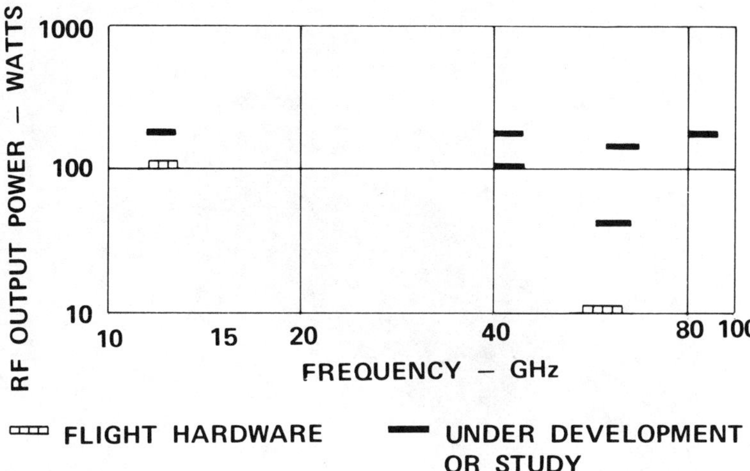

Fig. 7 Coupled cavity traveling wave tubes and amplifier.

will push the limit upward, widening the application of helix TWTA's.

Coupled Cavity Tubes and Amplifiers

Coupled cavity tubes have been developed for space applications only relatively recently. They are generally under consideration where more power is required and more weight is available. Also, as we have seen, there is an upper limit on helix TWT's, necessitating coupled cavity devices to fill the gaps even at relatively low output powers. Thus, at 40 GHz, for example, the coupled cavity TWTA at 10 W will be the approach taken to fulfill such a requirement. Figure 7 illustrates the variety of coupled cavity tubes and amplifiers for communication applications in space currently available or under development or study at Hughes.

Interest in millimeter wave power in space is on the increase, but we are only beginning to scratch the surface of applications and developments. The cost of these programs is higher than for helix tube amplifiers. The tight tolerances required in millimeter wave coupled cavity tubes are expensive. For example, at very high frequencies, tolerances of ±50 millionths of an inch in parts and ±0.0001 in. in critical cavity dimensions must be obtained. Higher operating voltage is also a cost. However, this is an area where considerable improvements in this young technology must be achieved in order to reduce these costs. These improvements must not be limited just to designing power supplied to stand off these voltages

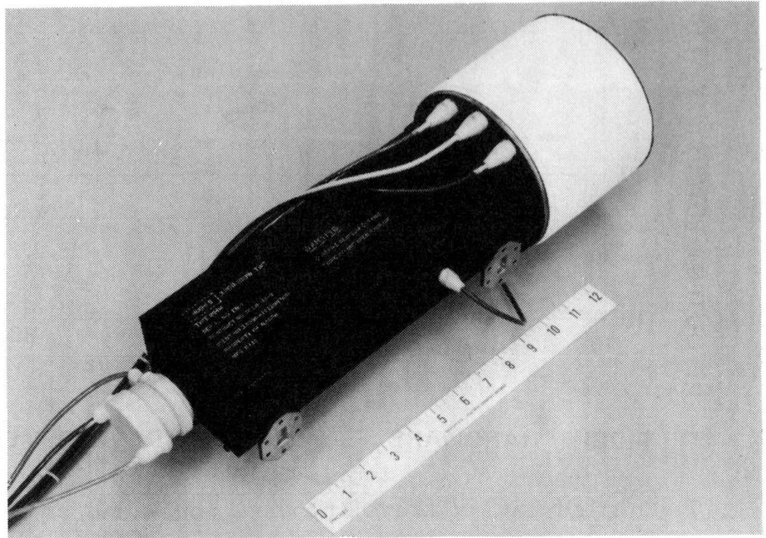

Fig. 8 294H traveling wave tube.

safely but must include the development of high-reliability, high-voltage electronic components.

A coupled cavity traveling wave tube, the 294H, currently is being supplied to provide greater than 100 W of power for use on the Japanese Broadcast Satellite (Fig. 8). This device is an offshoot of a tube, the 284H, providing greater than 200 W developed for NASA for the CTS program. It contains a velocity-tapered interaction circuit, a very effective efficiency-enhancement scheme. The velocity of the wave in the

Table 3 Characteristics of the 294H coupled cavity TWT

Frequency	11.95 to 12.13 GHz
Output power	>120 W
Gain	46 dB
Efficiency	>52%
Weight	15 lb
Cathode voltage	−8 kV
Cathode current	55 mA

Fig. 9 60-GHz TWTA.

coupled cavity circuit is slowed as kinetic energy is transformed to rf energy at the output end of the tube. Thus synchronism between circuit and beam is maintained over the full length of the tube in the saturated region of operation. It also employs a three-stage depressed collector to obtain further efficiency improvement as compared to a nine-stage depressed collector in the 284H. In order to eliminate the problem of thermal dissipation in the spacecraft, the collector is designed specifically to radiate directly into space. This tube employs an impregnated cathode gun operating at values to provide 30,000 hr of life. Characteristics of this tube are shown in Table 3.

Table 4 Characteristics of 60-GHz TWTA

Output power	13 W
Bandwidth	1%
Gain, saturated	40 dB
Noise figure	28 dB
AM/PM conversion	10 deg/dB
Efficiency	13%
Weight	50 lb

Fig. 10 50-W 60 GHz traveling wave tube.

Utilizing scaling techniques, the 284H, 294H designs can be redeveloped to provide about 600 W at 50% to 60% efficiency and bandwidths up to 250 MHz. With additional development and larger size and weight, output powers greater than 1 kW can be achieved.

Significant development activities have taken place in the area of 60 GHz. The first device (shown in Fig. 9) is a TWTA providing about 13 W of power. Because of the high operating voltage of near 13 kV, the high voltage is generated in an oil-filled chamber as shown in the foreground of the photograph. The choice of dielectric was based on earlier work at even higher voltage. The low-voltage module to the left contains circuitry that drives the high-voltage module plus timing, logic, and protection circuits. The tube is provided with an ion pump integrally constructed in the gun end of the tube to enhance life. The high-voltage areas of the gun and collector region are isolated dielectrically from the outer package with an insulating oil. A single depressed collector stage is used for efficiency enhancement. The tube operates at about 24% efficiency and the power supply at about 65%. This amplifier has been tested through qualification level tests. Some of the characteristics of the amplifier are shown in Table 4.

Based on this last amplifier, feasibility models have been developed for 50 and 100 W of output power. A photograph of the 50-W tube is shown in Fig. 10. Similar to the 13-W tube, it contains oil dielectric isolation for the gun and collector regions and an integral ion pump. With a power supply at improved efficiency, the TWTA operates at about 18% overall efficiency with 55 dB of gain. Physical characteristics of the amplifier are about the same as the 13-W TWTA, and the 50-W TWTA also has undergone a qualification test program.

The feasibility model of the 100-W tube contains a velocity-tapered interaction circuit for efficiency enhancement. For further efficiency enhancement, the tube also contains two depressed collector stages. This tube is 30% efficient at 1.2% bandwidth, with other characteristics similar to the 13-W TWTA. The overall TWTA efficiency for this device is approximately 24%.

Conclusion

A variety of helix and coupled cavity traveling wave tubes and associated power conditioners are currently available for communications spacecraft applications above 10 GHz. The technology is still very young, however, and future developments only await far-seeing spacecraft system developers. Higher frequencies, higher power levels, higher efficiencies, smaller size and weight, all commensurate with high space reliability and long life, are the avenues that will be opened.

Addendum - February 1977

Since the original writing of this paper, a number of significant advances have taken place above 10 GHz in the helix tube and coupled cavity tube areas.

Within helix tubes, developments are underway in the "grey area" in Figure 4. Notable of this activity is the 874H TWT under development for the Space Shuttle radar and communications system. Operating in the band of 13.4 to 15.1 GHz, it will provide 60 watts of CW and pulsed rf power in the communication mode and radar mode. This will be provided at 44% efficiency utilizing a multiple collector approach. The TWT will employ an impregnated cathode in order to achieve the current density needed at this power level and will have a design life of over 30,000 hours.

For a higher power option in the 20 to 26 GHz region, testing of a tube under development, the 250H, has demonstrated 30 watts at an efficiency of 34%. While the cathode current density at 30 watts is higher than that which is required for seven year life, this power is available for short durations to overcome rain attenuation without significant reduction in life when normally operated at a lower output level.

Another significant helix TWT in development in the past year is the 286H. It is designed to operate in the 10.9 to 12.2 GHz commercial space communication down link frequency band. This tube is available in a single collector design at 34% efficiency or a dual collector approach at 40%. We have developed techniques within the EPC designs that provide amplifiers for dual collector TWT's at no increase of size, weight, or complexity and which meet the stringent requirements of TDMA operation.

In the area of coupled cavity TWT's, development tests on the 100 watt, 40 GHz TWT, shown in Figure 7, have demonstrated a bandwidth in excess of 2 GHz. This tube design contains a single depressed collector stage and will have an efficiency approaching 25%. Also currently under development is a 30 GHz coupled cavity TWT. This tube will provide 200 watts of CW power.

It is apparent from the previous discussion that the activities and results in space traveling wave tubes (amplifiers) are extremely dynamic. As capabilities are demonstrated, requirements are formulated for even greater capabilities. New developments are continually pushing through old barriers and we can expect the current "grey areas" to move out in the future as it has in the past.

PASSIVE DEVICE INTERMODULATION ANALYSIS IN COMMUNICATION SATELLITES

Z.A. Sarkozy*

TRW Defense and Space Systems,
Redondo Beach, California

Abstract

In communication satellites and other applications where signal amplifications well in excess of 100 dB are required, it has been observed that intermodulation products (IMP's) are generated not only in active devices such as transmitters, but also in passive devices such as filters, transmission lines, and antennas. This paper describes an analytical, computer-aided method to determine the spectra of these passive device IMP's. Ultimately, we compute the IMP's spectra for complex multicarrier systems based on measurements of as little as the two strongest individual IMP's. These analytically predicted computations are compared to measured test data and excellent correlation is observed.

Introduction

Continuous advances in spacecraft technology have resulted in increasingly complex communication satellites. Increased channelization and high transmitter powers have led to the emergence of a hitherto unimportant phenomenon: the nonlinear behavior of passive components such as transmission lines, antennas, and bandpass filters. These components have been usually considered linear, but in the high capacity communication satellites designed for the late 1970's and for the 1980's, intermodulation product (IMP) interference from such "linear" sources represents a major frequency plan and equipment performance design consideration.

Presented as Paper 76-295 at the AIAA/CASI 6th Communications Satellite Systems Conference, April 5-8, 1976, Montreal, Canada.

*Member of Professional Staff.

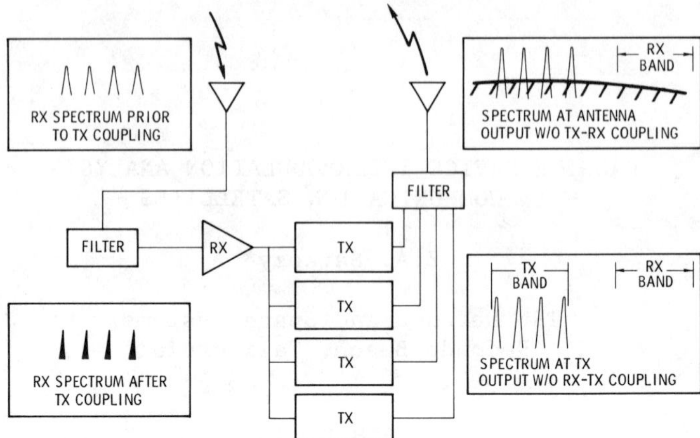

Fig. 1 IMP's degrade receiver sensitivity.

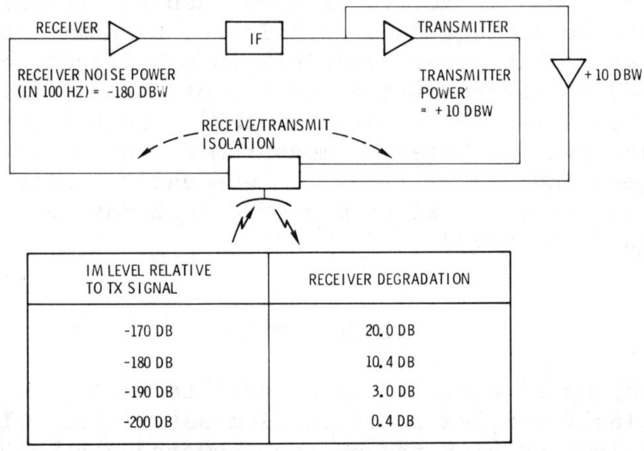

IM LEVEL RELATIVE TO TX SIGNAL	RECEIVER DEGRADATION
-170 DB	20.0 DB
-180 DB	10.4 DB
-190 DB	3.0 DB
-200 DB	0.4 DB

Fig. 2 IMP spurs nearly 200 dB below the generating signals degrade system performance.

The effects of intermodulation products on satellite communications have been studied by many authors.[1-8] These analyses are mostly concerned with the dominant IMP's, those generated by amplifiers, many of them by limiting amplifiers. These active device IMP's can be very harmful to a communication satellite. The dominant IMP's may be as little as 10 dB below the desired output signals; they can reduce the useful signal power, introduce signal distortion, and may degrade adjacent channels, but the receiver can be always protected from these IMP's by filtering.

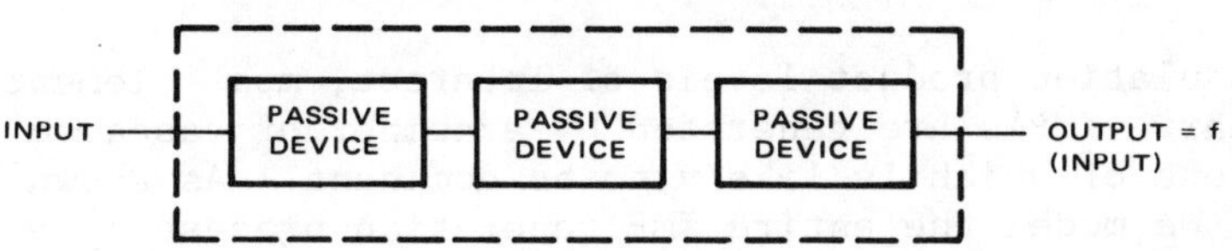

Fig. 3 Cascaded devices are modeled by a single transfer function.

In this paper we are concerned with a different class of IMP's: those weak products generated in or past the output filter. These passive device IMP's are, therefore, unfilterable and their spectra must be designed for in the frequency plan. Figure 1 shows how these potentially "weak IMP's" can reduce receiver sensitivity. The received spectrum is displayed in the upper left hand side. A high signal-to-noise (SNR) case is illustrated. The amplified output spectrum is shown in the lower right hand corner. The high SNR has been preserved. In the upper right hand corner, the effects of passive IMP's generated in the transmit filter are highlighted. These IMP's are at least 100 dB below the signal power. Therefore, they are far below the thermal noise level and remain undetected in the transmit band. They do not affect the quality of the transmitted signal. However, even weaker IMP's that fall in the receive band, if coupled to the satellite's receiver, may well exceed the receiver thermal noise floor, and reduce satellite capacity. The spectrum in the lower left hand corner shows that the quality of a subsequently received communication message is no longer a function of the receiver's noise figure; it is degraded by the IMP content of the receiver. Figure 2 shows a system where an IMP 190 dB weaker than the transmit signal significantly degrades the receiver sensitivity if it falls into the receive band. In this example the following parameters are assumed: 10 watt transmitter, 100 Hz signal bandwidth, and 4 dB satellite receiver noise figure. These effects can be particularly troublesome at VHF and UHF where the available RF spectrum is limited; therefore, transmitter IMP's as low as 3rd order may fall in the receive band.

In order to assess the effects of these IMP's from passive components, a computational tool, described here, was developed to predict the expected IMP spectra. This paper presents an analytical methodology for predicting and quantizing these spectra and discusses comparisons between data predicted by the analysis and actual measured data.

Analysis Technique

Mathematical Model

The analysis techniques used to predict IMP spectra are based on the following mathematical model. Recall that for the intermodulation product levels of interest, most elements are nonlinear. IMP's are generated by a number of cascaded elements, one of which is likely to be dominant. As shown in Fig. 3, we model the entire IMP generation process by a single transfer function. This approach is accurate because one of the cascaded nonlinearities is likely to be a dominant IMP source.

IMP measurements can also be made for an M carrier system. With arbitrary signal modulation, the input to the nonlinearity is

$$x(t) = \sum_{i=1}^{M} a_i \cos(\theta_i) \tag{1}$$

where

$$\theta_i = \omega_i t + m_i + \phi_i \tag{2}$$

a_i, ω_i, m_i, ϕ_i being the amplitude, frequency, modulation, and phase of the ith carrier. The output y(t) is

$$y(t) = f\left(x(t)\right) \tag{3}$$

There are a variety of mathematical functions that accurately describe the characteristics of nonlinear devices: Fourier series, power series, Volterra series, and a special function reported by Berman.[7] With a four-coefficient Fourier series model, Berman and Podratzky show excellent correlation between analysis and measurements.[2] They obtain the Fourier coefficients from single-carrier input-output characteristics. The higher harmonics in the Fourier series affect more significantly the magnitudes of the IM products than they do the signal components. Thus, the coefficient which has a negligible effect on the calculation of the output signal powers, may significantly contribute to the IMP's. For passive device IMP's, the single-carrier transfer characteristics are too linear to be useful for any Fourier coefficient measurements. Fourier analysis could be used, but more complex multicarrier analyses and measurements would be required. The highly linear nature of the IMP generating device suggests the use of the simpler, computationally faster, power series approach.

In our analysis, the IMP amplitudes are computed from a five-term power series nonlinearity model. Ignoring the even terms, which cannot appear in the fundamental zone

$$y(t) = \sum_{i=0}^{4} A_{2i+1}\, x(t)^{2j+1} \tag{4}$$

where A_j are complex coefficients describing the nonlinearity. The part of Eq. (4) multiplying A_{2i+1} can be expanded by repeated application of the binomial expansion

$$\sum_{m=1}^{M} x_m^j = \sum_{r_1} \cdots \sum_{r_2} \sum_{r_M} \frac{k!\, x_1^{r_1} x_2^{r_2} \cdots x_m^{r_M}}{r_1!\, r_2! \cdots r_M!} \tag{5}$$

with the constraint

$$r_1 + r_2 + \cdots + r_M = j \tag{6}$$

After rearranging, Sea computed the amplitude coefficients in a form suitable for digital computers[10]

$$V = \varepsilon_N \sum_{L=0}^{\infty} \frac{A_{N+2L}\,(N + 2L)!}{2^{(N + 2L)}} \sum_{q_1, q_2 \cdots q_M} \prod_{p=1}^{M} \frac{E_p^{|\alpha_p| + 2q_p}}{(q_p + |\alpha_p|)!\; q_p!} \tag{7}$$

where

V = the output voltage

$$\varepsilon_N = \begin{matrix} 1 \text{ for } N = 0 \\ 2 \text{ for } N = 1,2,\ldots \end{matrix} \tag{8}$$

N = the order of the IMP

α = the coefficient determining the intermodulation product frequency

$$\theta_{IMP} = \alpha_i\, \theta_i + \cdots + \alpha_M\, \theta_M \tag{9}$$

and

q's are nonnegative integers such that

$$q_1 + q_2 + \cdots + q_M = L \tag{10}$$

The summation can be more formally written

$$\sum_{q_1, q_2, \ldots, q_M} = \sum_{q_M=0}^{L} \sum_{q_{M-1}=0}^{L_{M-1}} \cdots \sum_{q_3=0}^{L_3} \sum_{q_2=0}^{L_2} \sum_{q_1=L_1}^{L_1} \tag{11}$$

This formulation requires us to classify the IMP's according to order. In the interest of computational speed, we also classify them according to zones. In communication satellites, the RF carrier frequency typically exceeds the bandwidth by an order of magnitude, in which case we only compute the frequencies of the IMP's falling into the fundamental zone

$$\alpha_1 + \alpha_2 + \alpha_3 \cdots \alpha_M = 1 \tag{12}$$

For a given IMP order and for M number of carriers, we first compute all possible combinations of coefficients, later to ignore those that cannot produce fundamental zone IMP's. For example N = 5 and M = 4, the absolute value of the coefficient α becomes

Type						
1	5	0	0	0	†	Ways to get fifth order products with four signals (13)
2	4	1	0	0	†	
3	3	2	0	0		
4	3	1	1	0		
5	2	2	1	0		
6	2	1	1	1		

In doing so we classified the IMP's of a certain order into types. In Eq. (13), we obtained six IMP types, the first two being absent in fundamental zone IMP's. For equal carriers, corresponding to the increasing type numbers, the spur amplitudes also increase. This classification is further useful to save computer time because the weaker types can be ignored.

We only compute the IMP frequency once it has been established that the product falls into the desired zone. Generally IMP products concern us if they fall into the fundamental zone. Once the IMP frequency is found, we evaluate its bandwidth, summing the individual bandwidths. We only evaluate the IMP amplitude if it falls into one of the requested output channels. In this process we preserve the sets of coefficients that generate IMP's in the zone of interest, and permute this coefficient set with the available number of carriers.

Because of the large number of IMP's found in most systems, it is important (1) to avoid any duplication in the IMP frequency generation process, (2) to utilize efficient permutational routines, (3) to provide the ability to select the dominant IMP types and orders, and (4) to evaluate amplitudes only for IMP's that are known to contribute to the desired output spectrum. For example, Fig. 4 describes a 12-channel system in which the signal band extends over only 3% of the RF

---

† Sign assignment is not possible to yield $\sum_i \alpha_i = 1$.

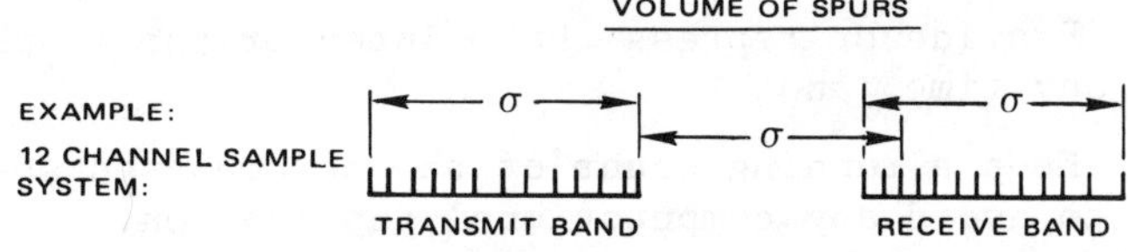

3 PERCENT OF ALLOCATED BAND IS OCCUPIED

SPUR ORDER	NUMBER OF SPURS IN RECEIVER
3	1
5	223
7	6,955
9	114,912

Fig. 4 122,091 IMP's fall within the receive channel bands.

bandwidth. Correspondingly, only the IMP's that fall in the used (3%) portion of the allocated uplink band are counted. Including up to 9th order spurs, this count exceeds 122,000.

Versatility was achieved by a control software that actually configures the computer program after execution. Such versatility would be extremely slow and annoying with what is commonly considered a "conversational language." We have developed a command decoder to understand a certain class of commands and structure the program for the desired execution. The preliminary evaluation of IMP effects, including some optimization tradeoffs, can be performed based on the location of certain dominant spur types without considering actual spur amplitudes. This method has two important advantages: the reduction of computer time/cost and the requirement of minimum input data. The only required data for this analysis is the identification of the candidate and/or some potentially permissible frequency plans. In this mode virtually no limitations exist concerning the program's capability of evaluating any system. This mode is particularly suited for RF frequency plan optimization at an early phase of system development.

For more detailed IMP analyses, the program computes the amplitude and the shape of each spur. For these computations, good accuracy and fast computational speed are important. In this mode, we:

1) Compute the shape of each IMP as a function of the shapes of the generating signals

2) Compute the complex amplitude of each IMP

3) Compute the sum of each IMP (in voltage or in power) in as small frequency increments as specified

4) Provide or repress the printed or the graphical display at any time, and

5) Keep a running count of the accumulated spectral density to avoid any computational duplication.

Computation of IMP Shapes

For carriers that are modulated by noise-like signals, the IMP spectral density is determined by convolving the spectral densities of the spur generating signals. IMP's generated by signals with Gaussian spectral densities and/or by unmodulated signals can be efficiently described because the spectral density of the resulting spurious product is known; that is, to calculate the IMP shape, the actual convolution need not be performed. Due to the hundreds of thousands of IMP's evaluated in a single run, it is critical that all IMP's spectra be calculated efficiently. Direct convolution is very inefficient.

A case of general interest is when the channels are thermal noise loaded or are used for spread spectrum communications. Good approximation can be obtained by precomputed piecewide linear approximations. This method eliminates the need for evaluating the convolution integrals or the alternative to the convolution integrals, namely conversion to the Fourier domain to perform the simpler arithmetics, followed by inverse Fourier transformation. For this increased efficiency we pay a small penalty of requiring that the bandwidths of the noise loaded channels be restricted to a finite number of ranges. With this restriction, all the convolutions of real, arbitrary phase spectra are approximated accurately.

Consider any number of channels that may contain either an unmodulated carrier, a carrier with Gaussian spectral density, or a bandlimited channel filled with either thermal noise or signal from a spread spectrum user. For this illustration, we consider noise bandwidths on the order of 25, 70, and 500 kHz. A rectangular shaped spectrum becomes Gaussian after many self-convolutions. More specifically, any rectangular spectrum is essentially Gaussian after three self-convolutions. Single self-convolution is similarly easy; it results in a triangular spectrum. The only remaining task is to precompute the IMP shapes resulting from any combination of signals having either one of the three rectangular spectral density widths or a Gaussian spectral density of any variance, σ^2.

When spectra of widely different bandwidths are convolved, the narrow bandwidth spectrum approximates a delta function relative to the wide spectrum. The output closely approximates the originally wide spectrum.

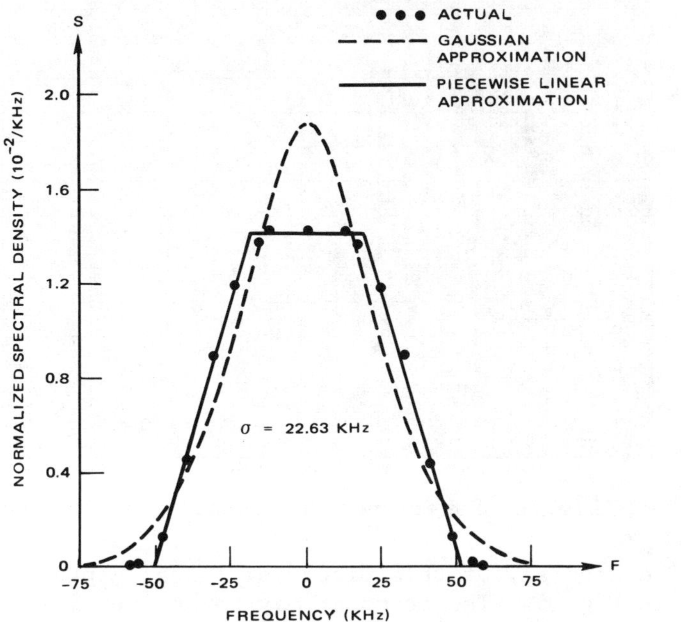

Fig. 5 Comparison of actual spectrum with approximations.

The 70 and 25 kHz spectra are convolved based on a logic diagram. Ignoring the Gaussian signal shape for the moment, we define the logic diagram parameter, x, to be the number of rectangular bandwidth signals, except in the presence of a 500 kHz rectangular signal, in which case the 70 and 25 kHz signals do not contribute to x. We determine the IMP based on x as:

1) For x = 0, 1, 2, and 4 the IMP shape will be a line of spectrum, rectangle, triangle, and Gaussian, respectively. The variance of the approximation will be identical to the actual variance

2) For x = 3 the IMP shape will be Gaussian provided all contributing channels are of equal bandwidth. The variance will be unchanged by the approximation

3) For x = 3 generated by the convolutions of a 25 kHz and two 70 kHz channels, the IMP assumes a triangular bandwidth. The variance will be unaltered

4) For x = 3 generated by the convolution of two 25 kHz and a 70 kHz channel, the shape will be obtained by a piecewise linear approximation.

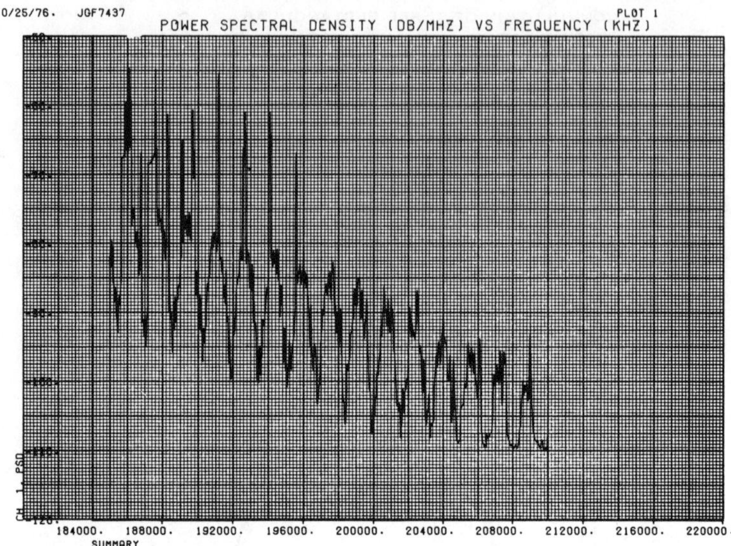

Fig. 6 Power spectral density vs frequency.

The accuracy of this piecewise linear approximation is shown in Fig. 5. The accuracy may be an overkill; however, piecewise linear approximations and the corresponding logic parameter growth is inexpensive relative to the point-by-point IMP density bookkeeping. We also observe from these approximations that reasonable variations about the nominal rectangu-

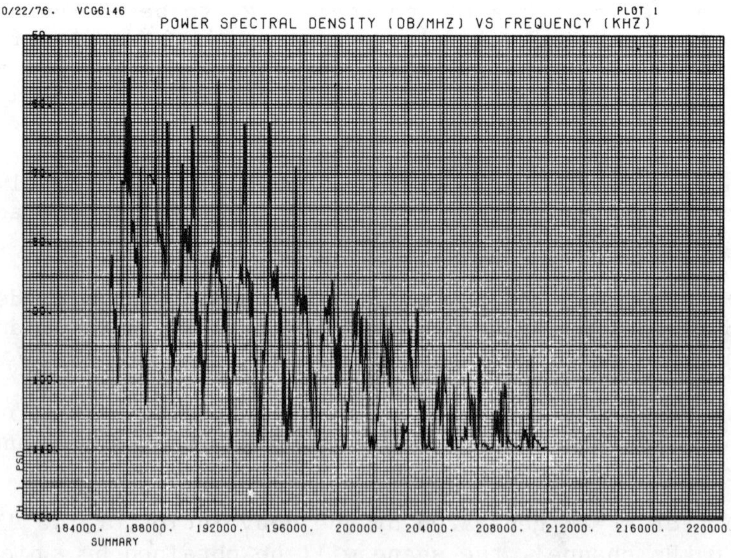

Fig. 7 Power spectral density vs frequency.

lar bandwidth do not degrade the accuracy significantly, provided the IMP shapes are computed based on the actual IMP variances.

We consider the convolutions of one or more Gaussian spur shapes to the above approximate spectral densities: rectangular, triangular, Gaussian, or piecewise linear. The resultant IMP is either assumed to take the shape of one of the generating IMP's or it is approximated by a precomputed piecewise linear shape, without performing the actual integral.

These precomputed convolutions are very accurate provided each IMP can be approximated by a real function of arbitrary phase in the frequency domain. By our assumption, signals with identical statistics must generate identical IMP spectra, except for an arbitrary phase difference. The phase of the IMP is a function of the nonlinearity and the phases of each carrier generating it. The phase is computed simultaneously with computing the complex amplitude of the IMP.

Computation of IMP Level

The amplitude of each individual IMP is given by Eq. (7). The coefficients A_j are determined by measurements of the individual spurs. No attempt is made to directly measure the transfer function of the nonlinearity. Instead, the dominant spurs are measured and the various coefficient A_3, A_5, A_7, and A_9 are subsequently computed, with A_1 set to unity. For sinusoidal input signals, the power series coefficients have been already computed in a form advantageous for digital computers.[9,10]

Table 1 Power series coefficients for Figs. 6, 7, 8, and 9

	Fig. No.			
Coefficient	6	7	8	9
A_3	$+2.5\text{x}10^{-4}$	$+2.5\text{x}10^{-4}$	$j2.5\text{x}10^{-4}$	$+2.5\text{x}10^{-4}$
A_5	$+5.0\text{x}10^{-7}$	$-5.0\text{x}10^{-7}$	$-5.0\text{x}10^{-7}$	$+5.0\text{x}10^{-7}$
A_7	$+1.0\text{x}10^{-9}$	$+1.0\text{x}10^{-9}$	$+1.0\text{x}10^{-9}$	$+1.0\text{x}10^{-9}$
A_9	$+2.0\text{x}10^{-12}$	$-2.0\text{x}10^{-12}$	$-j2.0\text{x}10^{-12}$	$+2.0\text{x}10^{-12}$
Spur addition	Voltage	Voltage	Voltage	Power

In theory, the highest power IMP's which are separated from all others in frequency are measured. Each such IMP is written in terms of A_1, A_3, A_5, A_7, and A_9 and the resulting equation is solved for the complex coefficients, A_j. If an overdetermined set exists, the "best" estimate is obtained according to some criteria, such as the minimum "least mean square" error.

Measuring the higher order IMP's generated by passive non-linearities is not a simple task. Often, most of the individual device IMP's, including those of the 5th order, are near the noise level. Therefore, measurements may lack the accuracy that is implied by this method. Solving an overdetermined set of equations, although straightforward in principle, can be time consuming. The solutions to an overdetermined set of equations can vary significantly on the selected criteria for the "best" fit. For these reasons, a simpler method of approximating the coefficients was developed: the strongest 5th order and the strongest 3rd order IMP's are measured and the power series is truncated to A_1, A_3, and A_5. A_1 is set to unity. The resulting two equations with two unknowns are readily solved, assuming A_3 and A_5 real. Finally, A_7 and A_9 are calculated by the assumption

$$\frac{A_3}{A_5} = \frac{A_5}{A_7} = \frac{A_7}{A_9} \tag{14}$$

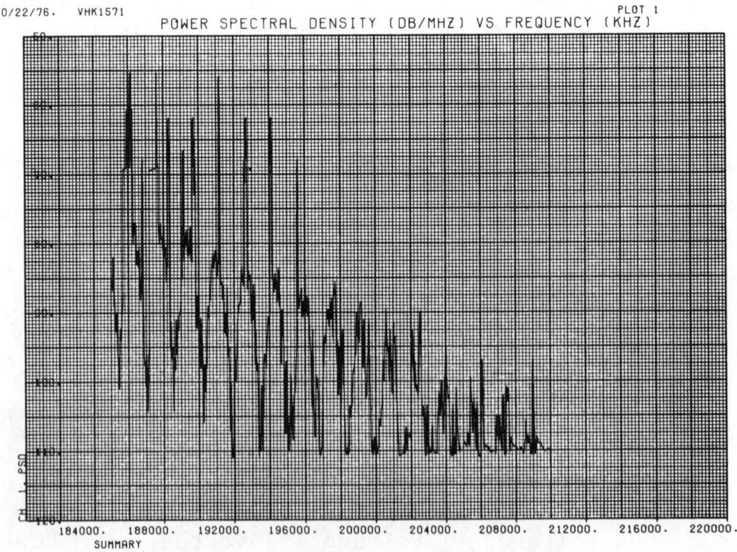

Fig. 8 Power spectral density vs frequency.

For these measurements, the passive device nonlinearity is subjected to a condition that closely approximates the operational environment. For example, it is driven during the test with a power level that nearly corresponds to that expected during operation. Nevertheless, this technique restricts the coefficients to real values. To qualitatively observe the corresponding error, we compare Figs. 6 through 8. These figures display the out-of-receive band IMP's computed for a 12-channel satellite system and for unequal power carriers of various bandwidths. The spectrum is pictured at the receiver input, riding on a -110 dBm/MHz noise floor. Table 1 displays the power series coefficients used in these figures. All coefficients are real and positive in Fig. 6. Here, all orders and types of IMP's add in phase. In the next figure, the alternating series ensures subtraction of each subsequent odd order. That is, the 5th order IMP's always reduce the overall IMP level when 3rd and 5th order IMP's are combined; similarly, the 7th order IMP's reduce the levels of 5th order IMP's. Complex coefficients are used in Fig. 8. The strong similarity between these figures suggest that using real coefficients introduces little error in the computations of the IMP spectra.

Our final task is the measurement of the relative phases of the signals generating the IMP's. If this measurement can be made, the respective phases are entered into the program; otherwise, to avoid a worst case estimate (all carrier phases equal), random carrier phases can be assigned, or power rather

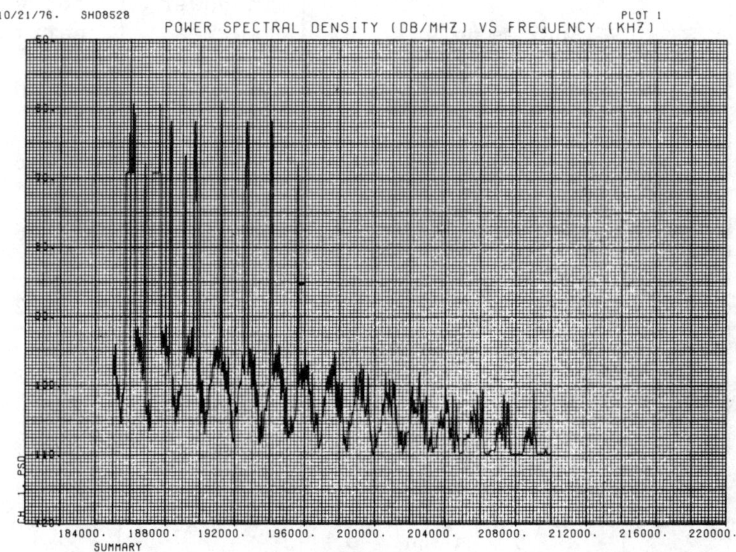

Fig. 9 Power spectral density vs frequency.

than voltage addition of IMP's can be selected. Figure 9 shows the spectrum calculated on the basis of adding each IMP in power; but using the identical coefficients to those of Fig. 6. The two spectra differ significantly, especially where many IMP's of like amplitude combine. A spectrum similar to Fig. 9 results when uniform random phases between $\pm\pi$ radians are assigned to all carriers.

Whereas setting all carrier phases to an identical constant provides a somewhat unrealistic worst case picture, either power addition or uniform random variable carrier phase assignments should be used with caution. It is possible to experience the effects of coherence between uncorrelated carriers.

Comparison of Measured Data with Predicted Data

In flight hardware the choice of materials and quality workmanship typically ensures passive IMP levels of 7th or higher order to be below the noise level. As a result the higher order power series coefficients A_7, A_9, A_{11}, etc. are extremely difficult to measure accurately. At UHF, for composite average powers in the range of 10 to 100 watts with as

Table 2 Highest IMP levels

Spur order	Spur type	Lead term in power series	Power in leading term for equal carriers*
3	$f_i + f_j - f_k$	$\frac{3}{2} A_3$	0 dB
3	$2 f_i - f_j$	$\frac{3}{4} A_3$	-6 dB
5	$f_i + f_j + f_k - f_\ell - f_m$	$\frac{15}{2} A_5$	-26 dB
5	$2 f_i + f_j - f_k - f_\ell$	$\frac{15}{4} A_5$	-32 dB
5	$2 f_i + f_j - 2 f_k$	$\frac{15}{8} A_5$	-38 dB
5	$3 f_i - f_j - f_k$	$\frac{5}{4} A_5$	-42 dB
7	$f_i + f_j + f_k + f_\ell - f_m - f_n - f_o$	$\frac{315}{4} A_7$	-46 dB

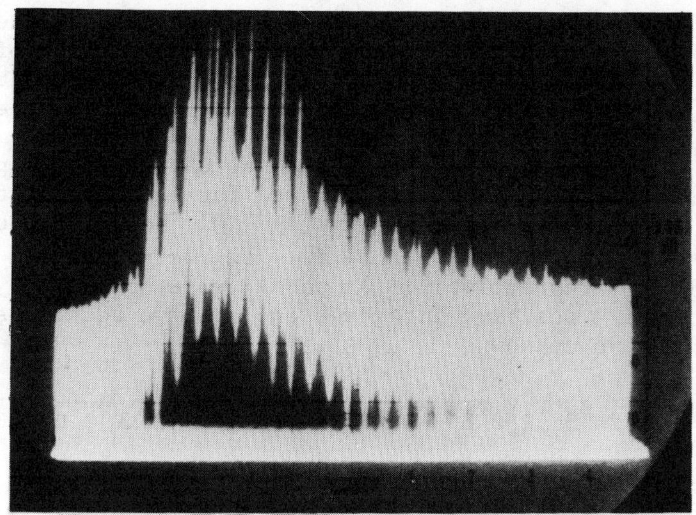

Fig. 10 IMP levels at the input to the reciever.

many as 12 carriers, we observed that in a power series representation the odd power series coefficients decrease approximately 30 to 40 dB per coefficient starting from A_3. Similar results have been reported at other frequencies.[11] That is

$$10^{-2} A_3 < A_5 < \sqrt{10}\, 10^{-2} A_3, \; 10^{-4} A_3 < A_7 < 10^{-3} A_3 \qquad (15)$$

However, the measurement of individual IMP's is extremely difficult at UHF at these IMP levels. Table 2 displays the dominant terms in the eight highest IMP's in a system of seven or more carriers. In a linear design, the strongest IMP may well be on the order of -110 dBm. In this case only four other individual products are likely to be measurable in a 100 Hz bandwidth. Measurement accuracy degrades fast near -150 dBm; therefore, accurate measurements can be hoped for only the two 3rd order and the two strongest 5th order IMP's. Although the IMP's in this particular system degrade system performance, spectrum analyzer measurements at UHF fail to show much of the IMP structure due to the wide bandwidth of the instrument. To compare our computations to measured data, a less linear system was selected.

Figure 10 shows a wideband receiver corrupted by nonlinear device IMP's. The spectrum analyzer was connected to the output of a wideband receiver. Subsequent to this receiver, the transponder was channelized and 12 unequal power and bandwidth transmitters were driven to saturation on thermal noise. The measurements were made at the output of the receiver in order to stay above the internal noise of the network analyzer. Using a linear receiver, however, allows calibration of the spectra to the receiver input. The labelling of the abscissa of Fig. 10 shows signal levels at the input to the receiver.

To compare our predictions with actual data, the power in two individual IMP's was measured for equal power transmitters. The most dominant 3rd and 5th order products, $f_i + f_j - f_k$ and $f_i + f_j + f_k - f_\ell - f_m$, were measured at -70 and -110 dBm, respectively. A_3 and A_5 were computed by setting $A_{5+2i} = 0$ (i a positive integer), and the two equations were solved for two unknowns

$$3.2 \times 10^{-4} = \frac{3}{2} A_3 V_1V_2V_3 + \frac{15}{4} A_5 \left(V_1^3V_2V_3 + V_1V_2^3V_3 + V_1V_2V_3^3\right) \tag{16a}$$

$$3.2 \times 10^{-6} = \frac{15}{2} A_5 V_1V_2V_3V_4V_5 \tag{16b}$$

$$A_3 \times 2.1 \times 10^{-4} \tag{17}$$

$$A_5 \times 4.3 \times 10^{-7} \tag{18}$$

$$\frac{A_5}{A_3} = 2.0 \times 10^{-3} \tag{19}$$

We computed A_7 at 5×10^{-9} from a measurement of the most dominant 7th order term. However, to illustrate the effect of our approximations, we ignored this measurement. Instead we set

$$A_7 = 2 \times 10^{-3} A_5 = 8.5 \times 10^{-10} \tag{20}$$

$$A_9 = 4 \times 10^{-6} A_5 = 1.7 \times 10^{-12} \tag{21}$$

For this coefficient, for the actual signal bandwidths and for unequal power transmitters, we computed the noise loaded spectrum shown in Fig. 11. To bound the IMP power, we have used voltage addition without introducing random phase errors among the transmitters.

The measured data corresponding to Fig. 11 is displayed in Fig. 10. For good resolution, the oscilloscope bandwidth was set to approximately 10% of the smallest channel bandwidth. Note the filtering effect on the lefthand side. From this fil-

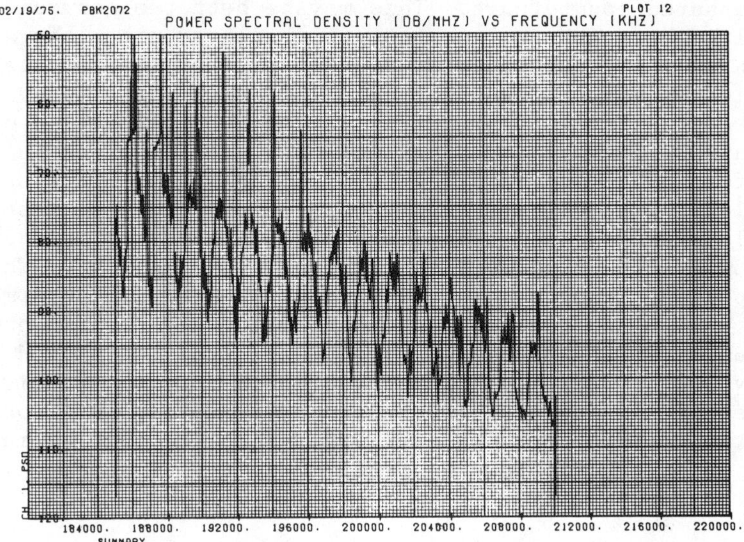

Fig. 11 Computed IMP spectral density.

tered spectra, we estimate the noise floor and the actual spur levels. At UHF, a thermal noise floor of -110 dBm/MHz is assumed.

The measured and the calculated results are in good agreement and appear to be approximately 5 dB apart. Almost every peak in the calculated sawtooth pattern can be correlated with

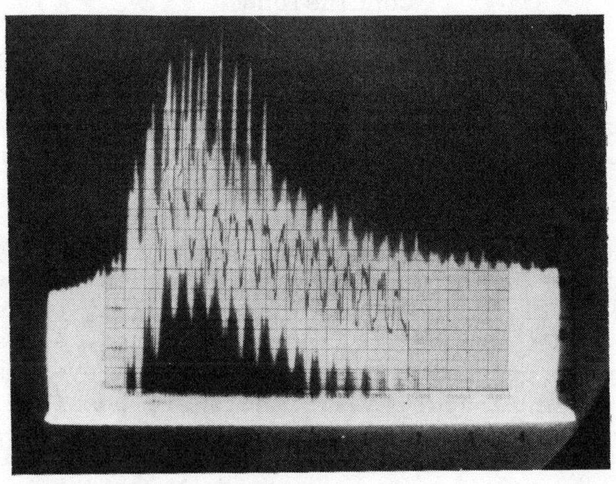

Fig. 12 Measured IMP spectral density.

its measured counterpart. This may be better observed from Fig. 12, where the two spectra are overlayed after the normalization of their frequency axes. In spite of the correlation, one discrepancy can be observed. The nulls in the computation are deeper. This difference is due to the combination of four unrelated effects: selecting less than actual values in analysis for A_7 and A_9, ignoring the 9th and higher order spurs, correlated transmitters, and imperfect oscilloscope resolution. In this example, the strongest individual IMP was measured at -70 dBm or 110 dB below the strongest output signal. The peak IMP spectral density measured at the receiver is -50 dBm/kHz. For a realistic receiver bandwidth of 100 Hz, the IMP power is -60 dBm or 90 dB above the noise power, implying a need of nearly 100 dB additional IMP reduction before the receiver will be free of this type of interference, resulting in a total IMP isolation of 200 dB. The same figure implies that as much as 25 dB less rejection may be required if the RF frequencies are selected optimally.

We have obtained this picture by computing all IMP's individually, summing them, and adding the sum to the thermal noise floor. Computationally any thermal noise floor or IMP level is acceptable. Of course, if an unrealistically low thermal noise floor is used in the computation, the IMP spectra are likely to exhibit a fine structure, never verifiable by measurement. Similarly, we cannot accurately predict IMP levels of many orders of magnitude below the thermal noise level, because all IMP computations start with the measurement of selected IMP's. In general, if we can measure the strongest IMP of a given order, all IMP's of that order will be computed accurately, but higher order IMP's will be increasingly less accurate.

Conclusions

This paper has described a method of computation for IMP's generated in passive components. Based on the measurement of the strongest IMP's generated by equal unmodulated carriers, the spectra can be computed for complex systems, for unequal carriers, and over a large class of modulations.

The accuracy of our approach depends on the accuracy to which measurements are made, and to which the phase information between the carriers is known. In general if the strongest spur of any order can be accurately measured, all spur types of this order will be computed accurately. In systems where accurate individual spur measurements are possible, high correlation between the measured and the computed spectra results.

The design implications of our approach are twofold. First the RF frequency plan can be optimized to reduce the system's sensitivity to passive device IMP's. Given complete freedom in the selection of the uplink and downlink frequency bands, the IMP threat can be eliminated. For an already existing allocation, optimum assignment can decrease the IMP impact by as much as 30 dB. The other important feature of our program is that it provides a good estimate of system performance prior to system testing.

References

[1] O. Shimbo, "Effects of Intermodulation, AM-PM Conversion, and Additive Noise in Multicarrier TWT Systems," Proceedings of the IEEE, February 1971.

[2] A. Berman and E. Podraczky, "Experimental Determination of Intermodulation Distortion Produced in a Wideband Communication Repeater," 1967 IEEE International Conf. Rec., Vol. 15, Pt. 2, pp. 69-88.

[3] D.P. Sullivan, "Intermodulation Distortion Study for FM Communication Satellite Systems," Space Technology Labs., Inc., June 15, 1962.

[4] P. Shaft, "Limiting of Several Signals and its Effect on Communication System Performance," IEEE Transactions on Communication Technology, December 1965.

[5] G. Hedin, "IM Interference Analysis," TRW Systems, Redondo Beach, CA, May 1971.

[6] S. Narayaman, "Application of Volterra Series of Intermodulation Distortion Analysis of Transistor Feedback Amplifiers," IEEE Transactions on Circuit Theory, November 1970.

[7] A. Berman and C. Mahle, "Nonlinear Phase Shift in Traveling-Wave Tubes as Applied to Multiple Access Communication Satellites," IEEE Transaction on Communication Technology, February 1970.

[8] S. Narayaman, "Intermodulation Distortion of Cascaded Transistors," IEEE Journal of Solid State Circuits, June 1969.

[9] R.G. Sea and A.G. Vacroux, "On the Computation of Intermodulation Products for a Power Series Nonlinearity," Proceedings of the IEEE, March 1969, p. 337.

[10]R.G. Sea, "An Algebraic Formula for Amplitudes of Intermodulation Products Involving an Arbitrary Number of Frequencies," Proceedings of the IEEE, August 1968, p. 1388.

[11]R.C. Chapman, J.V. Rootsey, I. Polidi, and W.W. Davison, "Hidden Threat — Multicarrier Passive Component IM Generation," AIAA/CASI 6th Communications Satellite Systems Conference, Montreal, Canada, April 1976.

HIDDEN THREAT: MULTICARRIER PASSIVE COMPONENT IM GENERATION

R. C. Chapman,* J. V. Rootsey,+ and I. Polidi≠
Ford Aerospace and Communications Corp., Palo Alto, Calif.
and
W. W. Davison§
U. S. Army Satellite Communications Agency, Ft. Monmouth, N.J.

Abstract

Passive intermodulation (IM) generation in microwave components has been identified as a serious threat to multicarrier communications systems. Particularly, in highpower systems with low-noise receivers, surprisingly high levels of passive intermodulation (PIM) can be encountered and can degrade system performance seriously. Because of geometrical and workmanship dependences to the generation mechanisms, it has been necessary to rely on semiempirical modeling to the phenomena. A phenomenological picture of the generation and discussion of the physics, power laws, and dependences are made. Procedures to implement sensitive test beds for obtaining actual PIM data and typical test results then are described. A number of ground rules and design guidelines for implementation of linear systems are presented and conclusions relative to the application to future systems made.

I. IM Generation Phenomena

Communications system design and subsystems specification for either Earth terminals or communication satellites conventionally are made based on ring-around, leakage, jamming, and receiver saturation calculations. Little account, however, is taken of passive component nonlinear IM or harmonic generation, since it usually is not recognized by the designer to be a real threat. Particularly, in the case of high-power multicarrier systems with low-noise receivers,

---

Presented as Paper 76-296 at the AIAA/CASI 6th Communications Satellite Systems Conference, April 5-8, 1976, Montreal, Canada.

*Sec. Supervisor Adv. Communications Systems Engrg. Dept.
+Microwave Engineering Specialist.
≠Senior Engineering Specialist.
§Chief, Phase II Terminals Division.

surprisingly high levels of IM can be encountered, and lack of sufficient design margin to accommodate for the generation can be catastrophic to system performance.

Passive component IM generation is a very real phenomenon. Although recognized early in troposcatter work and for a couple of decades by microwave engineers and mobile equipment operators, little has been published on the subject because of the rather unpredictable nature of the generation and the rather strong dependence on configuration and workmanship. Some of the first reports and attempts at comprehensive treatment of the phenomena were made by Illinois Institute of Technology Research Institute[1] (IITRI) in the early 1960's under the Hull Noise Program conducted for Naval Electronics Laboratory Center (NELC) and in study of discrete frequency intermodulation interference generated by nonlinearities within the launch service structure during the checkout of several Saturn vehicles.[2] Relative to satellite systems, note of the phenomena was made first in 1968 in the Massachusetts Institute of Technology (MIT) Lincoln Laboratories satellites (LES-5 and LES-6), where higher-order intermodulation products due to contacts in cavity-backed antennas were noted.[3] In early 1970, IM generation in waveguide components and at flange joints was reported by Cox[4] at General Electric, and a considerable amount of experimental data were presented. One of the first real attempts at modeling the phenomena responsible for the generation was made by Pound[5] in investigation of tunneling phenomena and ferromagnetic nonlinearities.

Detailed investigation of passive component IM generation at Aeronutronic Ford Western Development Laboratories (WDL) was begun in 1972 following isolation of the generation in the United States Army Satellite Communication Agency Heavy Terminal - Medium Terminal (USASCA HT-MT) antennas under development. Under a subsequent program sponsored by USASCA, study of the physics for the development of a phenomenological picture of the mechanisms and an experimental investigation of IM generation in X-band waveguide components were made.[6,7] Most recently, a number of different programs and investigations on passive generation have been conducted, and a number of reports have appeared in the recent literature. These have ranged from IM problems encounted on FLTSATCOM,[8] transmission lines, and connectors[9] to problems encountered in the deep-space network[10-12] and in specialized detection equipment.[13]

Basically, the problem arises because presumably normally linear components such as coaxial cable connectors, waveguide interfaces, or discontinuities in aerials or reflector panels in radio-frequency systems are very slightly nonlinear. For high-power systems with sensitive receivers transmit-to-receive isolations, the order of 150-200 dB typically are needed. For such systems, nonlinearities as little as the order of 1 part/10^{10} have been found to present a problem as is shown in Section III.

II. Nonlinear Processes

Although the exact microphysics responsible for the phenomena have not been resolved fully, three of the most predominant mechanisms suspected are 1) electronic tunneling and semiconductor action through thin oxide layers separating metallic conductors at metallic junctions; 2) microdischarge between microcracks, whiskers, or across voids in metal structures; and 3) nonlinearities associated with dirt, metal particles, and carbonization on metal surfaces. Each of these different mechanisms appears to manifest itself in the same power laws and similar levels of generation. As power levels and sensitivities are increased progressively, a hierarchy of different low-level mechanisms come into play, much like peeling off layers of an onion skin.

Physically, the nonlinearities responsible for the IM or harmonic generation are a result of the summation of many different microcurrent conduction processes. Microscopically, all surfaces are highly irregular. In addition, most metallic surfaces have an oxide layer several angstroms thick on their surfaces. When contact is made between surfaces, rupture spots through the oxide coating are formed, work-hardening and cold-welding of the metals occur at some spots, very thin oxide layers separate the metals at some spots, and thick oxide or void regions are formed in other spots.[14] The actual impedance at the contact then is determined by two factors: the current conduction through the finite number of discrete contact spots, and the displacement current through the oxide and void regions.

The effective nonlinearity of a contact will depend on the relative proportion of the conductive and displacement currents, and the net nonlinearity of the different conduction mechanisms at work. For metal surfaces separated by thin oxide layers less than 50 Å, nonlinear electron tunneling through the barrier occurs. For thicker oxide layers separating conducting materials, space-charge-limited semiconductor current flow can take place. At other microscopic contacts, small voids and impurity materials, such as small metal particles or carbon, may be present. As the microvoltage is increased at these positions, the resultant microcurrent becomes piecewise nonlinear as microbreakdown of trapped gases, punch-through of impurities, or current flow through the carbon material occurs, and the microcurrent takes different paths with increasing microvoltage. The total rf current flow, thus, becomes the summation of microscopic rf tunneling currents, shunt displacement currents, parallel discharge currents between whisker or through particles, discharge across trapped gas regions, punch-through currents through impurity materials, or actual microcurrents flowing in impurities such as carbon.

Contingent on the actual microgeometry, one or more of these different processes may be predominant. Experience at WDL has indicated that surface treatment and coating in many cases can influence the resultant IM generation level greatly in microwave components. Additionally, for most contracts, the IM level does not appear to be critically dependent on the actual contact pressure. Both of these observations tend to support microdischarge as one of the more predominant generating mechanisms in that it is not as subject to contact pressure and is highly dependent on the actual surface microgeometry and presence of impurities or whiskers, all of which can be influenced by surface treatment and coating. As power level or frequency is changed, a relative shift between the microprocesses undoubtly takes place. At high power levels, low-level water vapor, gaseous weak plasma, and nonlinear processes in the waveguide wall even can come into play.

The net macroscopic IM levels and power laws observed, however, do not appear to change markedly or to be associated with one particular process. This is, no doubt, because the observed IM currents are a result of the statistical summation of the microcurrents from many different nonlinear contacts.

III. Mathematical Relations For Third-Order Nonlinear IM Product

A nonlinear device is one that does not obey Ohm's law, and the relation of current and voltage for such a device is a curve that can be represented by a polynomial of degree n over a finite interval. The mathematical relation for this is

$$I = f(V) = C_o + C_1V + C_2V^2 + C_3V^3 + \text{-------}C_nV^n \qquad (1)$$

where V is the voltage applied to the nonlinear device, I the current through the device, and the coefficients C_o, C_1, C_2, C_3, etc., quantities that determine the extent of the nonlinearity. For the case in which V is the sum of two sinusoidal voltages, it can be shown through a binomial expansion using only odd terms that the third order intermodulation power delivered to the load is

$$P_{21} = (9/16)\ (C_3{}^2/C_1{}^6)\ R_e\ G_e{}^3\ P_T{}^3\ [R/(R+1)^3] \tag{2}$$

where G_e is the effective load conductance,

R_e equals $G_e/4C_1{}^2$

C_1 is the admittance of the junction,

P_T is the total carrier power,

and, R the ratio of the carrier powers.

If the total power in relation (2) is held constant then the third order intermodulation power, P_{21}, is found to vary with the carrier ratio R as

$$P_{21} \ \alpha\ [R/(R+1)^3] \tag{3}$$

Although these two relations show the dependency of intermodulation product on the total carrier power and carrier ratio, the question remains as to the relative order of nonlinearity that might present a problem. To get a rough order of magnitude to this question, it is possible to use a semiemperical approach and substitute test results and assumed values into equation (2). For a total power of 5 kW with a carrier ratio of 3 dB, the measured value of the IM was -164 dB. Assuming $G_e = 1/50$ and $C_1 = 0.2$ for a 40 Angstrom Cu-CuO tunneling junction the third order nonlinearity C_3/C_1 is found to be the order of 2×10^{-10}.

IV. Intermodulation Product Dependencies

Mathematically, passive component intermodulation products and harmonics are the result of slight nonlinearities in the relation of current and voltage. As just discussed, this deviation from Ohm's law is the result of a number of different microprocesses. In order to provide a means of predicting the expected IM generation from microwave components, a couple of different approaches can be taken. One method would be to model as accurately as possible the microprocesses responsible for the generation and their resultant current voltage characteristics. Using a polynominal of degree n over a finite interval, the various coefficients then could be calculated and, in turn, the resultant IM levels and dependencies relative to imposed carrier levels obtained. This method, although straightforward, suffers from the inability to model accurately the microphysics, the geometrical and configuration dependencies of the generation, and the basic statistical nature of the phenomena, that is, the fact that the macroscopic system test values depend on the summation of multiple sources. Therefore, at best, several orders of magnitude bounding to the levels could be expected using this approach.

Although less satisfying, an approach that somewhat alleviates these problems and the one chosen by the authors is to use a semiempirical approach for bounding the generation and obtaining information for predicting IM levels and their dependencies. In this, a phenomenological model first is developed for the generating mechanisms and their characteristics. This then is used as a basis for making the appropriate assumptions and deriving the power laws and the dependencies. For the case of the third-order intermodulation product, a summary of the results are given in Section III. The mathematical relations developed then, in turn, can be used in conjunction with test results for prediction of generation at different levels or conditions other than the actual test situations.

Relative to third-order passive component intermodulation production, which was selected in the testing because it is generally the highest IM and the one of most concern, three important conclusions can be obtained from the mathematics summarized in the previous section. First, the third-order IM should vary as 3 dB/dB with total power. Deviations from this in actual test results to slightly lower values are observed to occur. This is due to the second-order dependence on higher coefficients to the polynomial expansion, that is, the exact shape of the nonlinear I-V curve, and the dependence of the test result on the summation of multiple different processes and sources at a macroscopic level. If a system becomes very noisy and not well behaved, this breaks down completely, and such a relation then no longer has to be followed.

Second, for the case when third-order IM level is related to carrier ratio and total power is held constant, it can be shown from the limits of equation (2) that for the case that the IMP occurs adjacent to the weak carrier or in the region of weak carrier, the IM varies as 2 dB/dB with the carrier ratio, whereas in the strong carrier region, where the IM is adjacent to the strong carrier, the intermodulation product (IMP) approaches a 1 dB/dB dependency. The maximum IMP occurs at a carrier ratio of 2 to 1. These relations have proved to be very useful for correlation of experimental data and in predicting or scaling IMP. Methods to obtain the experimental data, test results, and correlation of the data are presented in the following sections.

V. Establishing a Test Bed

Figure 1 illustrates the typical WDL IMP test configuration, where all transmit frequencies are generated and stabilized in a common area, and subsequently switched to the roof for radiation tests or to a shielded room for component tests. The most critical components for IM measurement and suppression are the filters at the input and output of the test devices. The high-pass filter must remove all traces of receive band noise from the transmitter so that only spectrally pure carriers are present in the test region, and the bandstop/bandpass filters must suppress the transmit carriers so that spurious IMP's are not generated in the preamp or components external to the test sections. The frequency bands over which the WDL experimentation occurs create a particularly severe situation because the separation of the transmit and receive bands is significantly less than the individual receiver and transmit bandwidths. This means that there are many combinations of transmit carriers which will produce third-order IMP's in the receive band, so that the high-pass filter must have a rejection capability of between 180 and 200 dB with extremely steep skirts if a 10-kW to -140-dBM dynamic range is required. Fortunately, a high-pass filter is ideal for such a difficult task because it requires no tuning, and its characteristics can be calculated accurately instead of measured.

The amount of filtering required after the test device depends on the response of the low-noise amplifier to transmit carriers; typical saturation properties require about 100 dB of rejection over the 7.9-to 8.4-GHz band. A standard Chebychev bandpass design is recommended with the first model made without tuning (or with slight dent tuning) to avoid spurious IMP's. Once a quiet setup is achieved, subsequent tuned filter models can be examined for IMP's and, when quiet, can replace the earlier high-voltage standing wave ratio (VSWR) models. It is unfortunate that the linearity or IMP-producing nature of a component cannot be determined or improved without IMP-free components to put in the test bed. This initial quieting can be a difficult iterative process in which the various critical components are replaced alternately with improved quieter versions. The iteration involves not only the filters but also interconnecting waveguides and flanges.

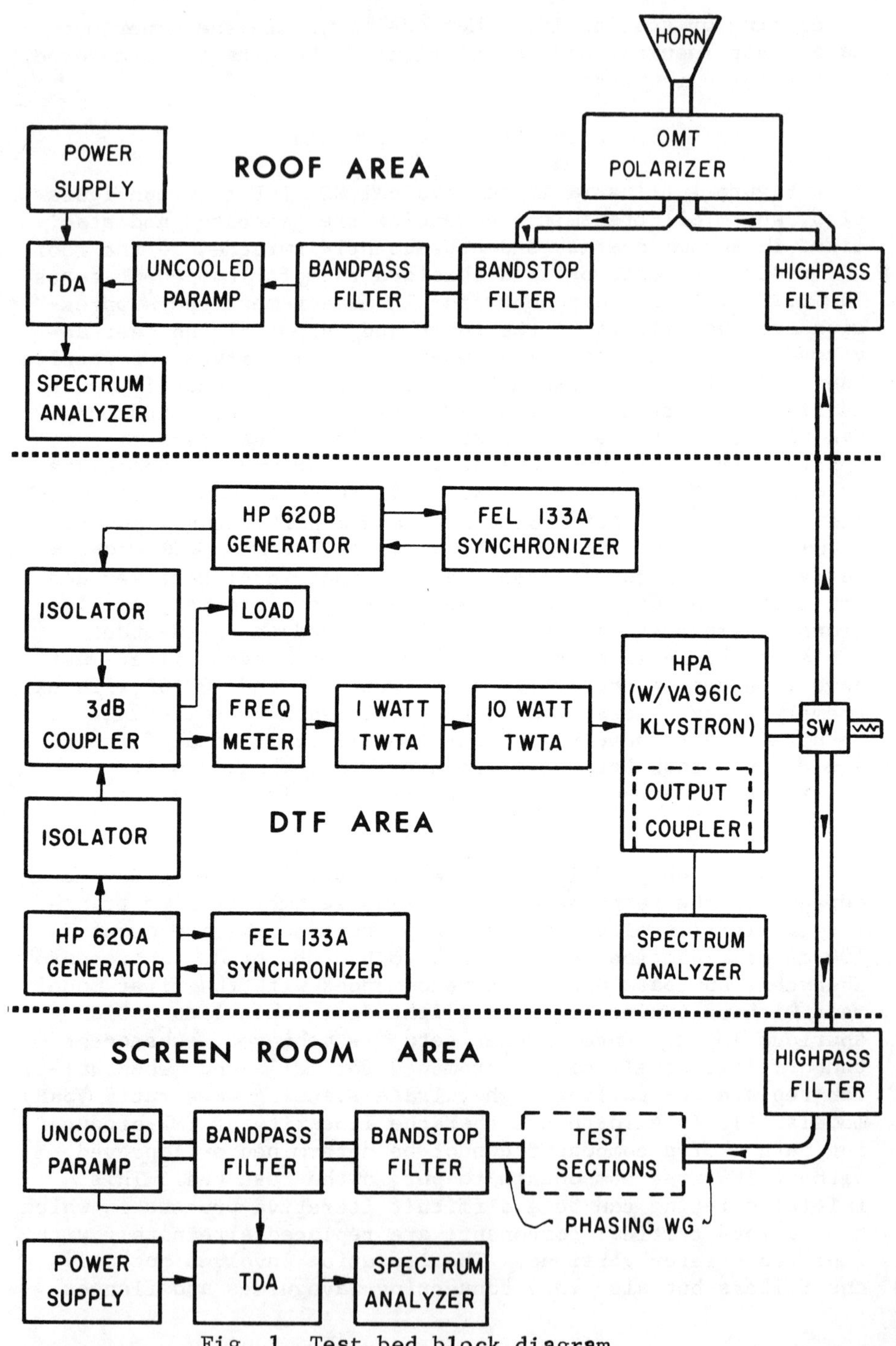

Fig. 1 Test bed block diagram.

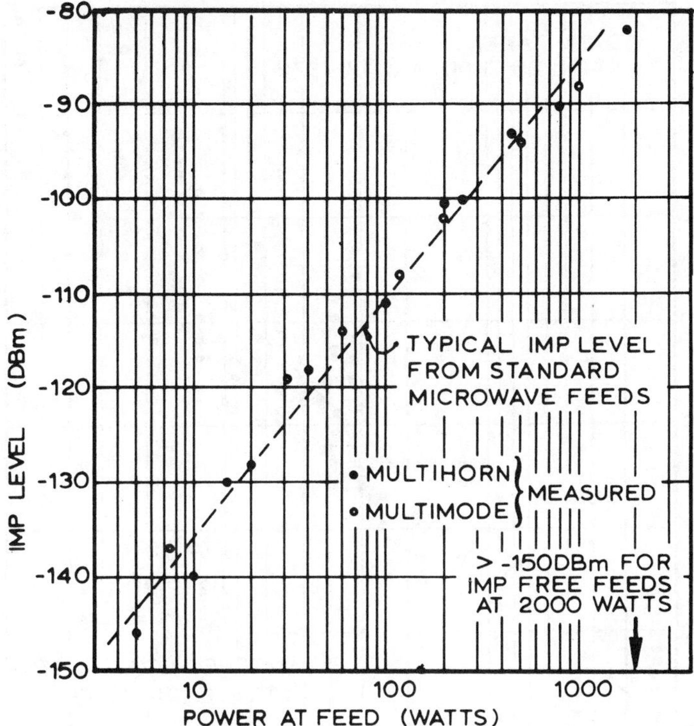

Fig. 2 IMP levels from typical standard microwave feeds.

The two carriers that excite the test device must be synthesized or stabilized to 1 part/10^9 or better for the short term. Instability greater than this will make IMP's very difficult to locate on the spectrum analyzer and may cause spreading of the spectral energy with an artificial reduction of the displayed IMP level. The high-power amplifying device can be a TWT or broadband klystron, but, if both carriers are amplified together, the test engineer must be sure that his filtering is sufficient to suppress tube-generated IMP's. On the receive side, a spectrum analyzer with a persistence feature (storage screen) and a 100-Hz predetection bandwidth preceded by a 6-dB noise figure amplifier will provide adequate sensitivity for recording at the -140-dBm level.

VI. Test Results

The significance of component nonlinearity became evident when Defence Satellite Communications System (DSCS) phase II terminals began highpower dual-carrier tests in the early 1970's. The high level of spurious signals falling into the sensitive receive band were blamed at first on the transmit-

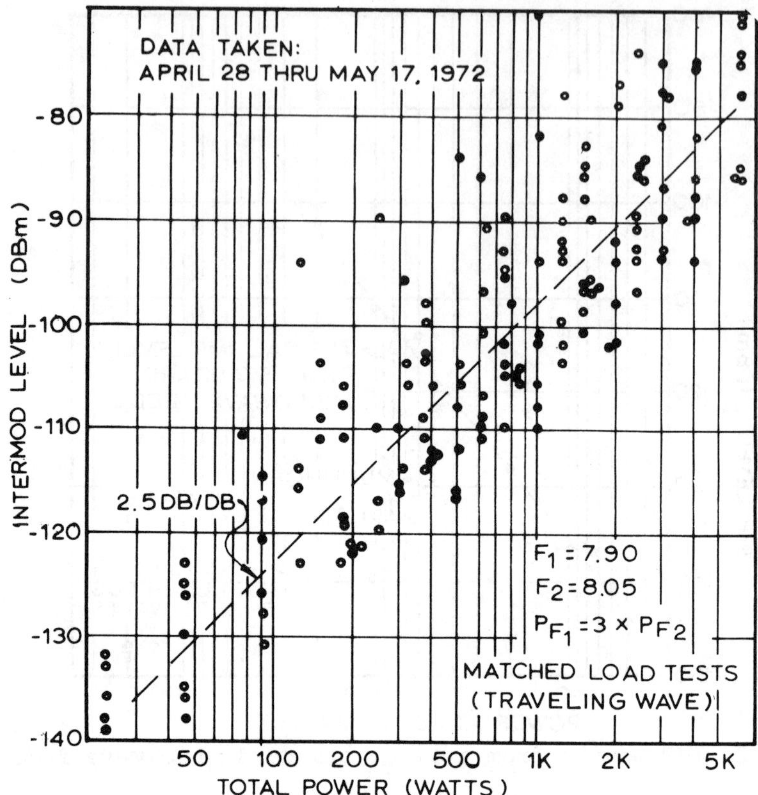

Fig. 3 Collection of component IMP levels.

ters, receivers, and inadequate filtering until extensive tests demonstrated that the IMP's actually were originating in the passive waveguide components. Subsequent tests were run on two general classes of feeds commonly used by the U. S. Army: multihorn monopulse tracking feeds, and multimode tracking feeds. The IMP levels generated within these feeds are shown in Fig. 2, where the recorded levels, even though taken at different times, were almost identical. These feeds were built by different manufacturers to appropriate military standards using commonly accepted microwave techniques, and both performed excellently in areas of gain, efficiency, insertion loss, and mode stability. IMP generation was an inherent hidden threat in even the best standard designs.

IMP generation of the type experienced in feeds appears to be independent of the kinds or combinations of materials used in the fabrication process. The test program found no correlation between IMP level and material during tests on gold, brass, copper, nickel, cadmium, silver, stainless steel, aluminum, and various dissimilar metal junctions. However,

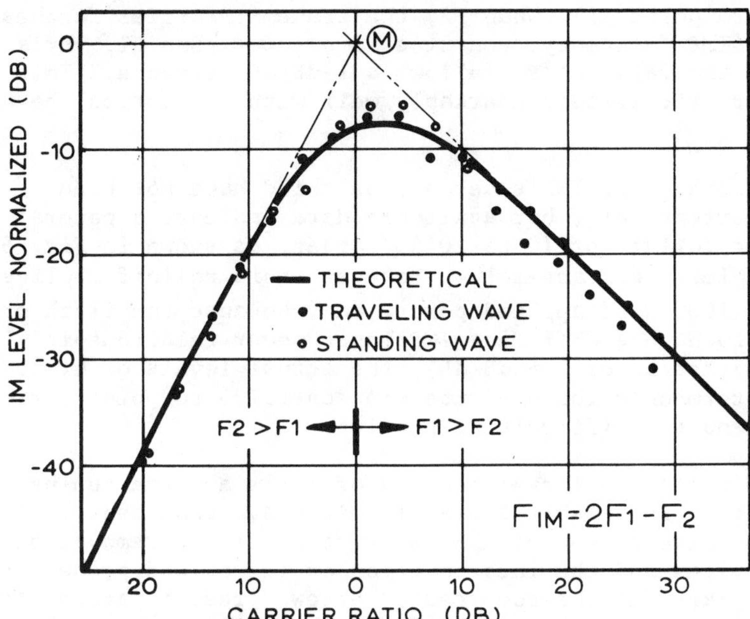

Fig. 4 Comparison of measured and theoretical responses.

the IMP levels were related directly to how these materials were joined together, the quality of the mating surface finishes, the pressure at the junction (from bolts or other fasteners), and the quality and cleanliness of the brazing or soldering process. During these measurements, IMP's varied both in absolute level and in their power-slope relationship. Single flanges produced IMP's in the -80-to -105-dB range, with IMP/input power relationships varying from 1 to 3 dB/dB, depending on the surface texture and cleanliness of the junction. After measuring many junctions in traveling wave (matched-band) situations, a trend in the behavior of nonlinear junctions was observed as shown in Fig. 3. The general slope of third-order data followed a 2.5-dB IMP variation for each decibel of input power variations. On the average, data from individual nonlinear junctions were about 10 dB below data from a complex assembly such as feed.

In an attempt to learn more about the generating mechanism, the third-order IMP was observed under various power ratios of F1 and F2 (the two transmit carriers) while the total combined power was held constant. As shown in Fig. 4, the strongest IMP for a given power is produced when the transmit carrier closest to the IMP frequency (F1) is about 3 dB higher than the other, in agreement with the previous derivation. (The three sets of data in this figure are normalized to point M.) When F2, the transmit carrier farthest from the IMP frequency, contains more power than F1, variations in the ratio F1/F2 follows a 1-dB/dB response. This characteristic agreed remarkably well with theoretical behavior.

The orders of IMP's beyond the third have not been studied extensively, but available data indicate a general monotonic rolloff of 10 to 20 dB/ order, as shown in Fig. 5 for a typical feed assembly. However, this rolloff applies only for 1000 W of applied power, and, because the fifth order appears to have a different IMP/input power relationship (4 dB/dB instead of 2.5 dB/dB), the actual levels of higher-order intermodulation products are sensitive to point current density and are difficult to predict.

The most significant sources of IMP's are the tuning methods that invariably appear in microwave components. The most effective method of IMP suppression is the removal of these devices and the incorporation of irises and probes as integral parts of the component. Screws present particular problems, as their threads often contain microscopic debris and make poor contact with the threaded hole.

Once a system is manufactured and is quiet to some specified level, is this a guarantee that the components will remain quiet? Not necessarily so, as experience with components has shown occasional degradation with time. Figure 6 illustrates WDL's experience with a bandstop filter, which, at the time of manufacture, was quiet to the -140-dBM level at 200 W input power. Two weeks later, an IMP first was ob-

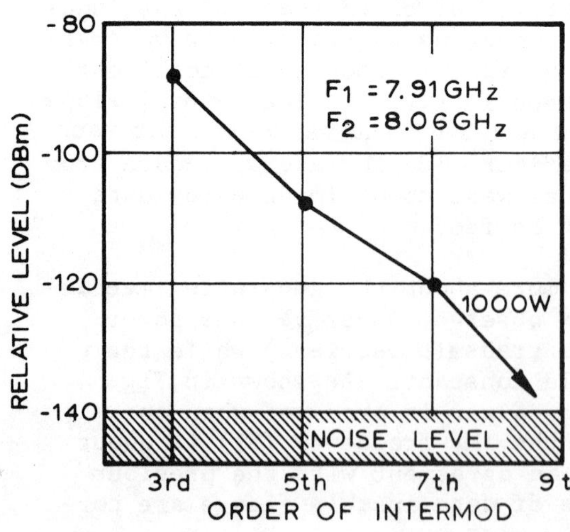

Fig. 5 Relative levels of high- order IMP's.

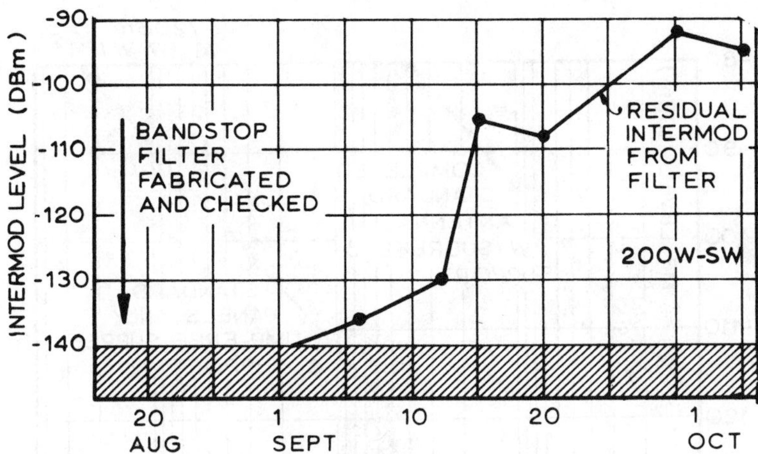

Fig. 6 A particular aging observation.

served, but the filter continued to function as a test device and was checked periodically for its IMP level. After a month, the IMP had grown to almost -90 dBm! Although such behavior is not typical, it does caution the designer to examine stress, humidity, and other conditions that may occur in the test system and components.

Even when IMP's from the feed are below the designer's specification, the antenna structure and environment can introduce strong IMP's into the system. Recent tests with medium and large antennas with quiet feeds produced the curves shown in Fig. 7. The data, being plotted as a function of power density on the main reflector surface, are relatively general and can be expected from standard antennas of all sizes. WDL currently is developing new fabrication methods to linearize the structure; however, data are too preliminary for publishing at this time.

VII. Design Guidelines

Although specific designs for low-IM-generating components cannot be made which guarantee low IMP's in all cases, a number of general design guidelines can be made which, if properly employed, help to minimize the resultant passive IM level. Briefly, and in order of importance, the guidelines and rules for producing the components are as follows:

1) First perform a frequency allocation if at all possible to minimize the effects of nonlinear generation.

2) Design the system to limit the nonlinear generating sources to as small a region as possible.

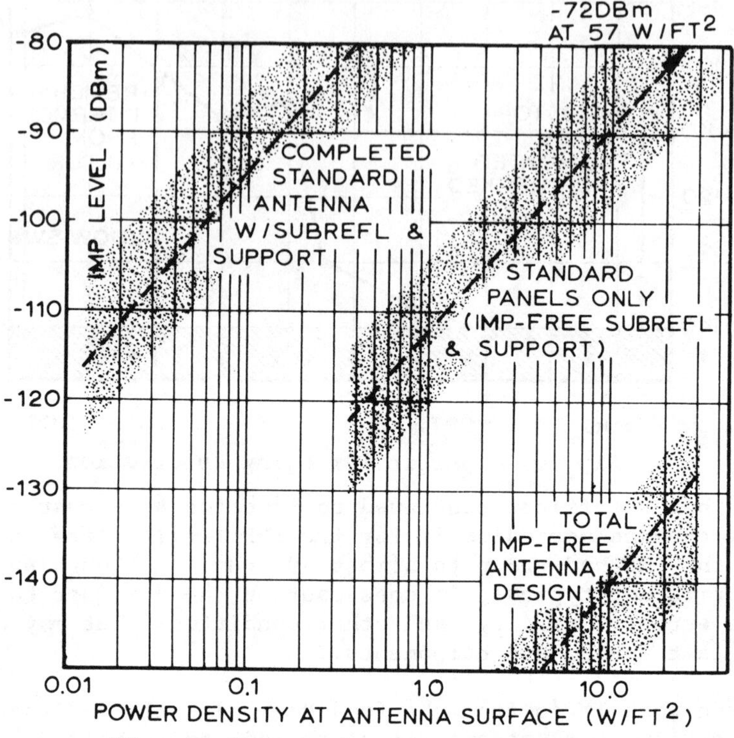

Fig. 7 Measured reflector IMP's.

3) Minimize the use of the high-Q resonant circuits in the resultant generating region. For example, use hybrid filter designs or band reject filters rather than bandpass filters.

4) Eliminate as many mechanical contacts as possible in the generating region.

5) Increase the dimensions of transmission lines and resonator components to reduce the rf current density.

6) Use extreme care in the design of mechanical components. For example, use unibody monolithic designs, lapped surfaces, welded contacts, or contacts under pressure near the Young's modulus.

7) Employ care in fabrication and cleaning procedures to avoid contamination with foreign particles at critical contacts. This is especially critical for the case of microdischarge generation.

These steps for linearized component design should be performed in the order given; that is, each design measure should be exercised, one by one, until sufficient linearity is reached to assure that the passive nonlinear harmonic or IM generation is below the desired level. As can be observed, each step becomes progressively more difficult and involves more careful and costly design.

WDL's experience with IMP's has shown that metallic junctions of all types are the most likely sources of system nonlinearity. This is not to say that junctions cannot exist in an IMP-free system, because most systems could not exist witnout them. However, a junction can be linear only if it is tight, smooth, flat, and clean. These, of course, are relative terms and are somewhat interrelated in that nonsmooth surfaces can be made smooth by sufficient pressure. In general, it is a good practice either to lap mating surfaces and tighten until full contact is achieved or to braze joints with high-quality techniques. The surfaces must be cleaned thoroughly with a suitable nonfilming solvent before assembling to remove oils, fluxes, etc. However, regardless of what method the designer chooses to unify parts and assemblies, if the procedure follows the general guidelines and includes precision manufacturing, tight assembling control, and extreme handling care, then it is likely that the system will meet equivalently a -140-dBm level of quietness for multiple carrier power totaling 10 kW. The biggest difference between a good linear design and a poor one will be in how long the component practically remains quiet.

Although most of the data and analysis presented in this paper have been for ground antenna systems and third-order intermodulation generation, it is applicable equally to satellite communication systems or other-order harmonics using appropriate transformations. The power laws and general levels presented in this paper then can be used to obtain rough order-of-magnitude estimates to the expected nonlinear generation.

References

[1]"Under the Hull Noise Program," TM 1-TM 18, 1962-1970, Illinois Institute of Technology Research Institute for Naval Electronics Laboratory Center.

[2]"Study Concerning Nonlinear Mixing of Radio Frequency Signals in Steel Structures," Final Rept. IITRI E 6047-7, Oct. 15, 1965, Illinois Institute of Technology Research Institute.

[3]"Lincoln Experimental Satellite - 5 (LES-5) Transponder Performance in Orbit," TN 1968-18, Nov. 1, 1968, Lincoln Lab., Massachusetts Institute of Technology.

[4]Cox, R. D., "Measurement of Waveguide Components and Joint Mixing Products in 6 GHz Frequency Diversity Systems," IEEE Transactions on Communication Technology, Vol. COM-18, Feb. 1970, pp. 33-37.

[5]Pound, T. R., Frazier, M. J., Dabkowski, J., Rudge, A. W., Bryant, G. W., and Ashford, N. A., "Research on Natural and Ferromagnetic Nonlinearities," Final Rept., Vol. 1, IITRI E6157, July 1971, Illinois Institute of Technology Research Institute.

[6]Chapman, R. C. and Darlington, J. C., "USASCA HT-MT Terminals Program - Intermodulation Generation in Normally Passive Linear Components," Final Rept. WDL-TR5242, Aug. 1973, Aeronutronic Ford.

[7]Rootsey, J. V., Gradisar, A. A., and Bordenave, J. R. P., "USASCA HT-MT Terminals Program - Intermodulation Study Final Test Report," WDL-TR5243, Aug. 1973, Aeronutronic Ford.

[8]Becker, N. L., Hannemann, T. W., and Kindorf, S. D., "FLTSATCOM IM Problems and Solutions," AFTRCC Spring Meeting, April 14-15, 1975, Las Vegas, Nev.

[9]Bayrole, M., and Benson, F. A., "Intermodulation Products from Nonlinearities in Transmission Lines and Connectors at Microwave Frequencies," Proceedings IEE (London), Vol. 122, April 1975, pp. 361-367.

[10]Bathker, D. A. and Brown, D. W. "Dual Carrier Preparations for Viking," TR 32-1526, Vol. XIX, 1973, Jet Propulsion Lab.

[11]Higa, W. H., "Spurious Signals Generated by Electronic Tunneling on Large Reflector Antennas," Proceedings of the IEEE, Vol. 36, Nov. 1975, pp. 306-313.

[12]Bathker, D. A., Brown, D. W., and Patty, S. M., "Single and Dual-Carrier Microwave Noise Abatement in the Deep Space Network," TM 33-733, Aug. 1975, Jet Propulsion Lab.

[13]Chapman, R. C., Polidi, I., and DeHaven, W. R., "Final Report METRRA Producibility Investigation," WDL-TR5880, June 1975, Aeronutronic Ford.

[14]Holm, R., Electric Contract Handbook, Springer-Verlag, Berlin, 1958.

Chapter VI – Special Topics in New Earth Station Technologies

CENTIMETER-WAVE LOW-NOISE AMPLIFIERS FOR SATELLITE COMMUNICATIONS

Robert Levin*

Comtech Laboratories, Inc., Smithtown, N. Y.

Abstract

The growth of satellite communications in the 4- and 6-GHz bands has resulted in the need to explore the higher frequencies for future satellite systems. The characteristics of parametric amplifiers used in satellite ground terminals operating above 10 GHz are a significant determinant of communications performance. At these frequencies, few, if any, alternatives to the parametric amplifier become viable considering the noise performance that must be achieved. Amplifier systems employing cryogenic cooling and other forms of temperature stabilization (e.g., thermoelectric devices and "above ambient" heating) must be evaluated in terms of overall system requirements. Improvements in GaAs varactor diode technology and solid-state millimeter-wave pump sources have enabled the design and implementation of parametric amplifier systems in communications terminals operating above Ku band.

Introduction

The increased usage of the frequency spectrum below 10 GHz has led to investigations of the feasibility of using higher frequencies for satellite communications. In particular, the performance of future downlinks, as characterized by a G/T, reflects the characteristics of the antenna and front end used at higher frequencies. This paper will discuss the parametric amplifier as used in the front end of ground terminals above 10 GHz, as well as some system considerations relative to components that affect G/T performance at high frequencies.

Presented as Paper 76-299 at the AIAA/CASI 6th Communications Satellite Systems Conference, April 5-8, 1976, Montreal, Canada.

*Assistant Engineering Manager.

General Considerations

The performance of a satellite communications downlink is related directly to the carrier-to-noise ratio C/N

$$C/N = C/KTB \tag{1}$$

where K is the Boltzmann constant = -228.6 dBW/°K/Hz, T is the receiving system noise temperature in degrees Kelvin, and B is the receiver predetection bandwidth in Hertz.

In order to compare systems without reference to bandwidth, performance can be specified by the C/KT. Given an EIRP, C/KT is

$$C/KT = EIRP - FSL - K + (G/T) \tag{2}$$

where FSL is the free space path loss in decibels, and G/T is the Earth terminal figure of merit in decibels per degree Kelvin. The G/T defines the Earth terminal's performance in terms of the receiving system sensitivity. Therefore, for a fixed satellite beamwidth and EIRP, G/T also determines the achievable C/KT and hence the maximum communications capacity of the link. The figure of merit G/T can be defined as

$$\frac{G}{T} = \frac{\text{gain of antenna at the receive frequency}}{\text{noise temperature of the receiving system}} \tag{3}$$

To determine G/T for a receiving system, the model of Fig. 1 can be used. The total G and T must be determined at the same point in the system and in the model; the reference point is 1 - 1'. Note that a mismatch contribution T_m is included.

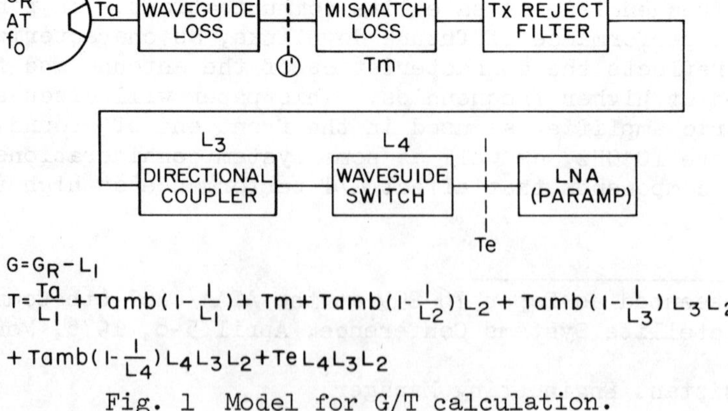

Fig. 1 Model for G/T calculation.

This is due to the resulting VSWR when the antenna and receiving system are mated, causing some fraction of the receive system noise to be reflected back into the system. This is explained in more detail below. The term T_e is the noise temperature of the receiver referred to the input of the low-noise amplifier (LNA). The T_e is determined almost wholly by the noise temperature of the LNA. The other elements shown in the model typically are found in Earth terminal receive chains.

There are four major components to consider in the model of Fig. 1, viz., the antenna, the waveguide components between the antenna and LNA, the mismatch contribution, and the LNA itself. The following section discusses the impact of these items at higher frequencies on receive system performance as characterized by G/T.

System Considerations at High Frequencies

As frequency increases, the change in gain of an antenna at f_o relative to a lower frequency f for constant diameter and efficiency is given by

$$G_R = 20 \log (f_o/f) \tag{4}$$

At higher frequencies, the antenna gain follows 20 log f_o/f, as shown in curve A of Fig. 2, which is normalized to 4 GHz. However, the free space loss also varies as 20 log f_o/f; there-

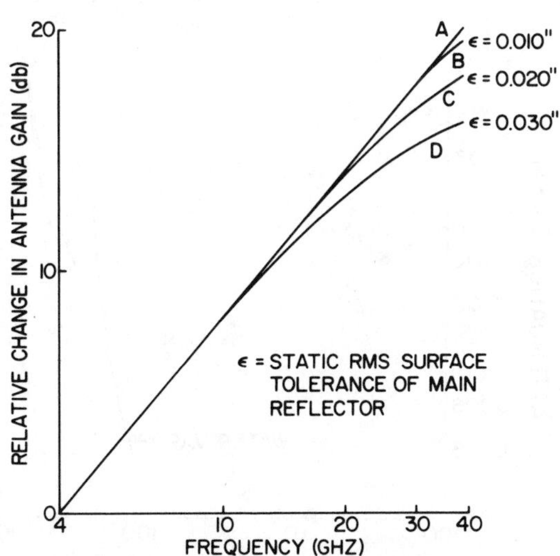

Fig. 2 Change of antenna gain with frequency for constant diameter and efficiency normalized to 4 GHz.

fore, it would appear that these effects would negate each other. But at high frequencies, the surface tolerance of large antennas will cause some gain degradation. Figure 3 shows the loss of gain vs frequency as a function of static rms surface tolerance of the main reflector.[1]

The curve for the 0.030-in. (0.762-mm) tolerance, for example, shows a loss of 0.04 dB gain at 4 GHz and 1.04 dB at 20 GHz. The data from these curves are combined with curve A of Fig. 2 and results in curves B, C, and D, which are the deviations from 20 log f_o/f as functions of ε (the surface tolerance) and frequency. The foregoing would indicate that the

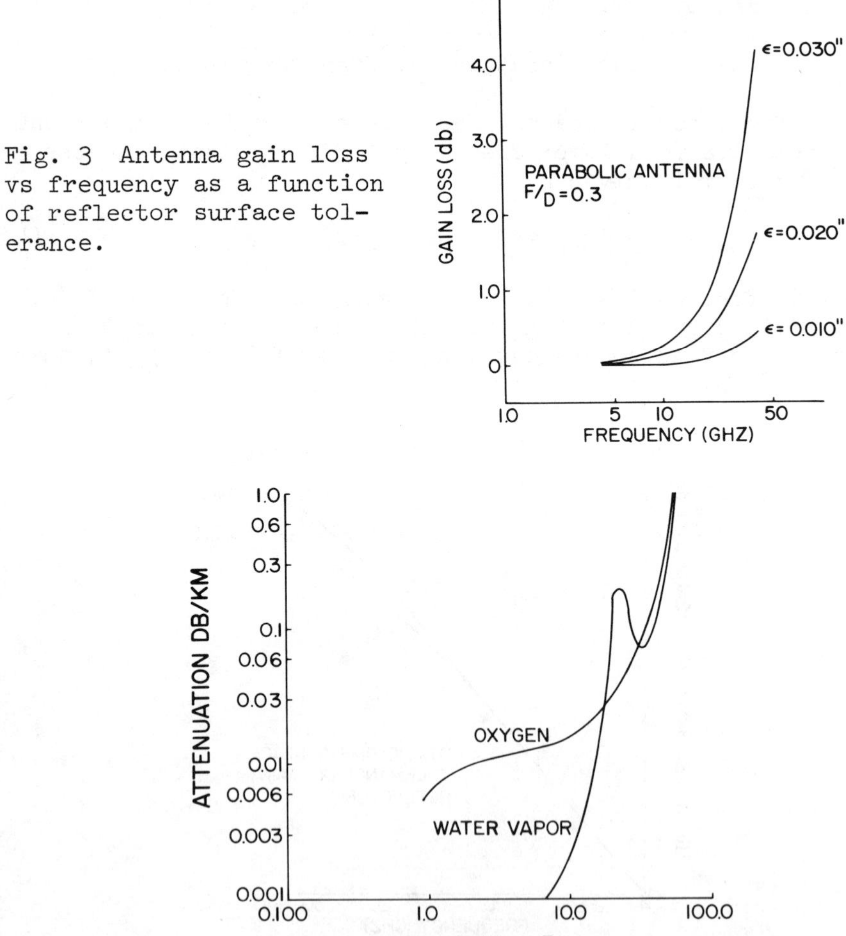

Fig. 3 Antenna gain loss vs frequency as a function of reflector surface tolerance.

Fig. 4 Oxygen and water vapor attenuation vs frequency.

trend to higher frequencies would result in smaller-aperture antennas being used because of the cost penalties associated with holding a small ε on a large structure.

To improve the G/T, one would require a reduction in the antenna temperature, losses, or the noise temperature of the receiver. With respect to the antenna temperature T_a, as the operating frequency is increased, T_a will tend to increase because of atmosphere attenuation. Above 10 GHz, oxygen and water vapor absorption, shown in Fig. 4 [2,3] will act as a source of noise. In addition, at higher frequencies the main beam and the side lobes of the antenna are looking at higher sky temperatures than at lower frequencies.[4] The preceding effects result in less than expected performance above 10 GHz.

The components between the antenna and LNA can be characterized as to their respective losses; hence their effect is well defined. Therefore, these components as well as the considerations just discussed will not be treated in any more detail. Two items, namely, the mismatch contribution and the receiver noise temperature, comprise the remaining items to be considered in the overall G/T. The first, the mismatch contribution, will be discussed below. The rest of the paper will be devoted to the parametric amplifier, which is used as the first active component in the receiver and which determines the receiver noise temperature.

Noise Temperature Contribution Due to Mismatches

A consideration in overall system noise temperature performance is the contribution due to the feed/LNA system mismatch. This occurs in LNA systems that incorporate an isolator before the parametric amplifier (varactor diode as shown in Fig. 5) and is a result of the isolator terminations' noise being reflected at the feed/LNA interface.

The analysis presented below is specific to the model in Fig. 5. However, for amplifiers not incorporating an isolator before the varactor, the same general results would apply because the isolator following the varactor would produce a similar effect. The thermal noise produced by the termination is a function of its absolute temperature. Referring to the model in Fig. 5, the interface 1 - 1' is defined as the point at which noise temperature measurements using a hot/cold source are taken to evaluate the system noise temperature performance. Therefore, assuming that the hot/cold source is well matched (i.e., VSWR < 1.05), the contributions of the transmit reject filter, test coupler, waveguide switch, and isolator termination are included in the overall T_e or excess noise

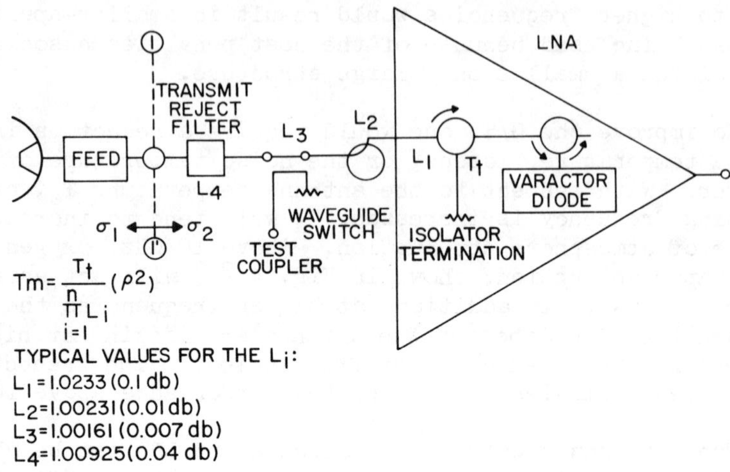

Fig. 5 Model for calculation of T_m.

temperature referred to 1 - 1'. However, when the system is mated with a feed whose VSWR typically is greater than 1.1:1, the isolator termination will cause a greater contribution than before.

The worst-case contribution to the system noise temperature due to termination reflected noise at the mismatched feed/LNA interface is given by

$$T_m = \frac{T_t}{\prod_{i=1}^{n} L_i}\left(\rho^2\right) \tag{5}$$

where T_m is the reflected noise contribution in degrees Kelvin, T_t is the absolute temperature of the isolator termination, L_i are the losses (expressed as a ratio > 1) between the termination and 1 - 1', and ρ is the reflection coefficient at 1 - 1' given by

$$\rho = (\sigma_1\ \sigma_2 - 1)/(\sigma_1\ \sigma_2 + 1) \tag{6}$$

where σ_1 is the feed VSWR, σ_2 is the VSWR at the input to the filter, and the worst case is represented by $\sigma_1\ \sigma_2$. This result neglects the effect on antenna gain and temperature, as well as any inherent change in LNA noise temperature due to the presence of the mismatch.

Figures 6 and 7 show T_m plotted vs σ_1, with σ_2 as a parameter for T_t = 300°K (noncryogenically cooled system) and T_t = 17°K (cryogenically cooled system), respectively. It can be seen that, for a cryogenically cooled system, this effect is negligible, but for a noncryogenically cooled system, a significant degradation in performance can occur if system VSWR's are not held as low as possible.

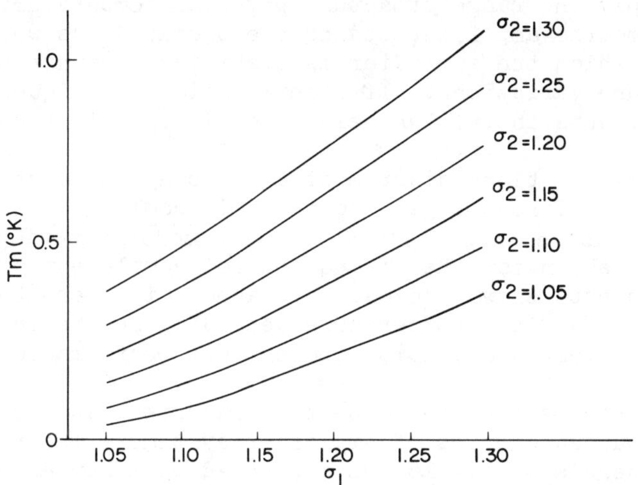

Fig. 6 Worst-case contribution to system noise temperature due to isolator termination reflected noise (T_t = 17°K).

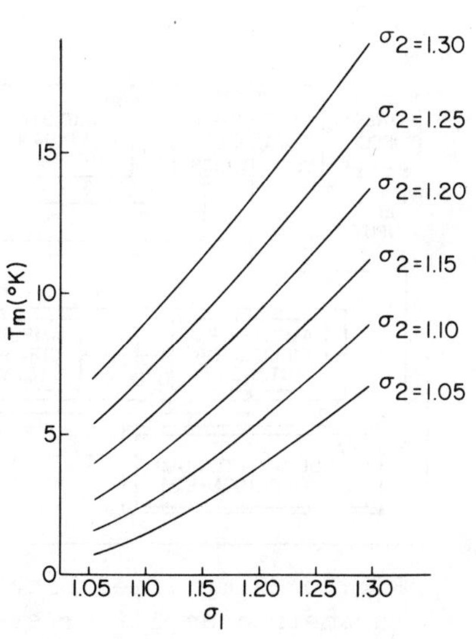

Fig. 7 Worst-case contribution to system noise temperature due to isolator termination reflected noise (T_t = 300°K).

High-Frequency Parametric Amplifiers

Once a target G/T has been determined for a particular system application and the effects of the previously discussed elements other than the LNA have been taken into account, consideration can be given to the attributes and achievable performance characteristics of various types of parametric amplifier systems. These fall generally into four categories, viz., cryogenically cooled, thermoelectrically cooled, thermoelectrically stabilized, and uncooled or heated. These designations imply the range of actual physical temperatures at which the parametric amplifier stages are operated, as well as the means by which the amplifier is stabilized against ambient temperature variations. If stabilization techniques were not employed, substantial LNA gain variations would result.

A general block diagram of a two-stage parametric amplifier system without regard to specific cooling or stabilization methods is shown in Fig. 8. The major functional components are shown for the rf and control complement of equipment. The components shown enclosed by dashed lines are those that are held at a fixed temperature depending on the input frequencies involved and noise temperature performance required.

Figure 9 shows the noise temperatures that can be achieved by the four categories of amplifier systems. The results shown for frequencies below 15 GHz are based on measured performance on systems that have been built. Above 15 GHz, the data shown are based on a noise temperature analysis using loss distribu-

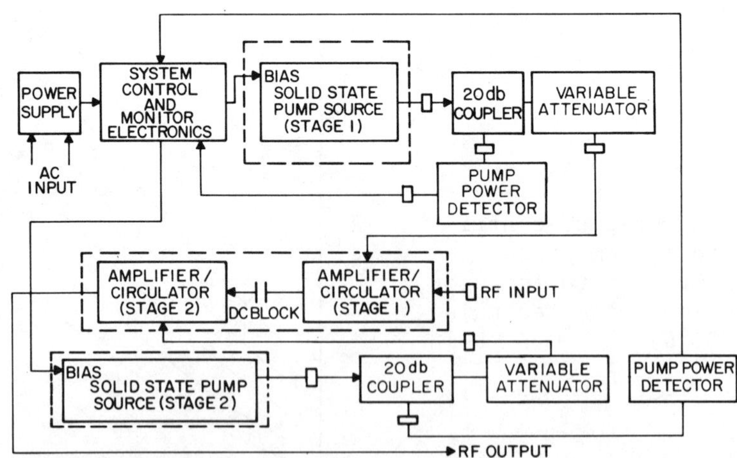

Fig. 8 Simplified block diagram of a two-stage parametric amplifier system.

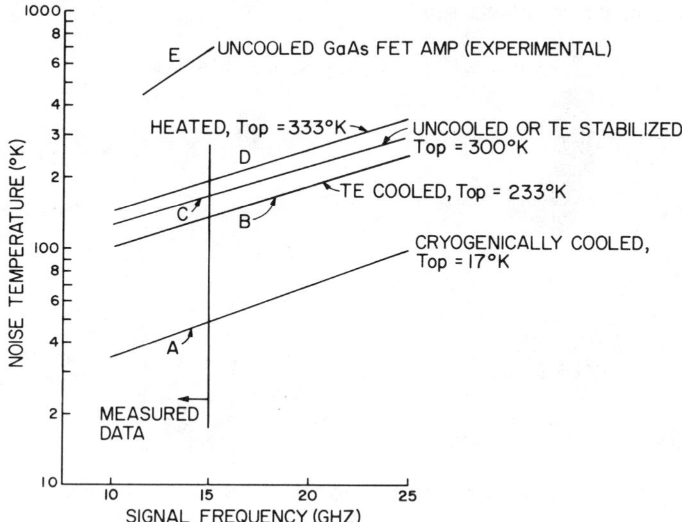

Fig. 9 Noise temperature performance of parametric amplifier systems.

tions that are practical and achievable at these frequencies. The results shown are for broadband (500-MHz minimum amplification bandwidths) amplifier systems operating at 20 - 30-dB gain. The noise temperatures are values readily achievable today using commercially available varactors with cutoff frequencies (at zero bias) ranging from 400 to 600 GHz and pump frequencies in the 40- to 80-GHz range. The following sections provide a description of the aforementioned categories of parametric amplifiers used in satellite communications terminals operating above 10 GHz.

Cryogenically Cooled Parametric Amplifiers

This type of amplifier employs a gaseous helium refrigeration system to lower the physical operating temperature of the amplifier components to 15° - 20°K. The amplifiers operate in a vacuum environment at a stable temperature. Figure 10 shows a complete 15-GHz cryogenically cooled parametric amplifier system including the antenna mounted unit, control units, and refrigerator compressor. The vacuum dewar contains three amplifier stages, each providing 10 dB net gain over a 500-MHz bandwidth.

The main advantage in using a cryogenically cooled paramp is the low noise temperature that can be obtained. Curve A of Fig. 9 shows the achievable system noise temperature for cryogenically cooled parametric amplifier systems between 10 and 25 GHz.

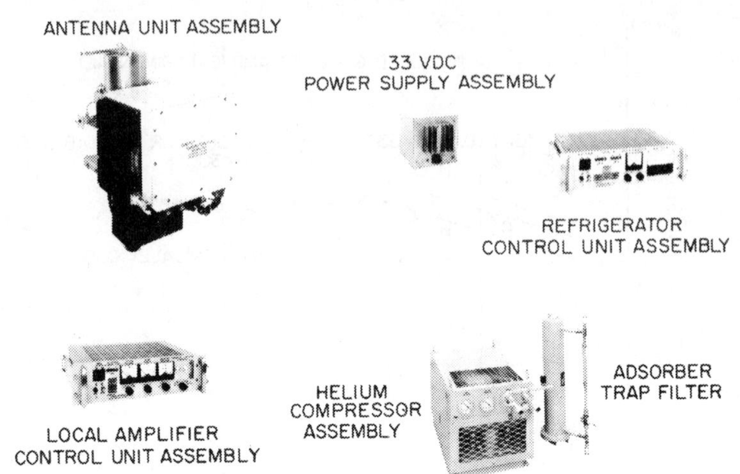

Fig. 10 15-GHz cryogenically cooled parametric amplifier system.

The major consideration in using a cooled paramp is, of course, to obtain a particular G/T required in the overall system. But other factors, such as installation and maintenance, must be considered. Cryogenically cooled parametric amplifiers require periodic maintenance of the refrigerator and compressor. For the CTI Inc. model 350 cryogenic system, which is used widely for parametric amplifier systems, 3000- and 6000-hr maintenance procedures are needed. A vacuum pump also is required as part of the maintenance equipment.

From a design standpoint, high-frequency, cryogenically cooled amplifiers embody special rf design techniques coupled with materials, vacuum, and packaging technologies. Material must be selected keeping in mind the temperature coefficient, specific heat, and mass of both metallic and nonmetallic (dielectric) substances. In addition, materials that outgas cannot be used, since a vacuum (<20 mTorr) must be maintained in the dewar. As examples, aluminum rather than copper or brass is used for amplifier and circulator housings because of its light weight (resulting in a smaller cooldown time for a given mass); stable dielectrics such as quartz are used which have very low temperature coefficients; low-vapor-pressure epoxy resins are used where it is necessary to bond materials together.

The input waveguide transition must be extremely low loss and yet provide a vacuum seal and thermal isolation. At higher frequencies, the tolerances to which components are made become more critical. Wavelengths are decreasing to the point where, in many cases, lengths of transmission line transformers and resonators must be held to within 0.0005 in. and surface flatness of mating parts to within 0.0002 in.

In general, the pump source for a cryogenically cooled parametric amplifier cannot be located physically close to the amplifier stages. This is because the pump circuit components (e.g., isolators, power splitters, attenuators, sampling couplers, and detectors) are external to the vacuum dewar. Therefore, care must be taken with the waveguide runs in the pump circuit to keep the losses as small as possible. Lower-frequency cryogenically cooled paramps generally can use one common pump source for the amplifier stages. But higher-frequency paramps pumped at frequencies in excess of 40 GHz may require a separate pump source for each paramp stage because of the power limitations of solid-state pump sources at these frequencies. As an example of performance of a high-frequency cryogenically cooled paramp, Table 1 gives typical specifications applicable to the 15-GHz amplifier system shown in Fig. 10.

Thermoelectrically Cooled, Thermoelectrically Stabilized, and Heated Parametric Amplifiers

These three types of amplifier systems have many features in common; therefore, they are discussed together. Figure 11 shows a heated system. Note that, in contrast to a cryogenically cooled amplifier, this unit is completely self-contained.

A parametric amplifier must be temperature stabilized, since the varactor diode is sensitive to temperature changes. In order to achieve stable gain over time and varying ambients, temperature control in the order of ±3°C is needed. The type of temperature stabilization used has an impact on the noise temperature as well as the packaging of the system. Size, weight, and power dissipation are factors to be considered when determining a temperature of stabilization. Since equipment used in small ground terminals usually is designed to operate over an environment ranging from 0° to +50°C, the parametric amplifier must be stabilized against these changes in outside ambient.

Thermoelectrically cooled (or TE cooled) amplifier stages operate at temperatures from -50° to 0°C. They are cooled by means of solid-state thermoelectric modules, which are capable of cooling and heating depending on the direction of current flow through the devices. Thermoelectrically stabilized amplifier systems also employ TE devices to maintain the amplifier stages and usually the pump source as well at temper-

Table 1 Typical specifications

Center frequency	15 GHz
1-dB amplification bandwidth	500 MHz
Gain response	30.5 ±0.5-dB system input to output
Gain slope	Less than 0.2 dB/10 MHz
Overall noise temperature	50°K maximum over a 500 MHz bandwidth
Input and output VSWR	Below 1.3:1
Gain stability	±0.5 dB/24 hr;±0.1 dB/1 min
Dynamic range	Less than 0.5-dB gain compression for an input level of -50 dBm
Intermodulation	Third-order intermodulation products at least 30 dB below carrier for two equal -50-dBm signals
Group delay	0.1 nsec/MHz linear; 0.01 nsec/MHz^2 parabolic; 1 nsec ripple, p-p
Pump source	Solid state
Spurious signals	35 dB below output signal level
Cooldown time	~2 hr

Fig. 11 10-GHz heated parametric amplifier system.

atures from 10° to 40°C. Heated amplifier systems utilize heaters to maintain the stages and pump source at a temperature of from 5° to 10°C above the highest expected operating ambient temperature. The latter three types of parametric amplifier systems are more desirable from a maintenance standpoint because they do not require periodic maintenance, are smaller in size, lighter in weight, require less power, and are less costly than cryogenically cooled systems.

The noise temperature performance achievable by thermoelectrically cooled, thermoelectrically stabilized, and heated parametric amplifier systems is shown by curves B, C, and D, respectively, of Fig. 9. Noncryogenically cooled paramps in the 10- to 15-GHz range generally will employ two stages of parametric amplification, and systems operating above 15 GHz generally would employ three stages of amplification. The gains per stage at these frequencies would be between 10 and 13 dB.

For rf gains greater than 30 dB, the additional gain required can be provided by a GaAs field-effect transistor (FET) amplifier. This type is preferable over a tunnel diode amplifier (TDA) because of the high intercept point and intermodulation product performance obtainable. Curve E of Fig. 9 shows the presently achievable noise temperature of the GaAs FET amplifier in the 12- to 15-GHz range. Gains of 20 - 30 dB with intercept points of +5 to +10 dBm can be expected. Higher-frequency GaAs FET amplifiers (up to 18 GHz) are under development by several manufacturers.

The discussion so far has emphasized performance on a system basis. We now consider some design factors affecting the parametric amplifier performance as a subsystem.

General Noise Temperature and Bandwidth Considerations

Parametric amplifiers can be characterized almost completely by their gain, bandwidth, and noise temperature. The consideration that a designer faces is one of meeting a noise temperature requirement consistent with a gain and bandwidth specification. The other operating parameters of a system, such as intermodulation, group delay, spurious outputs, etc., must, of course, be addressed. But these, with perhaps the exception of intermodulation products, are somewhat less a function of frequency than are gain, bandwidth, and noise temperature.

The noise contribution of the diode is used in the two-stage model shown in Fig. 12 to determine T_e, the overall parametric amplifier noise temperature referred to the input. Losses or gains shown above each component of the model refer to that loss or gain at the physical temperature shown below each component.

Once an overall system gain has been established, the noise temperature performance of a stage can be computed. This noise temperature will be a function of the signal and idler frequencies, the varactor diode characteristics, and the

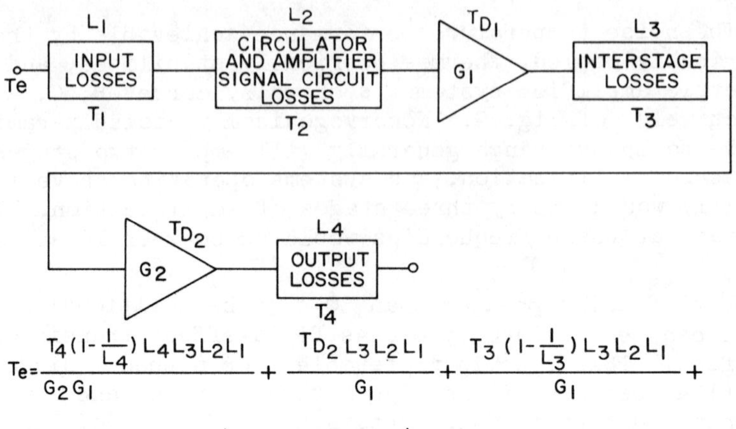

$$T_e = \frac{T_4(1-\frac{1}{L_4})L_4L_3L_2L_1}{G_2G_1} + \frac{T_{D_2}L_3L_2L_1}{G_1} + \frac{T_3(1-\frac{1}{L_3})L_3L_2L_1}{G_1} +$$

$$T_{D_1}L_2L_1 + T_2(1-\frac{1}{L_2})L_2L_1 + T_1(1-\frac{1}{L_1})L_1$$

NOTE: THE L'S ARE IN RATIO >1

Fig. 12 Model for determining T_e for a two-stage parametric amplifier system.

circulator and other losses in the stage. The noise contribution of the varactor diode in a finite-gain, nondegenerate one-port parametric amplifier is given by [5]

$$T_d = T_o \left\{ \frac{4S}{|1 - (M^2/f_s f_i) - S|^2} \left[1 + \left(\frac{M}{f_i} \right)^2 \right] \right\} \quad (7)$$

where T_d is the diode noise contribution in degrees Kelvin, T_o is the diode physical ambient temperature in degrees Kelvin, and

$$S = \frac{R_g}{R_d} = \left| 1 - \frac{M^2}{f_s f_i} \right| \cdot \frac{|\Gamma_o| + 1}{|\Gamma_o| - 1} \quad (8)$$

where R_g is the driving point impedance required at the varactor diode, R_d is the unpumped diode resistance, Γ_o is the stage reflection coefficient corresponding to the stage gain at the midband frequency, f_s and f_i are the signal and idler frequencies, respectively, and M is the diode figure of merit equal to

$$(C_1/C_o)f_{c_o}$$

where f_{c_o} is the diode cutoff frequency at zero bias, and C_1/C_o is the diode pumped nonlinearity factor. The value of T_d just found is used in the model of Fig. 12 to determine the overall parametric amplifier noise temperature referred to the input.

At the higher idler frequencies, it becomes more economical to use a better-quality varactor diode (i.e., higher cutoff frequency) rather than substantial thermoelectric cooling to achieve lower noise temperatures. Figure 13 shows the noise temperature improvement for a 12-GHz amplifier system as a function of the varactor diode cutoff frequency. The cost of the improved varactors that are available today (up to 1000 GHz) would be less than the costs involved in lowering the amplifier operating temperature.

Bandwidth

The bandwidth achievable from a parametric amplifier, whether cooled or uncooled, is a function of the varactor diode parameters, the idler frequency, and the broadbanding circuit used. In the 10- to 15-GHz frequency range, bandwidths of 500 MHz at the 1-dB points are achieved readily in amplifier systems of 30-dB gain using two or three stages, each with gains of 10 to 15 dB. This performance can be obtained using commercially available varactor diodes with cutoff frequencies of 400 GHz at zero bias and idler frequencies in the 30- to 40-GHz range.

At these as well as higher frequencies, greater bandwidths can be achieved. Figure 14 shows the gain bandwidth product plotted vs signal frequency for several different idler frequencies. The curves shown are the gain bandwidth product at the 3-dB points. The 3-dB gain bandwidth product is shown to facilitate the comparison between amplifiers operating at different signal and idler frequencies without reference to specific means of broadbanding these stages to obtain the bandwidth at the 1-dB points. It usually is possible by means of

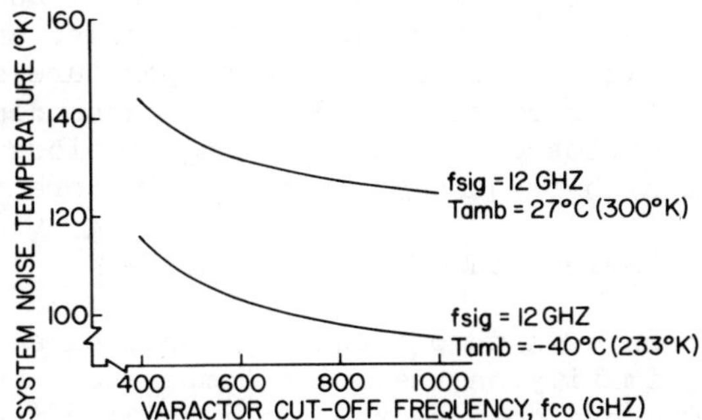

Fig. 13 System noise temperature vs varactor cutoff frequency for a 12- GHz parametric amplifier.

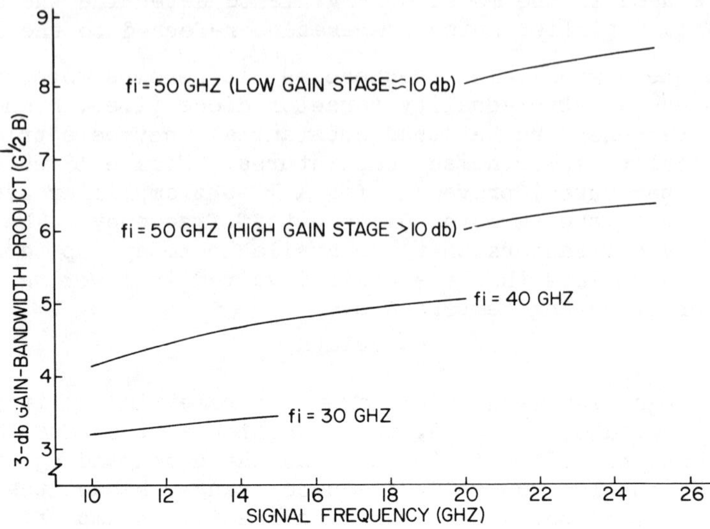

Fig. 14 Gain bandwidth product vs frequency.

broadbanding circuitry to achieve almost the same bandwidth at the 1-dB points as at the 3-dB points. The curves shown for idler frequencies of 30 and 40 GHz are for amplifiers stages operating at high gain: approximately 15 dB/stage.

Two curves are shown for the idler frequency of 50 GHz. One is for a high gain stage and the other for a low gain stage operating at no more than 10 dB gain. It can be seen, for example, that, at 20 GHz for a low gain stage, a 3-dB bandwidth of approximately 2.5 GHz is achievable. With suitable broadbanding, a 1-dB bandwidth in the order of 2.5 GHz also is achievable. As a practical matter, bandwidths of amplifiers operating in the 20-GHz range are limited to about 10% because achieving the theoretical limits would be accompanied by a degradation in noise temperature at the band edges and an increase in pump power required. Noise temperature degradation occurs because of the complexity of the broadbanding circuit and its associated losses. The following sections discuss the pump source for high-frequency amplifiers in more detail.

Selection of Pump Frequency

The selection of a pump frequency for a parametric amplifier is based primarily on the noise temperature performance to be achieved. One can compute the optimum idler frequency (and hence pump frequency) for a given signal frequency. For most high-frequency paramps, where $f_s/f_i<0.5$ and where a varactor

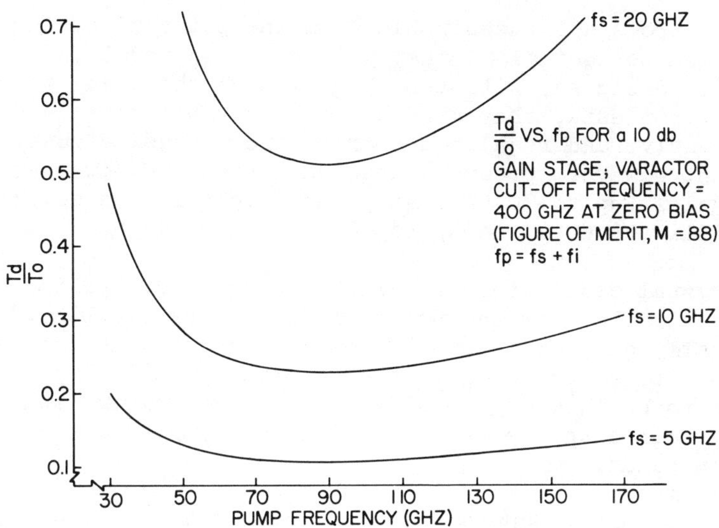

Fig. 15 Varactor diode noise contribution vs pump frequency.

with a cutoff frequency in excess of 400 GHz is used, the optimum idler frequency is fairly broad. Figure 15 shows the noise temperature contributed by the varactor, imbedded in the idler circuit, normalized to T_o (the actual operating temperature of the diode) as a function of the pump frequency f_p, for a 10-dB gain stage operating at 5, 10, and 20 GHz.

One would want to operate at the optimum pump frequency, but in many cases the ability to generate the required power at this frequency becomes the limiting factor. In such instances, a lower frequency can be selected with some degradation in T_e/T_o. The sacrifice in performance will be smaller for lower signal frequencies.

Pump Source Considerations

Up until the late 1960's, virtually all parametric amplifiers used klystrons as their pump sources. A few laboratory amplifiers did use crystal-controlled fundamental oscillators driving high-order multiplier chains, but these generally were pumped at frequencies no higher than Ku band. Long-life klystrons (>2000 hr) were available to 40 GHz. However, both the klystron and the high-order multiplier chain are undesirable today for pump sources that must be reliable, be free from field maintenance or adjustment, generate minimum heat, be low in cost, and have inherent temperature stability or lend themselves readily to temperature stabilization.

Klystrons are undesirable from the point of having to replace them periodically, they will not in general (unless specially designed) withstand high-impact shock and other severe environments, they generate considerable heat, and they are expensive and require a complex and expensive power supply. They can, however, produce large amounts of rf power at high frequencies (up to 70 GHz) at power levels in the several hundred milliwatt area and up to 50 MW at 200 GHz.

Crystal oscillators driving high-order multiplier chains are undesirable from the point of view of reliability. (They are usually quite complex devices.) Their output frequency is difficult to adjust even in the laboratory, and they are expensive to build and generally require temperature stabilization. Their output power is limited at pump frequencies above X and Ku bands.

Today, using Gunn and Impatt devices in oscillators, sources of microwave power are available which are small, lightweight, relatively inexpensive, reliable, efficient, rugged, stable, and require only a simple power supply. There are two techniques for constructing pump sources using Gunn devices or Impatt diodes for use up to 100 GHz. One is to use an X-band or Ku-band oscillator and drive a single-stage doubler, tripler, or quadrupler. The second is to use a fundamental oscillator at the desired pump frequency.

The choice between the multiplier approach and the fundamental oscillator approach is based on the required pump power. A paramp in the 10- to 20-GHz range requires from 10 to 70 MW/

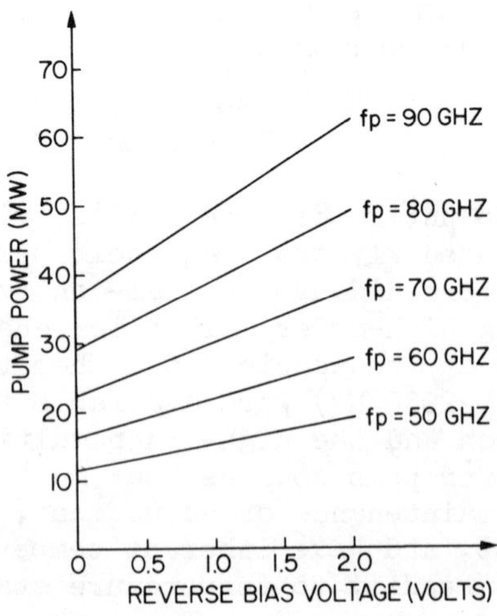

Fig. 16 Pump power vs reverse bias voltage for a varactor with f_{c_o} = 400 GHz, C_{j_o} = 0.25 pF, fully pumped.

stage at a gain level of 10 dB/stage, depending on pump frequency and bias level. Figure 16 shows pump power vs reverse bias voltage for the varactor diode for different pump frequencies. Note that the varactor cutoff frequency is constant at 400 GHz. As the cutoff frequency is increased, the pump power required will decrease proportionally.

The results shown in Fig. 16 are ideal in the sense that losses in the pump circuit are not taken into account. In addition, the amplification bandwidth may dictate increased pump power because of broadbanding circuit considerations. For example, based on experiment, a 12-dB uncooled amplifier stage in the 11.7-12.2-GHz band (500-MHz instantaneous bandwidth) pumped at 54 GHz requires 31 MW budgeted, as shown in Table 2.

The varactor at 0.5V reverse bias requires 16 MW. Therefore, the source must be capable of delivering at least 31 MW. A cryogenically cooled stage typically requires about 1.5 to 2 times the pump power of a noncryogenically cooled stage because of the additional waveguide and pump component losses encountered in such systems.

Paramps in the 10 - 25-GHz range used for communications systems applications generally are pumped in the 40 - 80-GHz range, and the fundamental or multiplier approach is based on generating the required power from available devices. Figure 17 shows the output power currently available from Gunn and Impatt oscillators vs frequency at an operating temperature of 40°C. These results are from actual performance measurements below 60 GHz and device data above 60 GHz. It should be noted that multiple Gunn and Impatt devices can be combined in a common oscillator structure for greater output power. This usually is done for driving multipliers.

Table 2 Example based on experiment

Pump waveguide loss (6 in. of WR-15), dB	0.26
Pump attenuator loss, dB	0.30
Mismatch loss (assumed worst case of 2:1), dB	0.40
Varactor mount loss, dB	0.20
Pump isolator loss, dB	0.40
Pump sampling coupler, dB	0.30
Margin, dB	1.00
Total, dB	2.86
	(1.93 ratio)

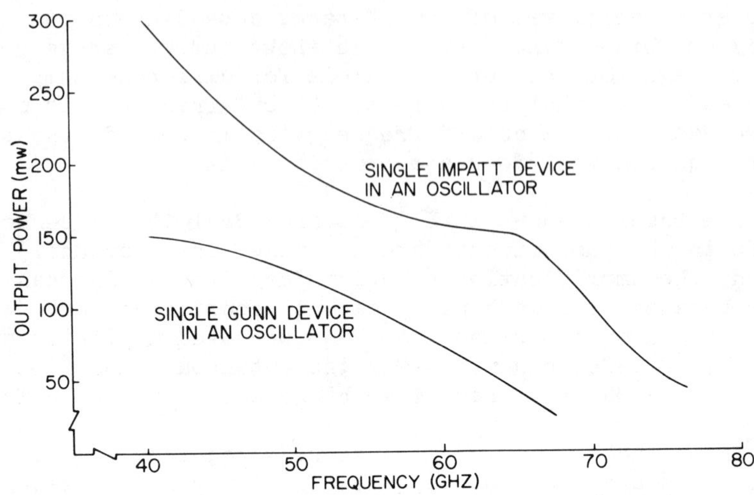

Fig. 17 Pump source output power vs frequency.

A parametric amplifier requires an extremely stable source of pump power. Typically, the gain of an amplifier stage operating at 10 dB will vary (in decibels) by about 10 times the change in pump power (in decibels). The factors that influence pump source stability are primarily temperature and bias voltage. Solid-state pump sources used for high-frequency amplifiers must be temperature stabilized (usually by TE devices) and be biased from a well-regulated power supply. A pump source (using a Gunn device) operating between 40 and 80 GHz should have (maximum) a frequency coefficient of bias voltage of 0.05 MHz/mV, a power coefficient of bias voltage of 0.002 dB/mV, a frequency coefficient of temperature of 2 MHz/°C, and a power coefficient of temperature of 0.03 dB/°C. These characteristics would be suitable for and result in acceptable gain variations in amplifier systems used for communications systems.

Summary

We have presented the primary performance characteristics, viz., gain, bandwidth, and noise temperature, that are associated with parametric amplifiers in the 10 - 25-GHz range. The use of a particular type of amplifier system is predicated by overall system G/T requirements. The parametric amplifier must be viewed as a system in itself with several interrelated subsystems, the most important of which are the amplifier stages themselves, the thermal arrangement, and the pump source. The characteristics of these items have been discussed, with emphasis on noise temperature performance.

Acknowledgment

The author wishes to thank D. Campbell and K. Everson for their significant contributions to this paper.

References

[1]Ruze, J., "Antenna Tolerance Theory - A Review," Proceedings of the Institute of Electrical and Electronics Engineers, Vol. 54, April 1966, pp. 633-640.

[2]Rice, P. L., Longley, A. G., Norton, K. A., and Barsis, A. P., "Transmission Loss Predictions for Tropospheric Communications Circuits," National Bureau of Standards, TN 101, Vol. 1, May 1965.

[3]Medhurst, R. G., "Rainfall Attenuation of Centimeter Waves: Comparison of Theory and Measurements," Institute of Electrical and Electronics Engineers Transactions, Antennas and Propagation, Vol. 13, No. 4, July 1965, pp. 550-564.

[4]Hogg, D. C. and Mumford, W. W., "The Effective Noise Temperature of the Sky," Microwave Journal, Vol. 3, No. 3, March 1960, pp. 80-84.

[5]Kliphuis, J., Neuf, D., and Albrecht, B., "Universal Performance Curves Speed Parametric Amplifier Design," Microwaves, Vol. 5, No. 7, July 1966.

HIGH-POWER TUBE AMPLIFIERS FOR SATELLITE COMMUNICATIONS ABOVE 10 GHz

R. Strauss*
COMSAT Laboratories, Clarksburg, Md.

Abstract

During the past decade, a variety of high-power cw communications amplifiers have been developed for uplink satellite communications service at frequencies below 10 GHz. Now, as the requirements at higher frequencies (14.0-14.5 GHz and 27.5-31.0 GHz) are coming into focus, the availability and capability of high-power amplifier tubes at or near these frequencies are of interest. This paper explores the advantages or limitations of the leading device types (traveling wave tubes and klystrons) currently capable of development. Solid-state devices are not being considered, since electron tubes still constitute the only practical source of power amplification (14 or 30 GHz) at output levels of the order of 1 W or more.

Introduction

Electron device technology is now a mature and established technology. Although innovations and new materials continue to be introduced, the basic physical electronic principles utilized in these high-frequency, high-power, high-gain amplifier devices all were established well over a decade ago. The line includes an impressive number of mature technologies[1,2]:

1) Converging beam-forming techniques using long-life dispenser cathodes (operating temperature ~ 1000°C) with current densities of 1 to 2 A/cm^2 for all but the lowest power levels, where oxide cathodes are used.

2) Beam-focusing techniques utilizing a periodic permanent magnet (PPM), a permanent magnet (PM), or the solenoid focusing method for the highest power levels.

Presented as Paper 76-300 at the AIAA/CASI 6th Communications Satellite Systems Conference, April 5-8, 1976, Montreal, Canada.

*Director Reliability and Quality Assurance.

3) Beam-collection techniques recently including shaped and depressed multistage collectors for higher efficiency.

4) Narrow-band "cavity" rf interaction with fixed or variable tuning within the desired band, as in klystron amplifiers.

5) Broadband slow wave rf interaction with helices in lower-power (~200-W cw or less at 14 GHz) traveling wave tubes (TWT's) or coupled-cavity/interdigital types of higher-power handling circuits in higher-power TWT amplifiers.

6) Ancillary techniques associated with maintaining a high vacuum for long periods of time in a high-power-dissipation-density or high-voltage environment (e.g., attached ion pumps, metal/ceramic technology for rf output windows, high-voltage standoff seals, and good thermal conductivity depressed collector isolation).

7) Materials techniques associated with achieving controlled frequency-selective or well-matched distributed rf loss inside the vacuum tube (e.g., pyrolytic carbon and hot sprayed kanthal), low-loss high-conductivity insulators (boron nitride), and stable metablized BeO (berylia) severs, or SiC (silicon carbide) wedges, all contributing to significantly improved higher-power tube characteristics.

8) Finally, new super-high-energy product permanent magnet materials (samarium cobalt and other rare Earth compounds),

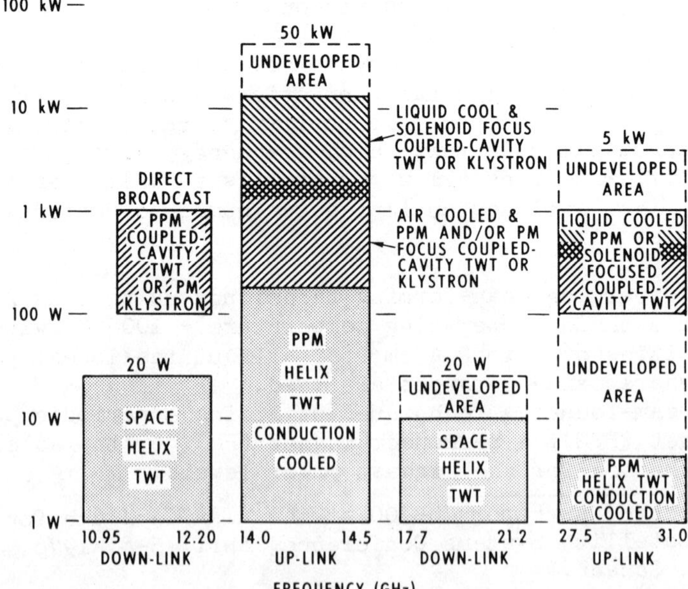

Fig. 1 Power tube amplifier type of designs at 11-31 GHz for satellite communications.

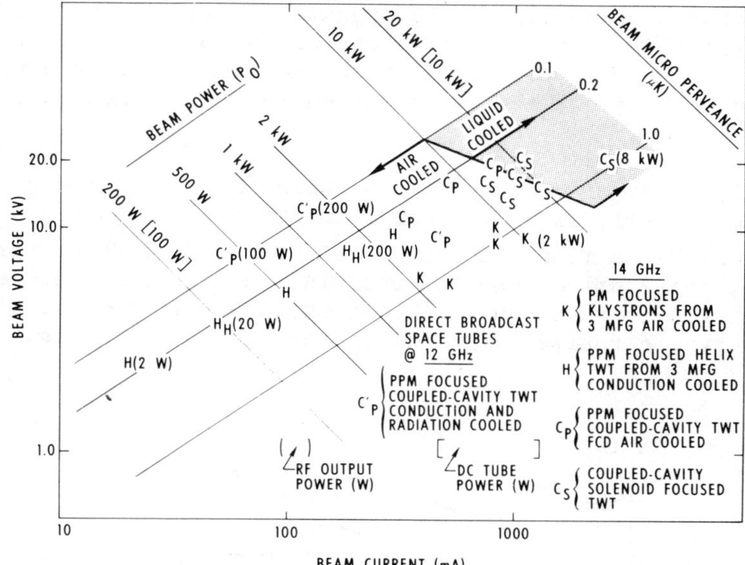

Fig. 2 Comparison of beam power and beam microperveance of TWT's and klystrons at 14 GHz.

making possible levels of small beam control not previously attainable.

This arsenal of technology has been applied extensively to the 6-GHz uplink frequency in each of its basic forms, i.e., klystron amplifiers to 20-kW rf output, 600-W helix TWT's, and 12-kW coupled-cavity TWT's. For conservative designs, lifetimes of the order of 20,000 hr are commonplace, although occasional quality control-problems mar such fine records. Earth terminals utilize redundancy-switchover features in their tube HPA installations to preserve high availability (greater than 0.999).

Helix Low-Power Amplifiers at 14 GHz

The helix-type TWT is a broadband device (typically an octave), and the relatively narrow band of 14.0 to 14.5 GHz presents an opportunity to exploit other features fully. As shown in Fig. 1, which includes other high-power devices for comparison, devices ranging from 2- to 20-W to 100- to 200-W saturated single-carrier output are available commercially. Periodic permanent magnet focusing is used exclusively in conjunction with conduction cooling to a baseplate. As indicated in Fig. 2, these tubes utilize beam perveances of about 0.2×10^{-6} at 14 GHz. Usually single-stage depressed collectors are available and operated in the vicinity of 50% helix/cathode voltage. Efficiencies in excess of 25% are obtained. Double collector tubes, now commercially available at the 20-W level, achieve efficiencies in the neighborhood of 40%.

In terms of communications characteristics, the gain slopes at full drive or backoff are typically less than 0.01 dB/MHz maximum with saturated gains of 50 dB or more. AM/PM, phase drive, and intermodulation values are comparable directly with values attained at the lower frequencies for helix-type circuits. Typical values are shown in Fig. 3.

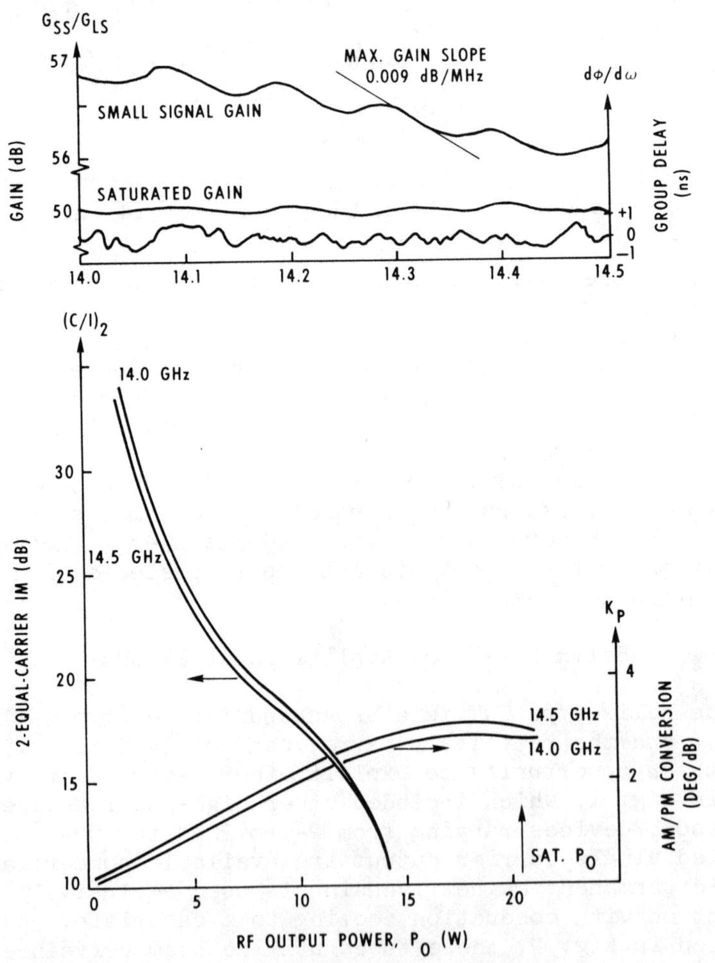

Fig. 3 Typical data, 14-GHz helix TWT.

As might be expected, group-delay values with broadband helix circuits are excellent, i.e., less than 1 nsec peak to peak for the various components over a typical bandwidth.

At the 2- to 20-W level, many excellent helix TWT's have been perfected, often as offshoots from space programs.[3] These devices have demonstrated lifetimes in excess of 20,000 hrs. At the 100- to 200-W level, less substantial information is available, particularly regarding life performance. The transition from helix-type structures to structures that can accommodate higher power through better inherent thermal conductivity (e.g., coupled-cavity slow wave structures) occurs at about these power levels. At 6 GHz, for comparison, the practical transition region appears to be between 400-600 W for operational devices, although 1-kW feasibility TWT's have been tested. If the theoretical scaling law

$$Pf^{8/3} = \text{const}$$

where

P = saturated rf power output, W
f = midband frequency, GHz

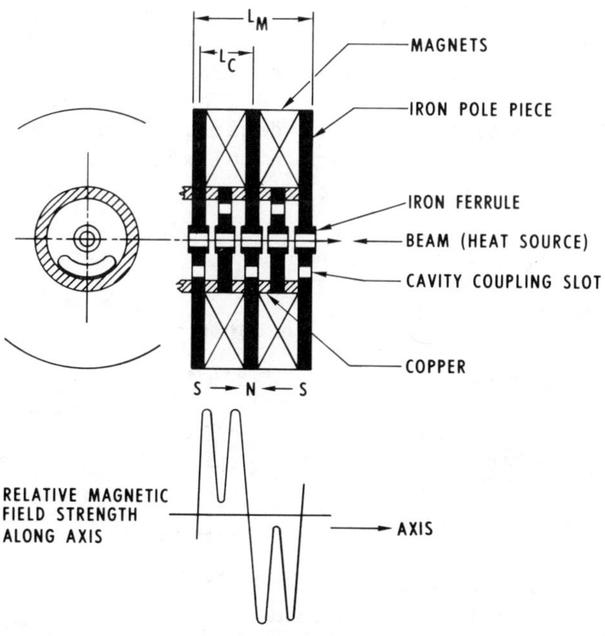

Fig. 4 Double-period permanent magnet (DPPM) focusing integral with the coupled-cavity slow wave structure.

for TWT's holds, then the comparative 6-GHz helix TWT capability would be between 1 and 2 kW, provided, of course, that the same design features are used.

From an applications standpoint, these helix tubes easily achieve sufficient gains for solid-state driving circuits. Typical weights range from 2-3 lb for the 2- to 20-W tubes and from 6-8 lb for the 100- to 200-W devices. Inputs are usually in coaxial format and outputs in waveguide format. Cooling is by conduction to a baseplate that must be maintained below a maximum temperature (typically 100°C maximum). Prime dc input power, as shown in Fig. 2, is approximately 50% of the beam power. Power supplies require tight helix voltage regulation and filtering due to the pushing factors (approximately 0.01 dB/V and 1 deg/V).

Helix Power Amplifiers at 30 GHz

Using the $Pf^{8/3}$ = const relationship and extrapolating to 30 GHz indicates helix structure TWT's in the range of 3-30 W. Tubes in the 1-W range have been available for several years. Recently, a 5-W (>40-dB-gain) design with an efficiency of 22% has been reported.[4] This design utilizes virtually all of the options available for a workable design: an integral pole-piece shrunk barrel, anisotropic boron nitride shaped support rods, a copper-plated molybdenum circuit, samarium cobalt periodic magnets, and a 45:1 beam compression ratio that yields a commendable 97% beam transmission with rf drive. To appreciate this TWT design, it should be noted that the circuit is 7 in. long with a helix tunnel of 0.025 in.

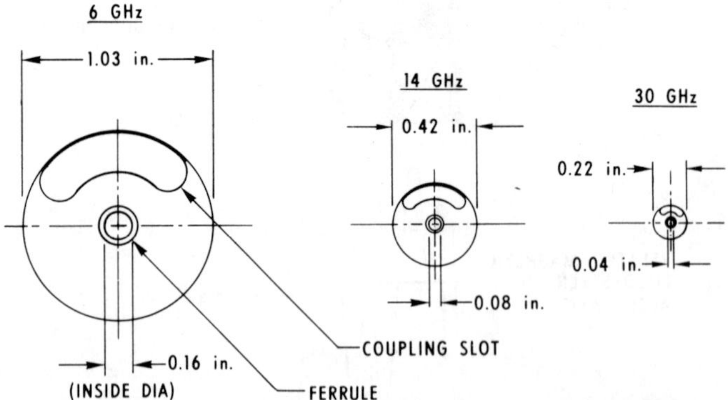

Fig. 5 Examples of coupled-cavity dimensions with each operating frequency.

The helix voltage is 5.5 kV, the cathode current is 16 mA, and the collector voltage depression is 85%. Life data are not available yet. However, the chosen cathode current density of 320 mA/cm^2 is somewhat high for long-life oxide cathodes.

As in the transition from 6 to 14 GHz, a similar evolution may be expected from 14 to 30 GHz. Satellite helix tubes will set the design pattern, and less expensive spinoff designs for Earth terminals will appear. During the next decade, it is projected that satellite helix tube work will be advanced at 20 GHz (satellite downlink) and at 60 GHz (intersatellite links).

Coupled-Cavity Medium-Power Amplifiers at 14 GHz

Like its lower-power helix counterpart, the coupled-cavity slow wave structure is capable of broadband performance (10 to 20%) within a wide range of power levels. At 14 GHz, the communications bandwidth for the uplink is only 3.5%. Medium power in this case represents a range from 100 to 2000 W, where periodic permanent magnet (PPM) focusing combined with conduction/forced-air cooling can provide an adequate design operating basis. The beam-forming, focusing, and collection principles are identical to those of helix TWT's, with the exception of the constraints imposed by the coupled-cavity structure. For instance, at this frequency a double-period permanent magnet (DPPM) focusing design is desirable for mating with the circuit. As shown in Fig. 4, the magnets are an integral part of the cavity structure in which the cavity walls serve as both field pole pieces and thermal conductors. Therefore, the magnet periodicity (L_M) frequently but not always is constrained by the coupled-cavity periodicity (L_c).

The operating frequency is the primary determining factor for the cavity dimensions. Figure 5 shows some typical cross-sectional dimensions at the three frequencies of interest. The extremely small "watchmaker" dimensions and tolerances at 14 and 30 GHz must be well controlled to minimize reflections (gain slope and group-delay distortions). The bandwidth (hot and cold) is determined primarily by the coupling hole angle, which is typically near 90° for communications tubes.

Figure 6 illustrates the relationship between hot and cold bandwidths for narrow-band and broadband cases operating at the same beam voltage (electron velocity) but different frequencies. The cavity diameter is adjusted to yield a midband phase angle of BL = 1.4π. In practice, a compromise between a low-gain-per-section (low-impedance), low-distortion, broadband design and a high-gain-per-section, greater distortion, narrower band design is attained. Variations on the klystron theme have been used in which each major section (typically 15- to 20-dB gain per major section with matching internal attenuators at the terminating sever) is centered at slightly different frequencies to maximize unit gain and bandwidth in a several-section tube. A typical 1-kW tube of this class may have a saturated gain of 45 dB (three sections), weigh 20 lb, and achieve an efficiency better than 25% with a single depressed collector. Reference 5 describes a similar 1.2-kW tube at X band with several unique forced-air-cooling features. Heat pipes are used to increase the axial thermal conductance, and the "contaminated" air flow is separated from the high-voltage insulation.

Instabilities

Unlike the helix structure, the coupled-cavity structure is subject to a greater number of possible instabilities. There are three potential sources: regenerative oscillations in the operating range, oscillations at the cutoff frequencies of the fundamental TM_{01} operating mode, and oscillations from higher-order modes. The first potential regeneration in-band

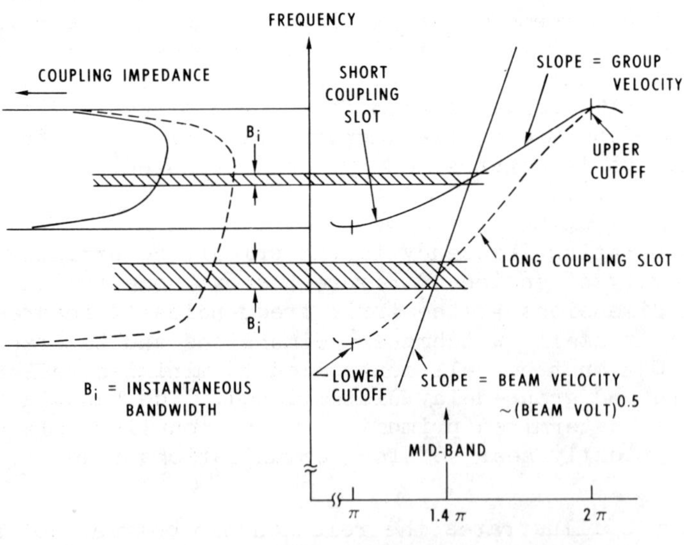

Fig. 6 Circuit characteristics of the coupled-cavity slow wave structure.

oscillations are caused by internal or external mismatches, which combine with a forward in-band gain component. This type of instability can be contained by limiting the section gain and using well-matched sever terminations, including some distributed loss. Indeed, all of the techniques for minimizing mismatches within a section or between cavities are also important in terms of reducing gain slope or delay distortion. Typical small signal gain values are 0.03-0.06 dB/MHz maximum and 2 nsec peak to peak for the group-delay components.

Oscillations at the upper cutoff frequency are more difficult to contain. Drive- or overdrive-induced oscillations

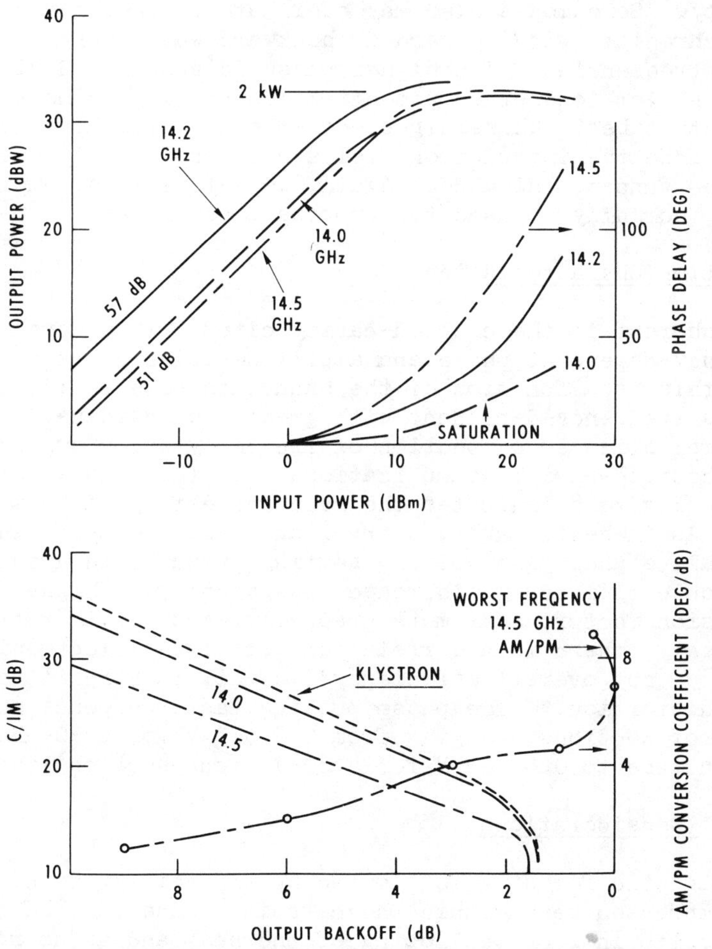

Fig. 7 Band-edge and midband coupled-cavity tube characteristics (10% velocity taper).

may occur when the average beam velocity is reduced and then becomes synchronized at the high coupling impedance cutoff frequency condition (see Fig. 6). For broadband communications tubes, tuned loss resonators or decoupling circuitry may be used to stop or shift these oscillations. The lower cutoff frequency oscillation is a less critical problem, since higher velocity election interaction is more unlikely. However, distributed loss with frequency-selective characteristics normally is used in the vicinity of the coupling slots.

Higher-order mode oscillations are also possible. Since the coupled-cavity structure is a filter-type circuit, higher-order modes may be propagated at the same frequency and group velocity. Some modes also may oscillate if the beam velocity is synchronized with forward or backward wave energy at the cutoff frequencies. The higher passband associated with the coupling slot is another potential instability in moderate-bandwidth tubes. These higher-order mode instabilities are weak, since the interaction fields are considerably smaller than the fundamental mode. Again, selective loss, empirically chosen, normally is used to suppress these modes.

Distortion Characteristics

Inherent in the coupled-cavity circuit is a more marked frequency-dependent phase and amplitude drive characteristic. Again this is a function of the bandwidth selected; i.e., it is less frequency-dependent with greater bandwidths. Typical characteristics as a function of frequency are shown in Fig. 7 for a coupled-cavity communications tube that has a velocity taper. Figure 8 indicates the relative effect of beam voltage. As in helix tubes, overvoltage (voltage above the maximum small signal gain voltage setting) results in greater electronic efficiency, increased phase delay, a higher AM/PM conversion factor, and, more gradually, reduced IM ratios. Conversely, there is a correlation with lower electronic efficiency (not overall efficiency), a uniform circuit, and lower distortion.[6] The phase pushing factor depends on the number of sections, ranging from 0.2 deg/V for a 30-dB two-section tube to 0.6 deg/V for a 60-dB four-section tube.

Thermal Considerations

Magnetic focusing is limited by the ability to contain the compressed beam within the circuit "tunnel." PPM focusing results in some scalloping of the beam and a "no rf drive" level of interception (about 3 to 5%) of the beam current. With rf drive, the interception increases (typically by a factor of 2) and, together with associated rf losses, results in increased ferrule tip temperatures, particularly at the output circuit sections (maximum dissipation density zone). These temperatures must be controlled and limited by the design to achieve reliable performance. Promising test results have been obtained for 2-kW PPM rf tubes realized by using the highest energy product samarium cobalt for magnet material and by providing low-thermal-conductivity paths from the ferrule heat source to the thermal sink while thermally isolating the temperature-sensitive samarium cobalt. These tubes utilize forced-air-cooled collectors but liquid-cooled "bodies" (approximately 0.5 gallon per minute at ground potential).

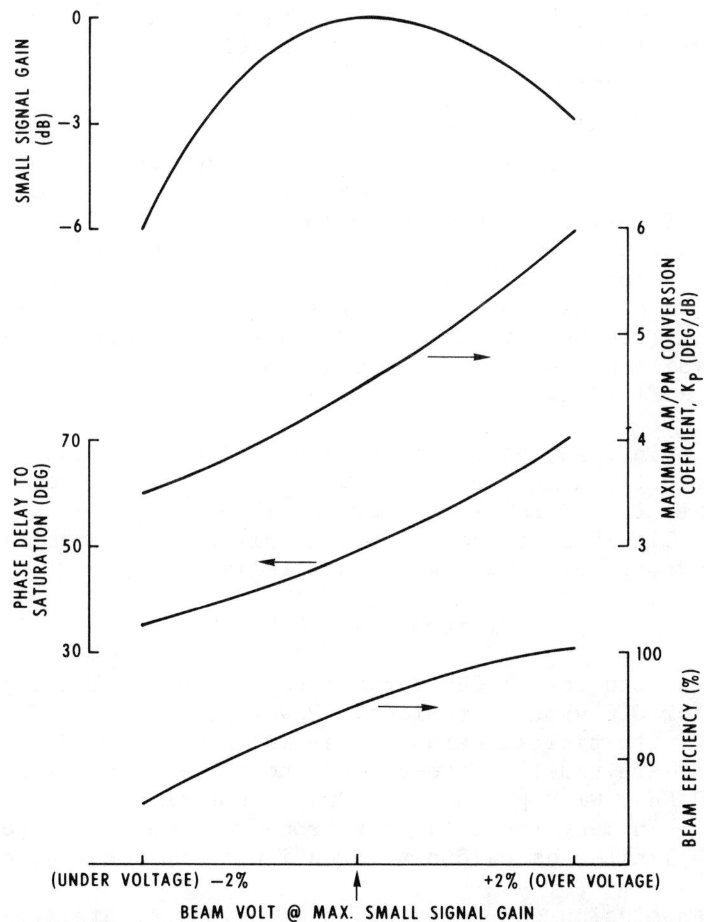

Fig. 8 Effects of beam voltage for a coupled-cavity tube with a 10% velocity taper.

An alternative approach is the use of a solenoid for focusing. Beam interception is reduced by at least a factor of 2. The resultant reduced dissipation (approximately 500 W less for a 2-kW tube) again permits less efficient forced air cooling for the whole tube.[7] Therefore, most tubes at the 2-kW level are solenoid focused and air cooled, reflecting the Earth station designers' preference for air-cooled tubes, particularly in unattended station applications. Thus, at 14 GHz, the transition from forced air cooling to liquid or vapor cooling is in the vicinity of 2.0-kW rf output power.

Coupled-Cavity High-Power Amplifiers at 14 GHz

At present, only a few designs are available at power levels between 3 and 8 kW. The prime application for such high-power tubes appears to be the multicarrier, high-capacity stations, possibly for operation with the INTELSAT V satellites.

These designs all use large solenoids for minimum body interception (typically less than 1% of the cathode current without rf drive). A combination of liquid and forced air cooling normally is used. The body, solenoid, and collector are liquid cooled, whereas the cathode is forced air cooled. The design of the coupled-cavity structures is unchanged, no different from that of their PPM counterparts except that the cavity walls are all copper and no longer require pole pieces made of soft iron.

In principle, power levels above 10 kW also could be obtained by utilizing the same design concepts.[8] These would be superpower tubes similar to earlier 100-kW tubes at C band. There is little likelihood that an uplink requirement for such devices and power levels will materialize, however.

Klystrons at 14 GHz

Klystrons for 14-GHz service have been developed primarily for CTS ground stations. Power levels range from 500 W to 2 kW, with typical gains in the vicinity of 40 dB. These klystrons are usually five-cavity tubes (1-dB bandwidth at 2 kW ~80 MHz) with permanent magnet focusing and forced air cooling. In many respects, klystrons offer a less expensive but more limited bandwidth alternative to uplink transmission.

Transmission characteristics of klystrons are acceptable for FM[9] or TDMA transmission. Once the klystron cavities are set correctly, the klystron distortion characteristics are typically slightly better and smoother (function of frequency) than those of comparable gain/power TWT's within approximately the mid-75% of the klystron 1-dB gain bandwidth (Fig. 9). The advantage is due primarily to the rf isolation of the cavities (no internal rf reflections) and the comparatively shorter electrical length of the klystron (one-third to one-fourth that of the TWT's). Other advantages resulting from these features, namely, less sensitive pushing factors, typically 0.005 dB of power (over perhaps a 6-dB range) per beam volt and 0.2° of phase shift per beam volt, simplify the power supply design.

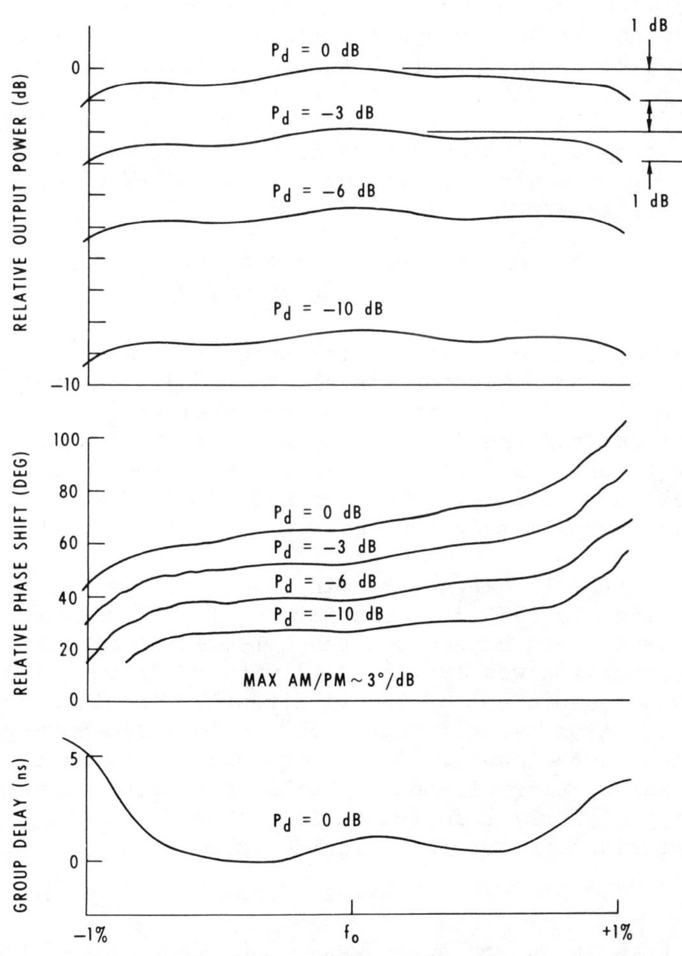

Fig. 9 Comparative data for a six-cavity klystron (see Fig. 7 for klystron IM).

The cavity temperature stability, which is of the order of 200 kHz/°C, may be a disadvantage for some applications. However, bimetallic cavity temperature compensation is possible so that this factor may be reduced to within 10 kHz/°C. The major disadvantage of the klystron is undoubtedly the basic narrow-band characteristic, although both broadbanding and tuning techniques have been developed.[10] Broadbanding by increased beam perveance and the addition, for instance, of a sixth filter cavity in the output which loads the output cavity are well-known techniques. Gang tuning or single "bellows" tuning, although possible, represents added mechanical complexity and a reliability disadvantage, which, for some applications, is undesirable. The effect of long-term, multicycle cavity tuning on the tube distortion characteristics is not known, nor particularly amenable to analysis. Therefore, klystrons parallel the power levels of coupled-cavity traveling wave tubes and in some select applications present a viable alternative to the TWT. This is evident in Europe, where a klystron design is competing with a coupled-cavity TWT for a high-efficiency (>50%), 500-W direct broadcast satellite application.[11]

Coupled-Cavity TWT Power Amplifiers at 30 GHz

The major impetus for uplink transmitters at 30 GHz is the Japanese Communications Satellite (JCS). Several related coupled-cavity TWT designs are being developed. The output power ranges from the 300- to 800-W level, with gains of the order of 40 dB. The latest designs use solenoid focusing and forced air cooling. The combined tube and solenoid weight is approximately 400 lb.

An unusual 1.2-kW/44-dB-gain tube developed in Europe was described in 1973.[12] Samarium cobalt PPM was used for a 25-kV/400-mA beam focused through a 1-mm-diam circuit tunnel. Body interception was typically 8% with rf drive. The PPM system was independent of the rf circuit. To minimize thermal electron velocity spreading, a low-beam-covergence and high-cathode-loading (3.6-A/cm^2) design was used. The cathode was a reservoir-metal capillary osmium-coated dispenser type. Early life indications from field tests appear to be satisfactory for this liquid-cooled tube.

Future Trends

At both 14 and 30 GHz, great power consumption inefficiency is evident. The first cause of inefficiency is the concept of active output backoff. HPA's are expected to incorporate active attenuator drive controls so that during heavy precipitation full uplink power (near saturation) is available. However, during perhaps 99.6% of the time, the 14-GHz output power amplifier is operated at levels 4-6 dB below saturation. At 30 GHz, comparative backoffs of about 10 dB are indicated. Since rain attenuation is not a fast-acting phenomenon (i.e., it has minimum time constants of the stop-order of tens of seconds), several power-efficient remedial steps are indicated. In TWT's, for example, much greater collector depression is possible, even to 10% of the body voltage during the 99% backoff operation. Both TWT's and klystrons can be backed off with their modulating anodes. Klystrons still can be operated quite efficiently at reduced beam voltage. For minute time constants, 3-dB power combining can be applied during power call-up by switching on the redundant standby tube.

The second more specialized type of power inefficiency is associated with single-carrier TDMA systems. The uplink burst from a given transmitter seldom is activated for more than 10% of the frame time. At present, the tube is always hot (beam on), and a PIN modulator is used to rf activate the tube with a four-phase PSK burst modulation. A triggered grid or modulating anode could turn the beam on and off in fractions of microseconds so that the rf burst is enclosed for stable rf amplification. The tradeoff between cost and lifetime (a somewhat shorter expected life is associated with a gridded tube) still must be considered after trial test data are available.

Uplink Domestic Transmitters

With the expected development of domestic TDMA systems such as SBS, there is an obvious need for an inexpensive, reliable power tube amplifier at 14 GHz. Hundreds of terminals are anticipated, and the advantages of tubes whose cost is reduced through quantity production should be realized. However, it is necessary to determine the 400- or 600-W tube type which best serves this application. Is it the more expensive, full bandwidth, coupled-cavity TWT, or is it the less expensive, narrow-band, fixed frequency klystron amplifier?

Uplink International Transmitters

Part of the evolving domestic uplink pattern already has been set by the CTS, OTS, MAROTS, JBS, and JCS satellites, but the uplink response at 14 GHz to the INTELSAT V satellite is still an open issue. Eight U.S. carriers are projected for three channels. Based on the devices described in this paper, several options are possible. All carriers can be processed with a single high-power tube (~6 kW), or three to four tubes handling multiple carriers (~2 kW) can be multiplexed. Finally, individual carriers can be multiplexed with 300- to 600-W power amplifiers.

Conclusion

With the advent of satellite communications at frequencies above 10 GHz, the uplink power amplifiers required for service are capable of development. At this time, klystrons have been installed for use with CTS. In Europe or Japan, coupled-cavity TWT's (both liquid and air cooled) in the 2- to 4-kW range at 14 GHz are being developed. These tubes usually are driven with a helix TWT. The general characteristics of these tubes and others being considered for development at higher frequencies have been described herein.

Acknowledgment

This paper is based upon work performed in COMSAT Laboratories under the sponsorship of the Communications Satellite Corporation.

References

[1] Mendel, J. T., "Helix and Coupled Cavity High Power Tubes," Proceedings of the Institute of Electrical and Electronics Engineers, Vol. 61, No. 3, March 1973, pp. 280-298.

[2] Staprans, A., "Coupled Cavity High Power Tubes and Hybrid Devices," Proceedings of the Institute of Electrical and Electronics Engineers, Vol. 61, March 1973, pp. 299-330.

[3] Marechal, J. C., "Traveling Wave Tube for Satellite Applications," Microwave Journal, Vol. 5, Nov. 1972, pp. 21-24.

[4] Sosa, E. N., "5W Ka-Band TWT for Space Applications," International Electron Devices Meeting, 1975.

[5] LeBorgne, R. H., "A Two-Level, High-Power, Air-Cooled TWT for Communication Systems," International Electron Devices Meeting, 1975.

[6] Tammaru, I., "Design of TWT's with Small Multi-Carrier Distortion Characteristics," International Electron Devices Meeting, 1975.

[7] Fleury, G., "A Coupled Cavity Traveling Wave Tube Delivery 2 KW CW at 14 GHz," International Electron Devices Meeting, 1974.

[8] Jones, W. G., "A 100 KW Peak 50 KW Average Coupled Cavity TWT With a 10% Bandwidth at 10 GHz," International Electron Devices Meeting, 1975.

[9] Bohlen, H., "FM Television Transmission Characteristics of a 12 GHz 1 KW Power Klystron," Eighth International Television Symposium, 1973, Montreux.

[10] Lien, E., "Wideband Turnable Klystron Amplifier for Satellite Earth Stations," International Electron Devices Meeting, 1974.

[11] "Comparative Study of TWT and Klystron Amplifiers For Use As Power Amplifiers in TV Broadcast Satellites," Final Rept. on European Space Research and Technology Centre [European Space Agency], Contract 1513/71/JS.

[12] Seunik, H., "A PPM Focused TWT for Upper Ka-Band With A CW Output Power of 1 KW," International Electron Devices Meeting, 1973.

WINDSCREENING EFFECTS ON ANTENNA TRACKING PERFORMANCE

R. Luik*
Aeronutronic Ford, Palo Alto, California
and
J. McCulloch+
Bell Telephone Laboratories, Holmdel, N.J.

Abstract

The effects of windscreening on the tracking and pointing performance of communications satellite tracking antennas were studied. Windscreens are structures that shield all parts of the antenna, except the reflector aperture, from windloads. Two typical windscreens were tested in a wind tunnel, and significant reductions in maximum force and moment components, and consequently in pointing and tracking errors, were noted. An analysis of the pointing and tracking errors for the tested configurations (assuming the structural elastic compliances of a typical 60-ft antenna and pedestal system used for synchronous satellite tracking) showed that the gain losses due to tracking errors in a 60 mph wind were only half those suffered in propagating the signal through a typical dry radome.

1. Introduction

Wind interaction with large antenna structures produces forces and moments sufficient to cause unacceptable antenna pointing errors and surface degradations at K-band frequencies. Classically, radomes have been used to protect an-

---

Presented as Paper 76-298 at the AIAA/CASI 6th Communications Satellite Systems Conference, April 5-8, 1976, Montreal, Canada.

* Since Mar 1976, Director of Engineering, TIW Systems Inc., Palo Alto, California.

+ Member Technical Staff.

tennas from wind and rain, but at K-band frequencies the attenuation of the signal through a wet radome is not acceptable.

One alternative to the radome is the windscreen, which reduces the wind-imposed forces on the antenna while providing a clear aperture. In concept, the windscreen structure completely surrounds the antenna except for a large fixed "window" through which the antenna can radiate without interference. Since antennas used in conjunction with synchronous satellites maintain essentially constant pointing angles ($\pm 5^{o}$), this type of windscreen can be used.

To determine the effectiveness of such a windscreen in reducing moments and forces on the antenna, a series of wind-tunnel tests was performed. Using the test results of bluff bodies and with an eye for ease of manufacture, two candidate windscreen configurations were chosen for testing. The results of these tests are the subject of this paper.

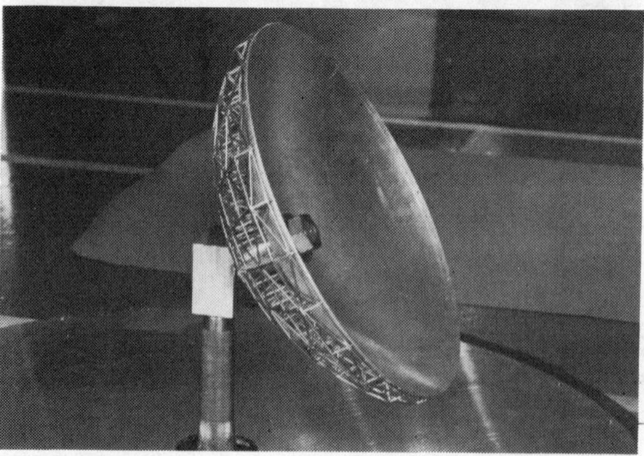

Fig. 1 Antenna model.

2. Wind-Tunnel Tests

The wind-tunnel tests were performed at the Kirsten Wind Tunnel, University of Washington Aeronautical Laboratory, which has a test section 8 ft high × 12 ft wide. Measurements were made of the component forces and moments on the antenna using the standard six-component balance. Tests were conducted for the antenna only and for the antenna in combination with each of the windscreen configurations. This section describes the antenna and windscreen models used in the wind-tunnel tests, together with the test procedures and test results.

Description of Models

The size of the antenna and windscreen models tested was based on blockage considerations in the wind tunnel, which

Fig. 2 Truncated elliptical windscreen model.

Fig. 3 Spherical windscreen model.

limited the maximum projected area to a 4.5 to 7.0-ft^2 range. Larger size models would have increased the distortion in the tunnel airflow to the level where the applicability of the test results to full-scale structures would be questionable.

Antenna Model. The antenna model is shown in Fig. 1. The configuration incorporates an 18-in. diam spherical reflector with a spherical radius of 13.7 in. This is equivalent to a parabolic reflector with an f/D of 0.38. The reflector was spin-formed from aluminum sheet 1/8 in. thick. A representative backup structure (not modeled exactly to scale) was included to simulate the effect of the actual backup structure on the wind forces. The ribs were fabricated from square 1/8 in. bars, whereas, the cross-bracing consisted of 1/8 in. diam circular members. No check was made on the Reynolds number effects for these members. Although these members deviate slightly from actual scale size of full-scale installations, their effect at least should be indicative of actual behavior.

Truncated Elliptical Windscreen Model. The truncated elliptical windscreen model is shown in Fig. 2. The model was fabricated from aluminum sheet. The major and minor diameters of the elliptical cross section are 24.0 and 12.0 in., respectively. These dimensions, which were necessary to accommodate the reflector and backup structure, resulted in an opening considerably larger than the antenna diameter. Thus, a 28.0 in. diam flange with an offset 19.3 in. diam hole was situated at the opening. This resulted in a clearance of 0.65 in. between the antenna and the windscreen window.

Spherical Windscreen Model. The spherical windscreen model shown in Fig. 3 was assembled by joining two plexiglass hemispheres together. A 19.3 in. diam window was cut in the model and reinforced by a 1/4 in.^{2} aluminum ring. The base is a cylindrical aluminum section 4.5 in. high.

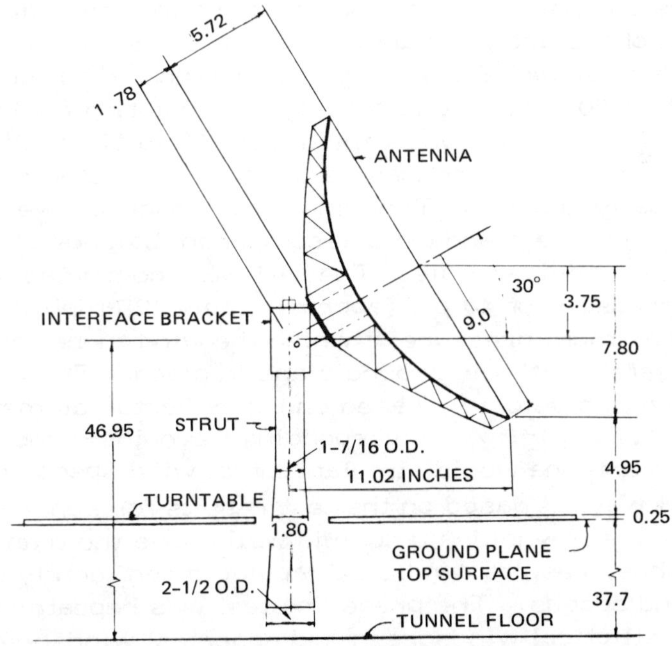

Fig. 4 Wind-tunnel test configuration: antenna model only.

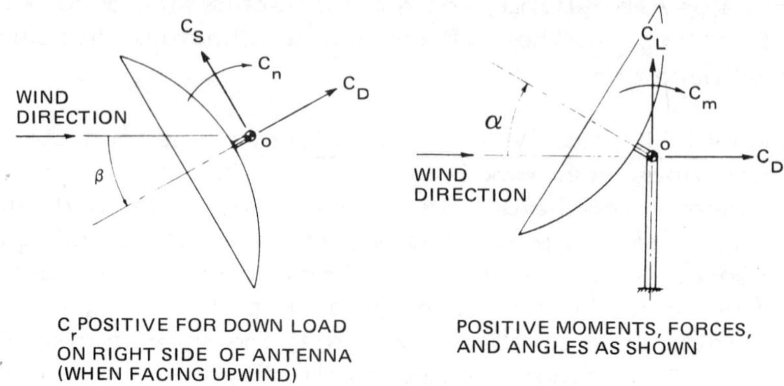

Fig. 5 Antenna stability axes.

Test Procedure

The purpose of the wind-tunnel tests was to determine the reduction in wind loads on the antenna due to each of the windscreen configurations. In this regard, an initial test with only the antenna model in place was made. The data from this test served as the baseline against which subsequent results were compared. The test configuration, including the placement of the ground plane, is shown in Fig. 4. Because of the limited scope of the study, the antenna was maintained at a constant 30° elevation angle (typical pointing angle). The azimuth wind direction was varied from 0° to 180° (0° is defined to exist when the concave side of the reflector is symmetrically upwind). The wind force reactions were recorded using the standard six-component balance at each 5° increment of the yaw angle. The test was conducted at a dynamic pressure of 40 psf (approximately 125 mph). This was the maximum pressure at which the wind tunnel could be operated safely with the ground plane in place. The Reynolds number for this velocity, based on the reflector diameter of 1.5 ft, is 1.7×10^6. A corresponding Reynolds number for a 60-ft-diam antenna would simulate actual wind speeds of only 3 mph. However, based on the test results for spheres and cylinders, the Reynolds number is well above the transition range. Thus, results can be extrapolated confidently to the higher wind speeds. The preceding test was repeated for the truncated elliptical windscreen and spherical windscreen configurations.

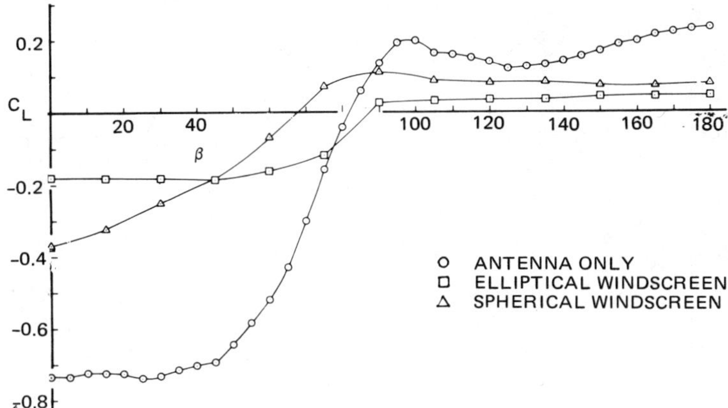

Fig. 6 Lift force: test results.

Tuft runs of each windscreen configuration also were made to determine airflow patterns around the antenna and windscreens. These runs were conducted for three yaw angles (at peak drag and yaw moment).

Test Results

The moment and force data obtained were reduced to coefficient form and transformed from the wind axes to the antenna stability axes. The force and moment coefficients are defined as follows:

$$C_D = \frac{\text{drag force}}{qA} \qquad C_m = \frac{\text{pitch moment}}{qAD}$$

$$C_L = \frac{\text{lift force}}{qA} \qquad C_n = \frac{\text{yaw moment}}{qAD}$$

$$C_S = \frac{\text{side force}}{qA} \qquad C_r = \frac{\text{roll moment}}{qAD}$$

where q is the dynamic pressure, A is the reflector frontal area and D is the reflector diameter. The stability axes to which these forces and moments are referenced are defined as an axis system that is parallel to the ground surface but is perpendicular to the model axis of symmetry. The positive directions for these axes are shown in Fig. 5. The moments are referenced to the intersection of the reflector surface

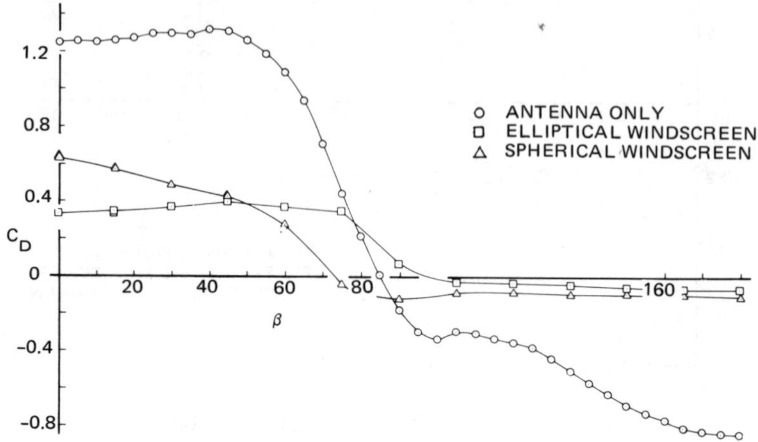

Fig. 7 Drag force: test results.

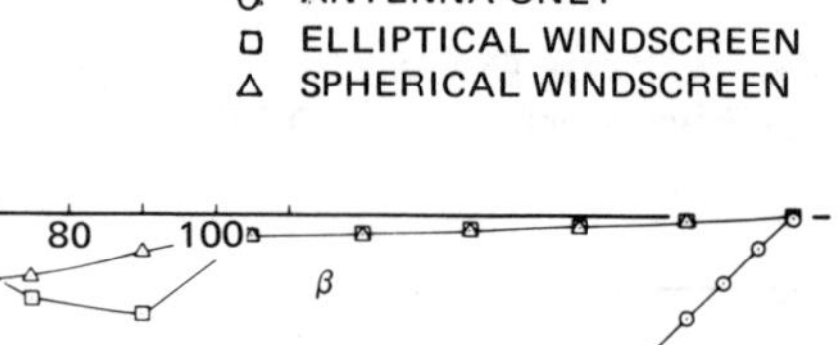

Fig. 8 Side force: test results.

generating centerline with the strut centerline (point 0 in Fig. 4).

The results are shown in coefficient form in Figs. 6–11 for the following configurations: antenna only, elliptical windscreen, and spherical windscreen. Photographs of the tuft runs are shown in Figs. 12 and 13.

Discussion of Results

The test results given in Figs. 6–11 show that both the elliptical and spherical windscreens substantially reduce the wind forces acting on the antenna. Measurements were taken only at 15° intervals for the windscreen configuration instead of at 5° for the test with the antenna only. Thus, there exists the possibility that the peak values may be slightly higher

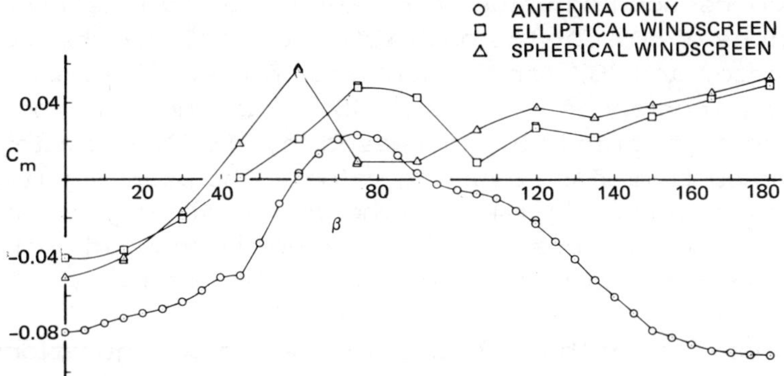

Fig. 9 Pitching moment: test results.

than measured, especially for the pitch and yaw moments, which experience sharp peaks (Figs. 9 and 10). No attempt was made to extrapolate between the measured values.

As shown in Figs. 6–11, the elliptical windscreen reduces the peak lift force by 75%, the peak drag force by 70%, and the peak side force by 66%. The peak negative pitch moment is reduced by 56%, but at the expense of increasing the peak positive pitch moment by 110%. The absolute value still is reduced by 47%, however. Reductions of 42% and 72% occur for the peak negative and positive yaw moments. Similar reductions of 78% and 20% occur for the peak negative and positive roll moments.

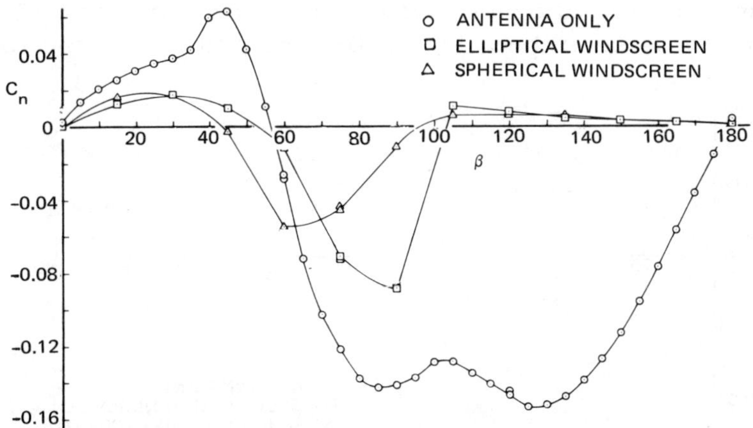

Fig. 10 Yawing moment: test results.

The corresponding values for the spherical windscreen show reductions of 50% for the peak lift force, 51% for the peak drag force, and 76% for the peak side force. The peak negative pitch moment is reduced by 45%, with the peak positive pitch moment again being increased (by 150%), resulting in a 38% reduction of the absolute value. The peak negative yaw moment is reduced by 64% and the peak positive yaw moment by 73%. The peak positive roll moment is reduced by 18%, whereas any negative roll moment is eliminated.

A comparison of these values for the two configurations shows that the elliptical windscreen provides the largest reduction in the lift force, drag force, and pitch moment, whereas the spherical windscreen provides the largest reduction in the side force and yaw moment. The wind force reductions are larger than would be expected to result from the elimination of only the windflow on the convex side of the antenna. It thus becomes obvious that the aerodynamic effect of the windscreen shape is also an important factor in reducing the wind forces on the antenna. In this regard, it is somewhat surprising that the elliptical windscreen provides the largest reduction in the lift and drag forces, since the spherical windscreen is felt to be a slightly better aerodynamic shape. Evidently, the flange used on the elliptical windscreen is a factor in reducing the wind forces. Although its exact nature is not understood completely, some inferences can be drawn concerning the effect of the flange.

Examining the tuft run photographs shown in Figs. 12 and 13, it can be seen that, for $\beta=0^{\circ}$, the primary direction of

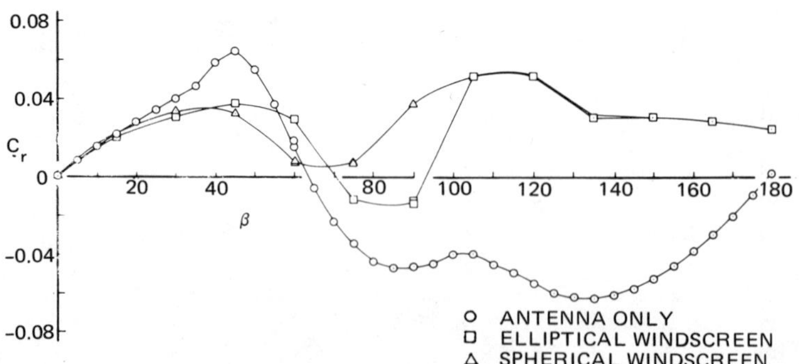

Fig. 11 Rolling moment: test results.

flow is into and upward across the antenna for both the elliptical and spherical windscreen configruations. However, for the elliptical windscreen, the flange tends to direct the air so that the angle of flow is in a direction more parallel to the antenna face, thus reducing the pressure perpendicular to the antenna face. The result is reduced values of the antenna wind forces for low values of the yaw angle ($\beta < 30^\circ$ to 45°). This effect is especially noticeable for the lift and drag forces (Figs. 6 and 7). As the value of the yaw angle increases, the wind diversion begins to cause a detrimental effect on the wind forces acting on the antenna. Instead of directing the airflow outward away from the antenna, as occurs for the spherical windscreen, the flange again directs the flow in a direction more parallel to the antenna face. The resultant increase of pressure on the antenna face increases the values of the resultant forces, especially for the yawing moment (Fig. 10). It also is felt to be the cause for the slight shift in the yaw angle at which the peak value of the wind forces occurs. Figure 12 shows that at $\beta=90^\circ$ a clockwise vortex is created along the flange surface. The reason for this flow or its effect is not understood presently.

As the yaw angle increases past 90°, the airflow at the antenna face becomes more turbulent. The wind forces acting on the antenna thus are reduced to fairly low, uniform levels. The large flange for this region ($\beta > 90^\circ$) causes a slightly larger reduction in the wind forces than occurs for the spherical windscreen.

3. Pointing and Tracking Accuracy

Wind forces and moments have a major deleterious effect on pointing and tracking, with consequent link gain losses. This section describes the effects of windscreening on the pointing and tracking performance of a typical antenna system. The following definitions and notations are used:

1) Rf axis: that direction in space in which the antenna has maximum gain at any frequency of interest and any instant of time.

2) Indicated direction: that direction in space defined by the outputs of the pedestal shaft angle indicating devices (e.g., encoders, synchros).

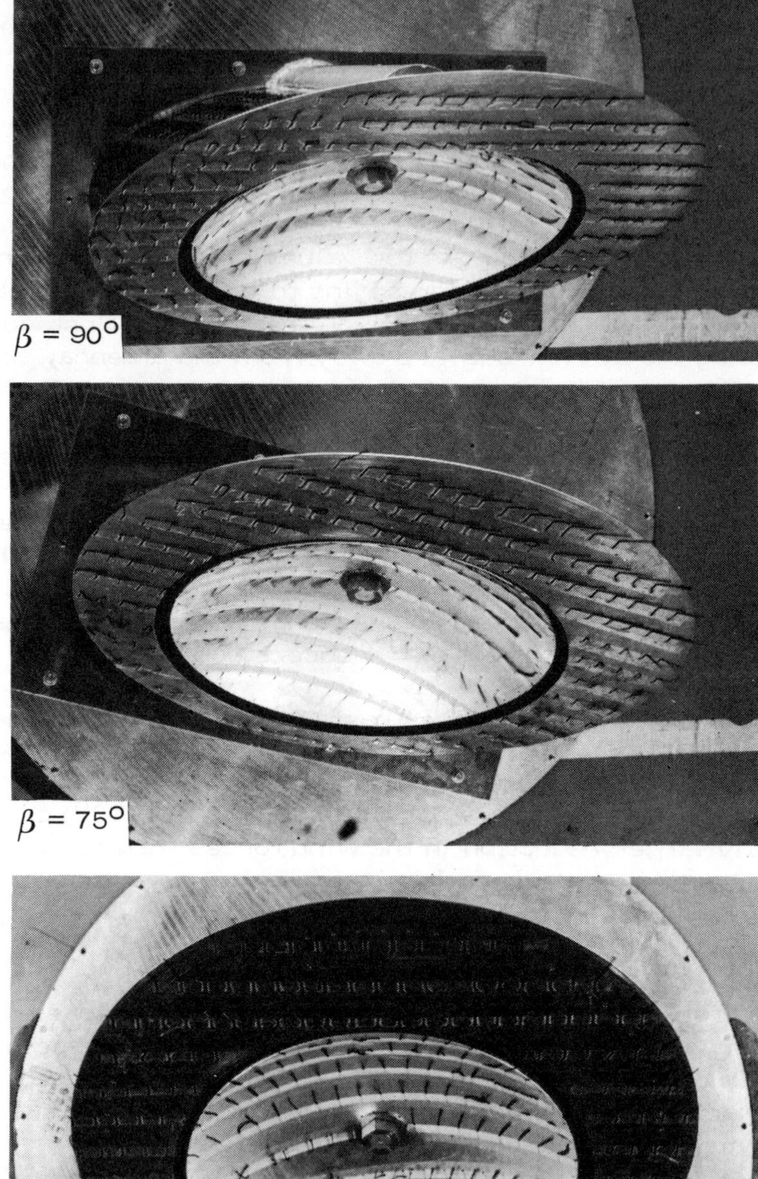

Fig. 12 Tuft run: elliptical windscreen.

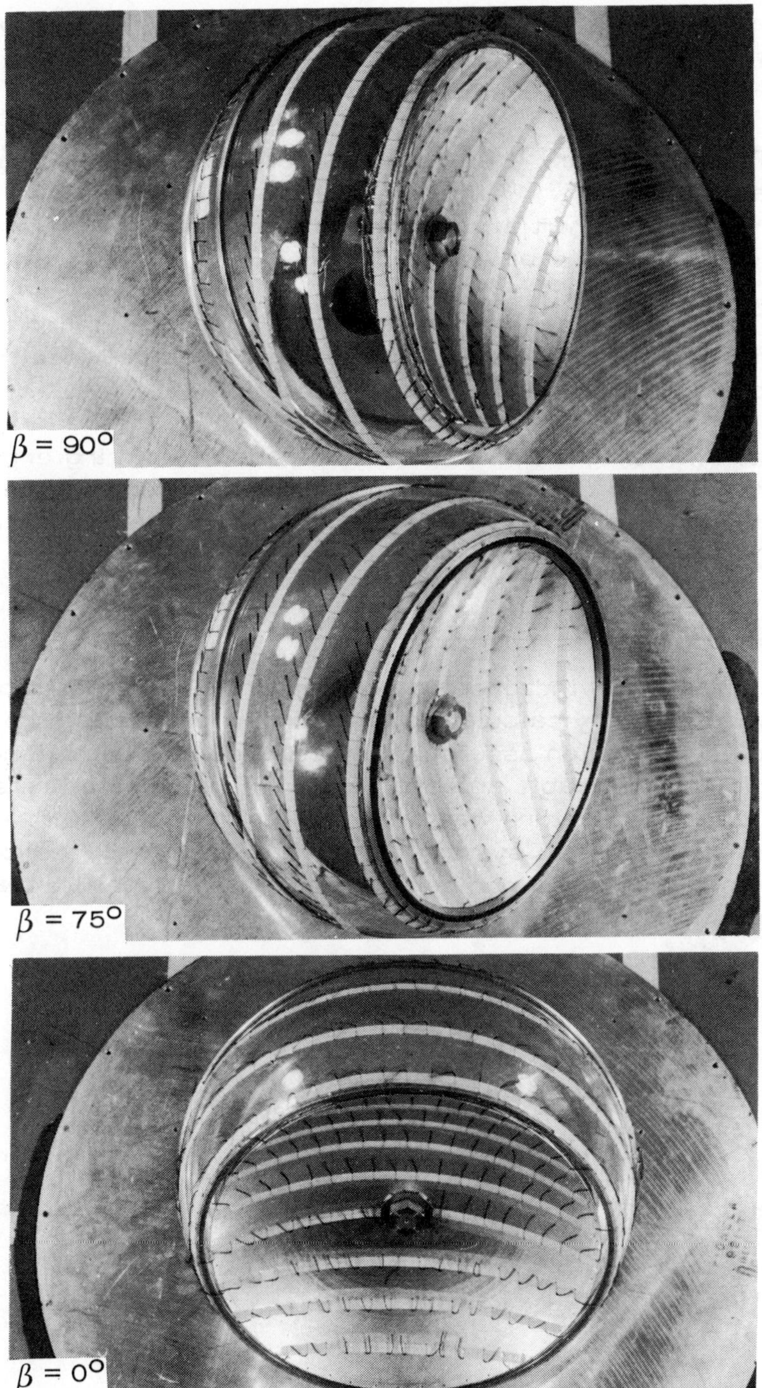

Fig. 13 Tuft run: spherical windscreen.

3) Apparent target direction: that direction in space which is normal to an isophase plane of incoming radiation from a distant source in the vicinity of the antenna aperture.

4) Pointing error: the angle between the rf axis and the indicated direction. Pointing error is a measure of the accuracy of the shaft angle indicators as instruments to measure the rf axis orientation.

5) Tracking error: the angle between the rf axis and the apparent target direction. Tracking error determines the loss of antenna gain in the target direction.

In summarizing the effect of wind loads on pointing error, it is convenient to divide the errors into components in the elevation direction (ε_y) and the cross-elevation direction (ε_x), as shown in Fig. 14.

Pointing Error Evaluation and Results

The effects of windscreening on pointing error reduction were evaluated by calculating the pointing errors for the tested configuration assuming structual elastic compliances of a typical applicable antenna design. The selected reference antenna structure was an Aeronutronic Ford 60 ft antenna and pedestal system, which has been analyzed extensively and built for a number of synchronous satellite tracking applications.

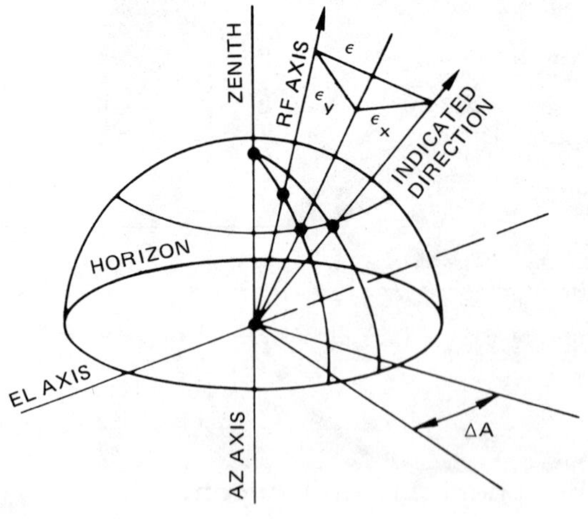

Fig. 14 Angular error components.

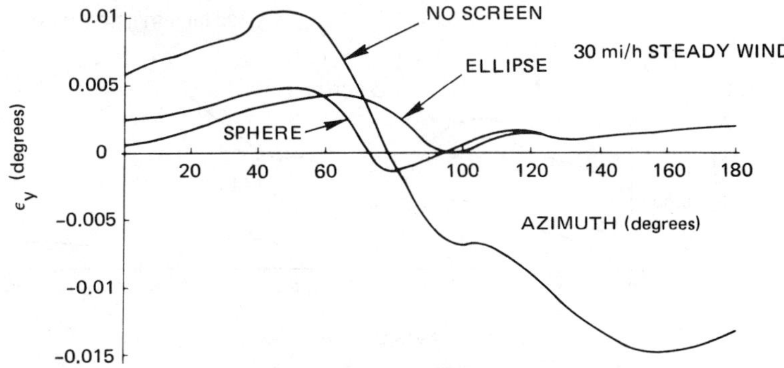

Fig. 15 Elevation pointing error.

The elevation, cross-elevation, and total pointing errors are shown in Figs. 15–17 as a function of wind angle of attack for 30 mph winds. It is apparent that the windscreens have reduced the effects of azimuth moments and headon and rear drag forces considerably, particularly when blowing from the back.

Figure 17 indicates that the worst-case total pointing error will be reduced by about a factor of 3. This plot also shows that there is no significant difference between the two windscreen configurations. Pointing errors for other wind speeds can be evaluated by velocity-squared proportioning.

Autotracking Error and Tracking Loss

The normal mode of operation for a high-performance communications antenna consists of proportional control

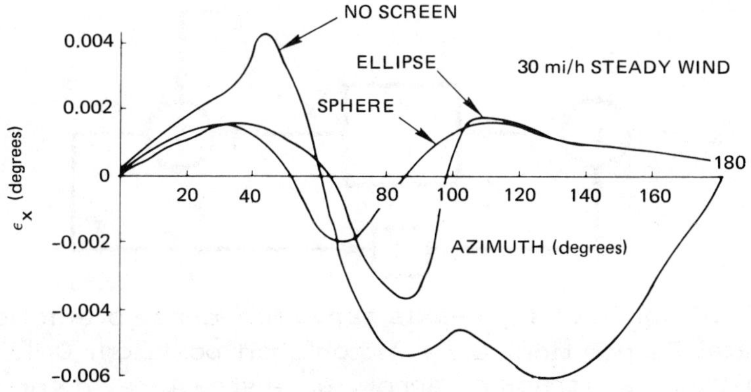

Fig. 16 Cross-elevation pointing error.

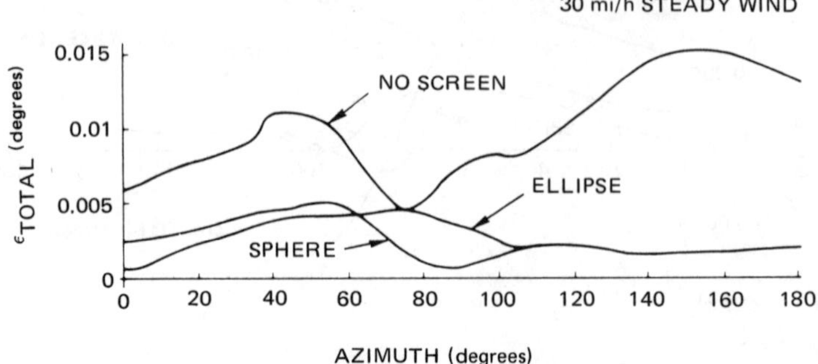

Fig. 17 Total pointing error.

automatic tracking with rf axis error feedback from a tracking feed and receiver system. The rf beam deviations from the satellite direction (tracking errors), when autotracking, will be significantly less than the absolute pointing errors, discussed previously for the following reasons:

1) An appropriately configured autotrack servo loop will have sufficient integrations to "track out" any constant structural deflections; only the gusts (windspeed variations from the mean) will have any effect on tracking error.

2) The servo system can sense the gust deflections and provide some attenuation within its bandwidth capability.

Analysis Procedure. The error produced by wind gusts is a result of load fluctuations due to windspeed variations. Wind velocity varies continuously and randomly in time due to

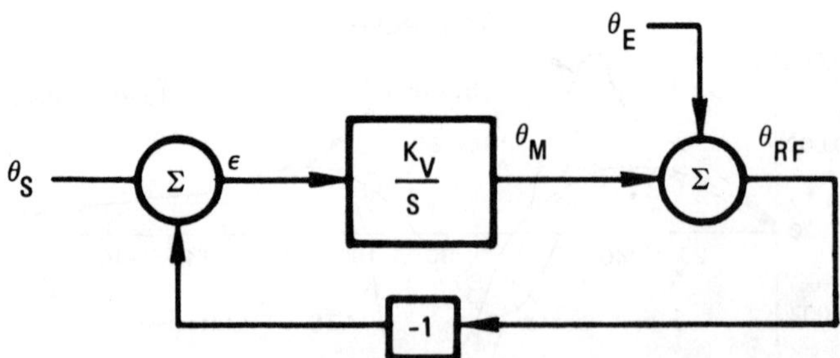

Fig. 18 Simplified single-axis servo for error prediction (Θ_S = satellite position, Θ_M = motor shaft position, Θ_{RF} = rf axis position, ϵ = tracking error, Θ_E = structural deformation due to wind, and K_V = position loop gain).

turbulence effects and is described quite well as a nonzero gaussian process.[1] The random component of wind velocity has a narrow-band, one-sided spectral density function defined by

$$S_V(f) = \frac{2\sigma_V^2 f}{3f_W^{2/3}\left(f^2 + f_W^2\right)^{4/3}}$$

where

f_W = wind velocity spectrum characteristic frequency, Hz

σ_V = wind velocity standard deviation

By correlation of a number of wind spectra, Davenport determined that the frequency f_W, which is near the spectral peak, is directly porportional to mean windspeed and is given by

$$f_W = 0.0008V_W$$

where V_W is the mean windspeed in meters per second. For 30 mph (13.5 m/sec), f_W will be about 0.011 Hz.

In high-gear ratio systems where the reflected inertia of the drive motors is much larger than the antenna inertia, the motor shafts are virtually immovable by external wind disturbances. The motors can be moved only by torque commands generated as a result of errors due to elastic deflec-

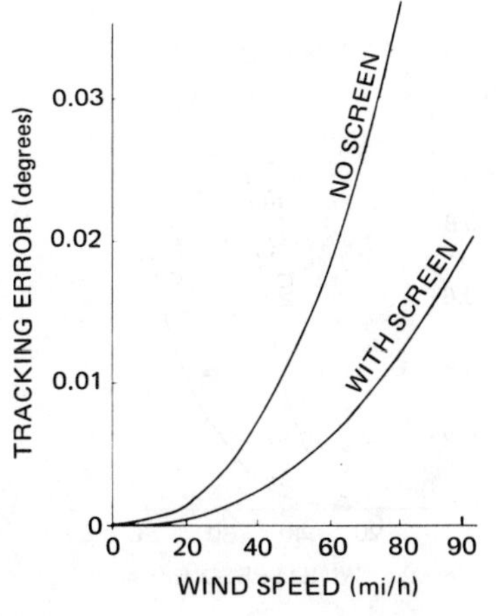

Fig. 19 Rms tracking error due to wind gusts.

tions of the structure. These deflection disturbances are processed by the position servo, as indicated by the low-frequency model of Fig. 18. In fact, the structural deflections have the same exact error-producing effect as position command variations. The time variation of the position disturbances is essentially proportional to the windspeed variations and hence can be represented as a gaussian process with a spectrum similar to the velocity spectrum. The standard deviation of the gust component of wind deflection about either axis can be estimated by noting that all instantaneous loads, and hence deflections, are proportional to velocity squared, i.e.,

θ_E = structural deflection (pointing error) for either axis

$$= C\,(\mu_V + \delta_V)^2$$

$$= C\mu_V^2\ [1 + (2\delta_V/\mu_V) + (\delta_V^2/\mu_V)]$$

$$= \mu_\theta\ [1 + (2\delta_V/\mu_V)]$$

where

μ_V = mean windspeed

δ_V = gust speed deviation from mean

μ_θ = structural deflection at mean windspeed

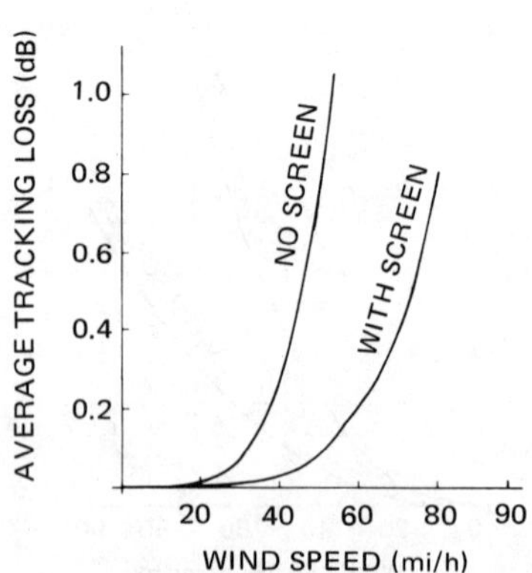

Fig. 20 Tracking loss at 20 GHz.

The standard deviation of the gust deflections then is approximated by

$$\sigma_\theta = (2\sigma_V/\mu_V)$$

The mean deflections were evaluated in the previous section.

The rf axis response of the system to wind can be analyzed relatively simply using a model based on the following assumptions:

1) The electrical bandwidth of the drive system (internal tachometer loop) is large compared to the gust spectra bandwidth; this is valid for most antenna applications, where the velocity loop bandwidths are typically 2–10 Hz and the spectral bandwidth is less than 0.1 Hz.

2) Structural resonances are not excited significantly. This is a reasonable approximation if both the gust spectrum and the servo bandwidths are significantly lower than the lowest structural natural frequency.

These assumptions are embodied in the block diagram of Fig. 18, which shows a simplified servo loop model for either of the axes and the nature of wind torque input disturbances. The total rf axis motion due to wind gusts is the sum of the motor motion and the elastic deformation between the motor shaft and the rf axis.

It also has been assumed that a type 1 position control loop will be used, since this configuration has been found to be the most effective for wind gust rejection. The transfer function deflection disturbance to tracking error for this model is

$$-\mathcal{E}/\mathcal{E}\theta = s/(s + K_V)$$

The error spectrum due to wind gust effects is the product of the gust deflection spectrum and the transmission vs frequency characteristics of the preceding transfer function. The mean square tracking error due to wind gusts for either axis is then

$$\sigma_W^2 = \int_0^\infty \frac{(2\pi f)^2 \; S_E(f)}{(2\pi f)^2 + K_V^2} \, df$$

This integral has been evaluated numerically; for servo bandwidth (proportional to K_V), large compared to the wind spectrum bandwidth, the rms tracking error for either axis is

$$\sigma_W = 2.2\left(\frac{f_W}{K_V}\right)^{1/3}, \quad \sigma_\theta = 2.2\left(\frac{f_W}{K_V}\right)^{1/3}\left(\frac{2\sigma_V}{\mu_V}\right)\mu_\theta$$

The two-axis tracking error is the vector sum of the azimuth and elevation components and so the total rms space angle tracking error can be evaluated by using the total two-axis structural deflection in the preceding relationship. For the final evaluation, the following two additional assumptions are made:

1) The standard deviation/mean windspeed ratio is assumed to be 25%; this is an intermediate typical value for rural environments.

2) The servo loop gain K_V was taken as 1.0/sec; this is a value easily obtainable with an antenna in the 60- to 100-ft diam range which provides a sufficiently low noise bandwidth to attenuate the effects of tracking receiver thermal noise adequately.

Discussion of Results. The rms tracking errors were evaluated by the procedures just described as a function of mean windspeed for the unprotected and screened configurations. The results are shown in Fig. 19. These and all subsequent results were evaluated for the worst-case antenna

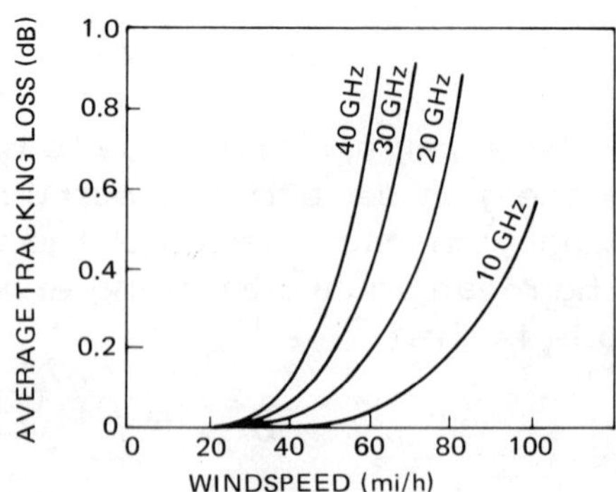

Fig. 21 Tracking loss with windscreens.

azimuth relative to the wind (150° unprotected, $50^\circ - 70^\circ$ screened). As is the case for the pointing error, the addition of screens reduces the errors by a factor of 3.

The effect of tracking errors on communication link performance can be characterized by the tracking loss, which is the reduction in gain due to rf axis boresight offset from the target direction (tracking error in the autotrack mode). This loss (in decibels) is estimated by the standard quadratic approximation, which matches the 3-dB points on the secondary pattern, i.e.,

$$L = 12\,(\varepsilon/\theta_{hp})^2$$

ε = instantaneous tracking error

θ_{hp} = half-power beamwidth

Since this loss is a time-varying function, some statistic must be selected to characterize the overall performance. The time average tracking loss is a good general metric of the general performance characteristics in gusty wind conditions and will be used in subsequent data presentation. For the quadratic pattern model, the average loss is

$$L_{av} = 12\,(\sigma_W/\theta_{hp})^2$$

The average losses with and without screening are shown in Fig. 20 for an intermediate K-band frequency (20 GHz). The loss reduction with wind screening is proportional to the

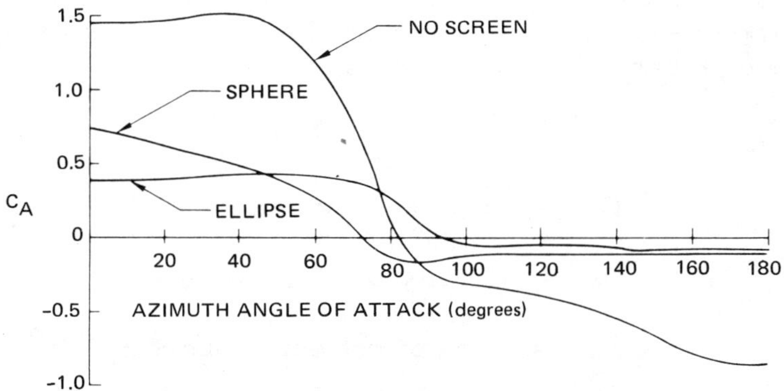

Fig. 22 On-axis force component.

square of the tracking error reduction (factor of about 9, or about an order of magnitude). The average loss will be about 0.5 dB for a 45 mph mean wind for an unprotected antenna, whereas this same loss level occurs at about 70 mph for the screened configurations.

The worst-case tracking losses for the screened configurations are shown as a function of communication frequency in Fig. 21. This figure simply shows the increased loss with increasing frequency and correspondingly decreased beamwidth.

Comparison of System Gain Losses for Exposed and Protected Antennas

The previous section has defined the losses in gain caused by tracking errors for screened and unscreened antennas. To quantify the communications performance of the windscreened antenna, it also is necessary to consider surface degradations. Unfortunately, the scope of this study did not allow the taking of pressure data on the reflector; therefore, the surface degradation due to wind can be estimated only roughly from the on-axis force resultants. For this analysis, it will be assumed that the rms surface degradation will be given by

$$\sigma_W \propto F_A$$

where σ_W is the rms surface degradation due to wind loads and F_A is the resultant force parallel to the axis of the reflector acting through the vertex.

The constant of proportionality K will be estimated from the wind deformations of an existing 60-ft antenna design by ratioing the on-axis force coefficient C_A for the exposed and screened antenna configuration as follows:

$$\sigma = KF_A = KC_A \ \rho AV^2$$

where

C_A = on-axis nondimensional force coefficient

ρ = density of air, lb-ft-sec^2

A = area of reflector aperture, ft^2

V = wind velocity, fps

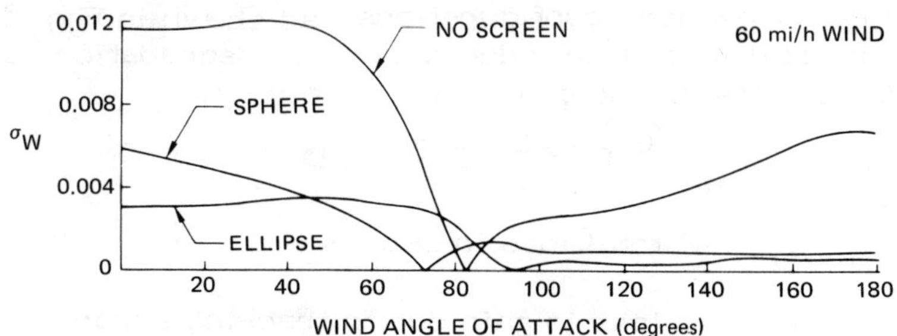

Fig. 23 Rms surface degradation.

The maximum wind-induced surface error for an existing Aeronutronic Ford 60-ft diam antenna is 0.012 in. rms in 60-mph winds:

$$\sigma_W = C_A \bar{K} V^2$$

where $\bar{K} = K\rho A$.

Figure 22 gives the on-axis force component C_A as a function of the azimuth angle of attack. As can be seen from this figure, the maximum coefficient is 1.5, and so

$$\bar{K} = 0.012/(88)^2 (1.5) = 1.0 \times 10^{-6}$$

Now the rms surface degradation can be estimated as a function of the wind angle of attack for the exposed, as well

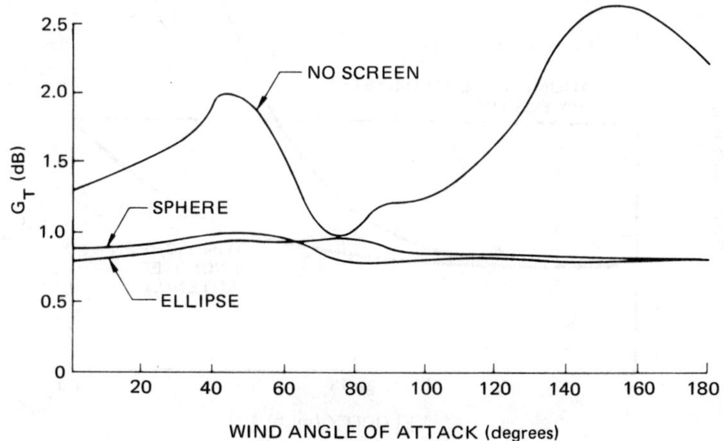

Fig. 24 Total gain loss.

as the two screened configurations, as shown in Fig. 23. Then the total loss of gain due to surface degradations and tracking errors can be calculated as follows:

$$G_T = G_{TE} + G_{SD}$$

where

G_T = total loss in gain

G_{TE} = loss in gain due to tracking error

G_{SD} = loss in gain due to surface degradation

The loss due to the tracking error can be approximated as a function of the wind-induced tracking error and half-power beamwidth as follows:

$$G_{TE}(\alpha) = 12\,(\varepsilon\alpha/\theta_{hp})^2$$

The gain loss due to surface degradation is given by the well-known Ruze's formula as

$$G_{SD} = 685\,(\sigma_T^2/\lambda^2)$$

where σ_T is the total rms surface distortions, and λ is the wavelength.

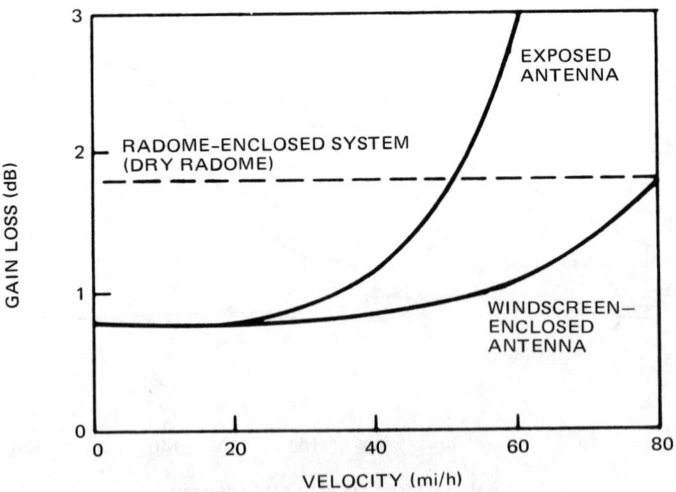

Fig. 25 Maximum gain loss.

For the system loss degradation, the frequency of interest will be considered to be 20 GHz. The wavelength at this frequency is approximately 0.60 in. In addition to the wind-induced surface distortions, the surface distortions due to manufacturing and alignment imperfections also must be considered in the analysis. Because of the state-of-the-art today, manufacturing distortions (σ_M) for 60-ft antennas are typically 0.020 rms. Therefore, the total surface degradation is given by

$$\sigma_T^2 = \sigma_W^2 + \sigma_M^2$$

The total gain loss as a function of the angle of attack is given by

$$G_T(\alpha) = 12 \left\{ \varepsilon(\alpha)/\theta_{hp} \right\}^2 + (685/\lambda^2)[0.020^2 + (\alpha)]$$

Figure 24 gives the estimated total gain loss G_T as a function of the angle of attack of the wind for the exposed and screened 60-ft antennas in 60-mph winds. Figure 25 gives the maximum gain loss as a function of peak windspeed for exposed, windscreen-enclosed, and radome-enclosed antennas. For the purposes of estimating the radome-enclosed system gain losses, the loss in a dry radome is estimated to be approximately 1 dB. For wet radomes, this loss can be as high as 8 dB, as shown by tests at Bell Telephone Laboratories.[2] For the radome-enclosed system, the effect of reflector manufacturing errors must be considered, in addition to the radome itself. As can be seen from the results, the total system gain loss is the lowest for the windscreened configuration for all wind speeds below 80 mph.

4. Conclusions

It can be concluded that, for operation in the K-band frequenty region, a windscreen minimizes the gain losses due to a combination of aperture blockage and tracking errors. As was shown in Sec. 3, the losses due to tracking errors in a 60-mph wind were only half the losses suffered in propagating the signal through a typical dry radome. When effects of weathering and wetness are considered in the radome performance, the advantages of the windscreen appear even greater. No attempt was made to optimize the performance of the selected windscreens, and it is conceivable that even greater improvements are possible.

Acknowledgement

The authors wish to express their gratitude to J. Bockholt of Aeronutronic Ford Corporation, Western Development Laboratories Division Staff, who performed much of the work associated with the design of the windscreening models and conducted the wind-tunnel testing.

References

[1]Davenport, A.G., "The Buffeting of Large Superficial Structures by Atmospheric Turbulence," Annals of New York Academy of Science, Vol, 116,

[2]Anderson, I., "Measurements of 20 GHz Transmission Through a Wet Radome," G-AP International Symposium, Aug. 22-24, 1973, University of Colorado, Boulder, Colo.

A MULTIPLE-BEAM TORUS REFLECTOR ANTENNA

R. Kreutel*
COMSAT Laboratories, Clarksburg, Md.

Abstract

The proposed implementation of regional satellite communications systems operating in the 20/30-GHz frequency band necessitates investigation of high-gain, multiple-beam antennas for Earth station application. An n-satellite system, for example, may utilize m Earth station antennas, each capable of generating A beams, to result in a total of n x m possible two-way communications paths. From both economic and operational points of view, the advantages of utilizing a multiple-beam Earth station antenna rather than n single-beam antennas are evident, especially for n > 2. This paper describes a version of the torus reflector antenna, specifically designed for use as a multiple-beam, high-gain Earth station. The general geometric configuration is discussed, and the results of rf evaluation of an experimental model are presented.

Introduction

In the examination of various antenna configurations for application as a satellite communications ground terminal antenna, the following performance factors have been considered: 1) multiple-beam-forming capability, 2) field of view, 3) wide-angle sidelobe envelope, 4) access to equipment for maintenance and operation, 5) production cost, and 6) compatibility with high beam gain (G > 60 dB). The first two items are related in that the multiple-beam-forming capability must be maintained over a wide field of view (5°–10° or larger). In general, this eliminates from further consideration antennas such as parabolic reflectors, which have beamwidth-dependent beam-forming limitations (e.g., a parabolic reflector). It also eliminates configuration that are angle-limited to a very small field of view (e.g., the Cassegrain reflector).

Presented as Paper 76-302 at the AIAA/CASI 6th Communications Satellite Systems Conference, April 5-8, 1976, Montreal, Canada.

*Manager, Antenna Department.

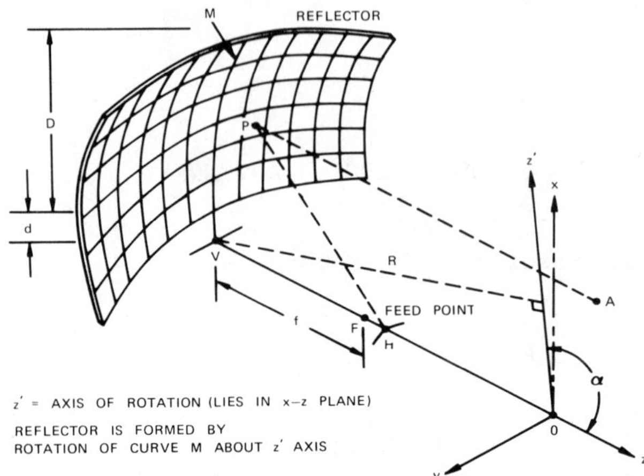

Fig. 1 Torus antenna geometry.

The wide-angle sidelobe envelope is of particular importance. In a system in which a number of satellites and many Earth stations share the same frequency band, interference can be a serious problem.[1] Since the antenna sidelobe is the performance factor that determines interference levels, it is especially important to maintain very low sidelobe levels.

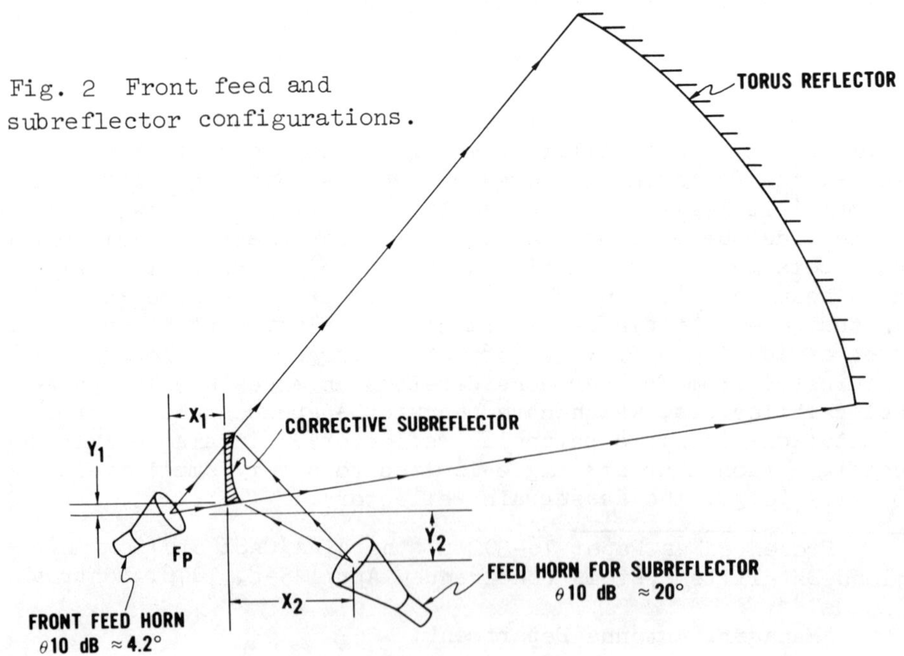

Fig. 2 Front feed and subreflector configurations.

Antennas normally employed as Earth terminals, namely, Cassegrain-type antennas, have wide-angle sidelobe envelopes that are limited by electromagnetic scattering by obstacles (feed, subreflector, and support members) contained within the aperture. To avoid this problem, a reflector antenna must be offset-fed. Offset feeding permits the feed system to be removed completely from the reflector aperture.

Consideration of the foregoing factors has resulted in the selection of the offset-fed torus antenna as an attractive configuration. Design and performance detail that supports this selection is presented in the following section.

Torus Reflector

The general geometry of the torus reflector configuration is shown in Fig. 1. The reflector surface is generated by rotating a parabolic arc M about the axis. Note that the z' axis forms an angle α with the plane containing the focal axis of the parabolic arc. In a conventional torus design, $\alpha = 90°$, and the field of view of the antenna is planar. In the present case, α is chosen as 95.5°, and the field of view is conical.[2] The 95.5° angle provides a beam-forming surface that closely conforms to the geostationary orbit as viewed from Earth station sites in the range of 30° to 50° lat.

Fig. 3 Experimental model of the phasecorrected torus antenna.

When the parabolic arc is rotated about z', its focal point also is rotated, hence generating a focal arc. The multiple-beam-forming capability of the reflector is realized by distributing feed elements along the focal arc. Each feed element generates a beam, and, as is evident from the geometry of Fig. 1, the focal arc is offset from the reflector, thus eliminating the sidelobe problem associated with aperture blockage.

It is recognized that the advantage of simple multiple beam forming is realized at the expense of the imperfect focusing provided by the torus reflector. It easily can be shown that the phase errors that result from imperfect focusing can be controlled by proper selection of the reflector f/D, where D is the effective azimuthal direction used for forming a beam. As f/D is increased, the focusing properties improve, and the reflector area to provide a given field of view also increases. It follows that a good cost-effective design involves a tradeoff of phase error and related performance degradation against reflector surface area and f/D. It has been found that, for a multiple-beam torus antenna with an antenna gain of about 55 dB or less, an f/D selection of about 1.25 and the use of a simple point-source-type feed system result in economical antenna system design. Reference 3 provides a complete treatment of a multiple-beam torus falling in this category.

Fig. 4 Feed system of the phase-corrected torus antenna.

For antenna gain requirements greater than 55 dB (which is the usual requirement for the 20/30-GHz system), the f/D of a reflector which allows tolerable phase error is exceedingly large. Consequently, the reflector area required for a given field of view becomes excessive. For this reason, it is desirable to employ an aberration-correcting feed system.

The aberration-correcting feed system generates wavefronts that are distorted to compensate for the aberration caused by the reflector. There are a number of techniques that can be used for this purpose. For the application described herein, a subreflector has been chosen because it is relatively simple. In addition, because it is designed to provide aberration correction on a path length equalization basis, it has a wide bandwidth and consequently can effect correction over the full 20/30-GHz band. Figure 2 is an elevation view of the torus reflector with the aberration-correcting feed system.

An experimental model of the 60+-dB phase-corrected torus antenna shown in Fig. 3 has been constructed and evaluated. The illuminated aperture was 10 ft in diameter. The testing was done over the frequency range of 12 to 42 GHz. The feed system, shown in Fig. 4, consists of the subreflector illuminated by a corrugated horn. To demonstrate the bandwidth capability, efficiency, and effectiveness of this aberration-correcting feed system, the demonstration model has been evaluated experimentally over a wide frequency range. In Fig. 5, calculations of gain vs frequency are plotted for the reflector with and without aberration correction. As can

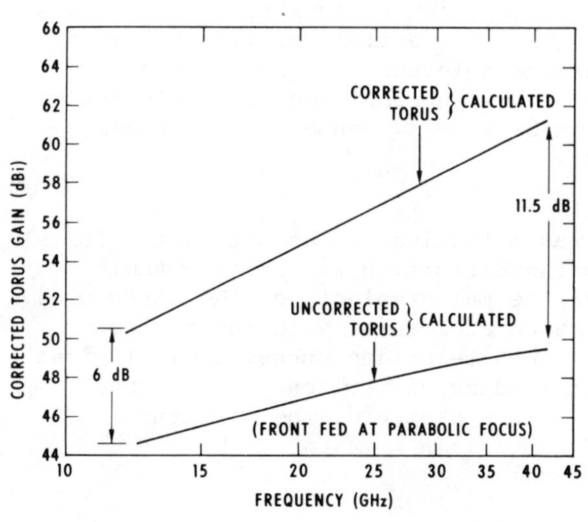

Fig. 5 Calculated gain vs frequency relationship for the corrected and frontfed torus.

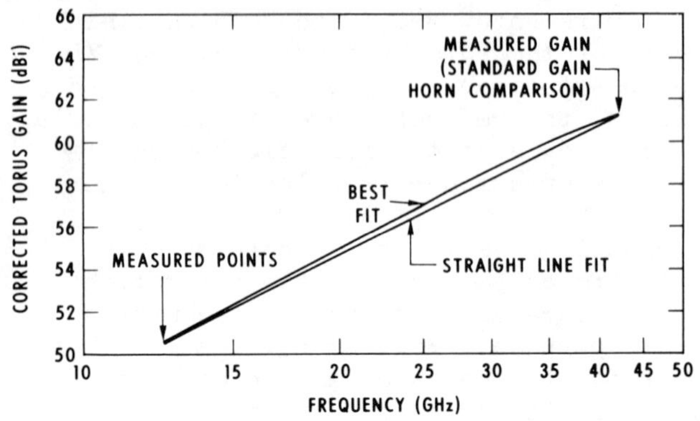

Fig. 6 Measured gain vs frequency for the corrected torus.

be seen, the improvement resulting from correction is substantial.

In Fig. 6, the measured data exhibit substantial agreement with the calculation shown in Fig 5. The data clearly demonstrate the frequency-independent characteristics of the correcting feed system. It should be noted that the wide-band testing was accomplished using a set of corrugated horns, each having essentially identical radiation characteristics.

Finally, Fig. 7 shows a typical radiation pattern measured at a test frequency of 42 GHz. The pattern is indicative of sharp focusing and fast sidelobe decay with angle.

Structural Design

A structural design concept for a torus reflector with an effective aperture of 30 ft and a field of view of about 20° was derived. The surface tolerance was prescribed as 0.015 in. (rms). The panel design consisted of an aluminum skin over grillage ribs on 6- x 6-in. center. Each panel was about 5 x 5 ft square.

Thermal distortion was determined to be the most critical design problem. To keep the distortion within acceptable bounds (i.e., to minimize thermal gradients on the structure), the structure was built as an enclosure, with the reflector panels forming one wall. This technique successfully limited the combined thermal/wind loading distortion to less than 0.006 in. (rms). Figure 8 is a sketch of the structural configuration.

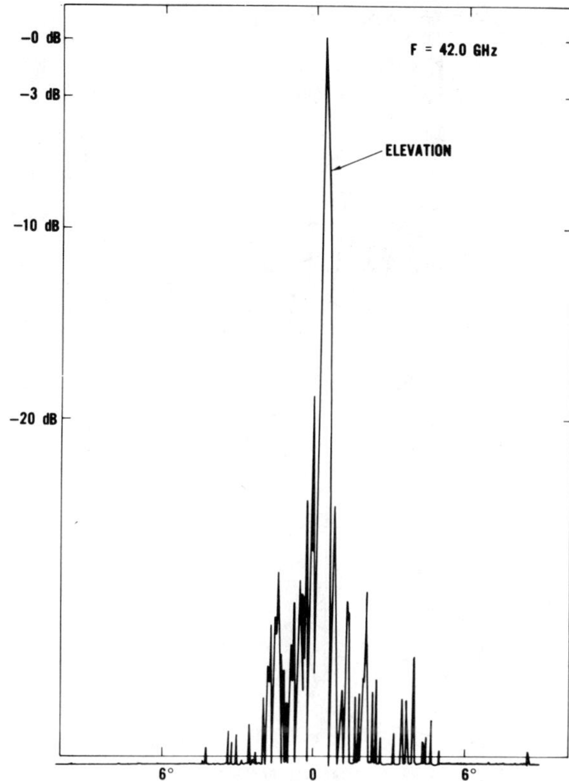

Fig. 7 Corrective torus.

Additional Features

An important feature of the offset-fed torus antenna is the incorporation of an enclosed structure to house the feed system and all associated equipment, including transmitters and receivers. This arrangement permits ease of access and maintenance for all equipment and generally allows direct connection of transmitters to the antenna feed. Such an arrangement is, of course, made possible by the offset feed configuration.

Beam positioning generally is accomplished by positioning the feed system. When satellite position is maintained to within small tolerances, no active beam pointing may be needed. When a small amount of diurnal pointing is necessary, it can be accomplished by moving the feed. The mechanism for positioning the feed can be based on a step track, program track, or conventional autotrack system.

The final factor is economy. Since the antenna is a fixed structure employing no heavy bearings and drives, it

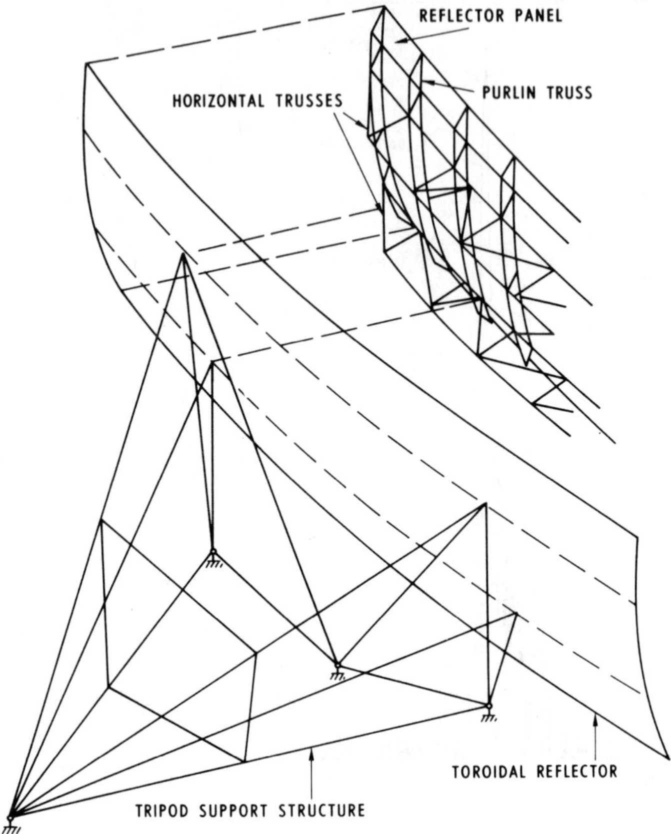

Fig. 8 Support structure for the fixed Earth station toroidal antenna (enclosure not shown).

is inherently an economical structure. Furthermore, the use of overlapping portions of the reflector surface to form multiple beams contributes substantially to the overall economy of the Earth station.

Acknowledgment

This paper is based upon work performed in COMSAT Laboratories under the sponsorship of the Communications Satellite Corporation. The experimental evaluation of the corrected torus antenna was carried out by R. Gruner and A. Williams. The correcting subreflector was designed by G. Hyde. The structural design study was conducted by Simpson, Gumpertz and Heger, Inc.

References

[1] Tillotson, L., "A Model of a Domestic Satellite Communication System," Bell System Technical Journal, Vol. 47, Dec. 1968, pp. 2111-2137.

[2] Kreutel, R. W., and Hyde, G., "Torus-Type Antenna Having a Conical Scan Capability," U.S. Patent 3,852,763, Dec. 3, 1974.

[3] Hyde, G., Kreutel, R. W., and Smith, L. V., "The Unattended Earth Terminal Multiple-Beam Torus Antenna," COMSAT Technical Review, Vol. 4, Fall 1974, pp. 231-262.

A NOVEL EARTH STATION ANTENNA FOR GEOSTATIONARY SATELLITES

M. Akagawa* and H. Yokoi+
Kokusai Denshin Denwa Company, Ltd. Tokyo, Japan

Abstract

The Earth station antenna system can be economized by limiting the operative angular range, since moving angular range of geostationary satellites has become very small. This paper describes a modified polar-mount antenna with a semifixed main reflector fed by a beamguide-type primary radiator. The longitudinal drift of a satellite is tracked by controlling the main reflector around the quasipolar axis, which is shifted slightly from the polar axis. The movement perpendicular to the longitude is tracked by controlling small beamguide reflectors with a simple drive subsystem. The antenna is able to track geostationary satellites, such as Intelsat IV, within longitudinal position of as much as $\pm 5°$ on the geostationary arc. The performance of this antenna and system parameters also are presented.

I. Introduction

Moving angular range of geostationary satellites has become very small owing to the great improvement in launching skills and attitude control technique for such satellites. In the near future, the moving range will be constrained much narrower than that of its present state to make efficient utilization of the geostationary-satellite orbit. The opera-

---

Presented as Paper 76-303 at the AIAA/CASI 6th Communications Satellite Systems Conference, April 5-8, 1976, Montreal, Canada. This study has been conducted by Radio Transmission Laboratory of KDD Laboratories, and the authors wish to express their appreciation to Dr. Nakagome, Dr. Kaji, Dr. Satoh, Mr. Ono and Mr. Mizuguchi for continuing guidance and helpful suggestions.

*Research Engineer, Radio Transmission Laboratory.
+Deputy Manager, Research and Development Laboratories.

tional angle of the Earth station antenna, then, may be limited within a very narrow angular range, and the Earth station antenna would not necessarily be a fully steerable type, but a limited steerable one would suffice. From this point of view, Cassegrain antennas with a movable subreflector and beam-scanning spherical antennas have been studied by several authors.[1-4]

In this paper, typical examples of the movement of geostationary satellites such as Intelsat IV are shown first. Then a new concept to compose such a limited steerable Earth station antenna accessing geostationary satellites is presented. Calculated steerability and electrical performance of this type of antennas also are described.

II. Location and Movement of Geostationary Satellites

Figure 1 shows the location and drifting direction of the Intelsat satellites as of Oct. 31, 1975. In the figure, the operating satellites are IVF-3 and IVF-7 (Atlantic Ocean region), IVF-1 (Indian Ocean region), and IVF-8 (Pacific Ocean region), and spares are IVF-2, IVF-5 and IVF-4, respectively. These operating and spare satellites are arranged to be located within 10° of longitude for the convenience of operating Earth stations in each ocean region.

If a new satellite is to be launched or an operating satellite fails, the beam directions of Earth station antennas

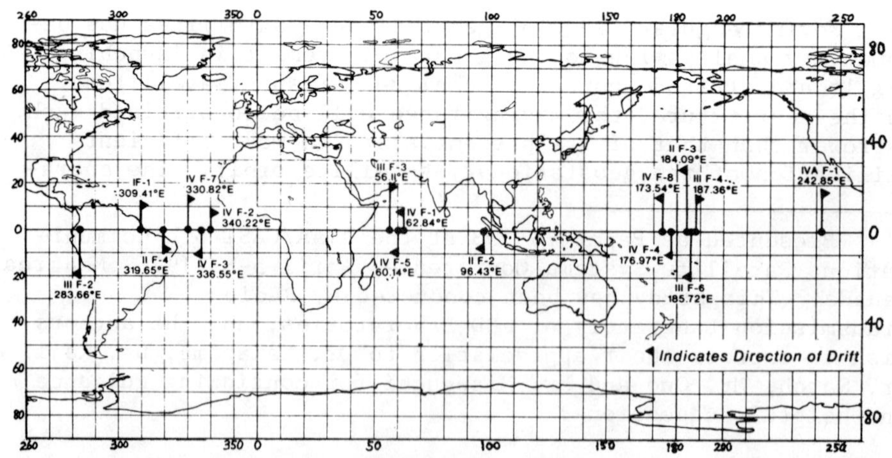

Fig. 1 Location of Intelsat satellites in orbit as of October 1975.

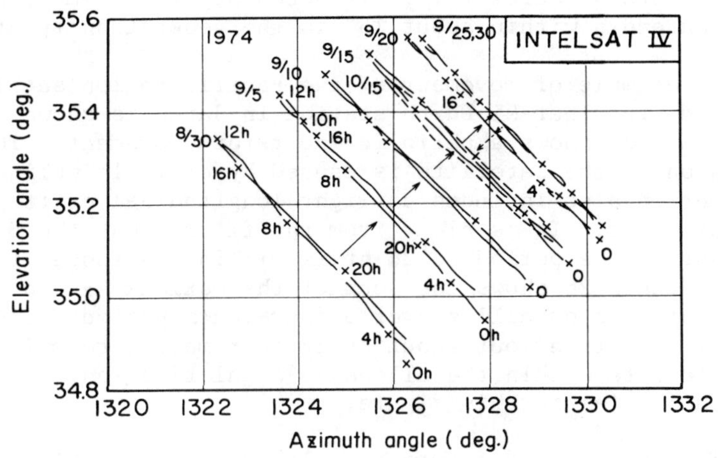

a Short-term movement

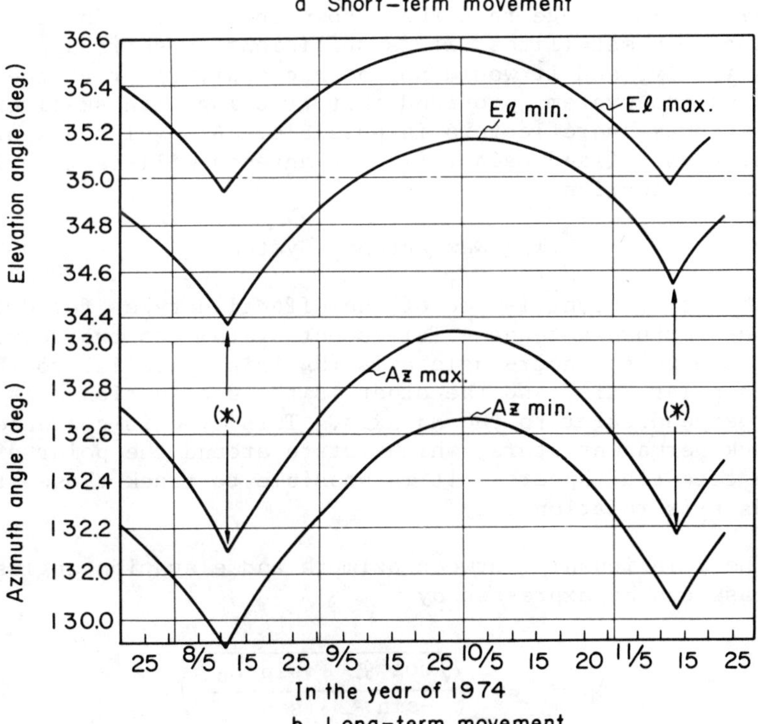

b Long-term movement

Fig. 2 Movement of satellite.
a Short-term movement
b Long-term movement

must be changed slightly in order to access the new satellite or the spare. Therefore, it is necessary for them to move their antennas within about ±5° in the geostationary arc.

An example of movement of the Pacific region satellite observed from Ibaraki Earth station in Japan is shown in Fig. 2. Figure 2a shows a periodic short-term movement. The daily excursion of the satellite is caused by orbital inclination. This motion also includes a slight longitudinal drift perpendicular to it. It is clear from the figure that the daily movement of the period of 24 hr is within the range of about 0.6°. Figure 2b shows the loci of the maximum A_z and El angles of this satellite over a four-month period. The orbit control is carried out about every four months to maintain the satellite within the allowable angular region, causing sharp nulls (*) in this figure.

The movement of future satellites will be limited in smaller angular range than that shown in Fig. 2. In case of tracking such satellites, it is sufficient to shift an antenna beam slightly, and it would not be necessary to drive the whole antenna system, provided that tracking with small elements such as subreflectors is possible. A novel configuration of such a fixed main reflector antenna will be described in the next section.

III. New Antenna System

The polar mount is one of the effective types for driving a large-aperture antenna. This mount system has two rotation axes: one is the hour-angle axis (H_a axis) parallel to the Earth's polar axis, and the other is the declination axis (Dec axis) perpendicular to the Ha axis. This type is convenient to track permanent stars, which rotate around the polar axis on the celestial sphere. It is possible to track those stars with Ha axis rotation only.

The relationship between azimuth and elevation angles in this case can be expressed by

$$A_p = \tan^{-1}\left(\frac{\sqrt{\cos^2\varphi_T - \sin^2 E_p}}{-\sin\varphi_T \cdot \sin E_p}\right) \tag{1}$$

where

Ap, Ep = azimuth angle and elevation angle of the pointing direction of the polar-mount antenna,

φ_T = latitude of Earth station.

The dot-dash curve of Fig. 3 shows this relationship for the

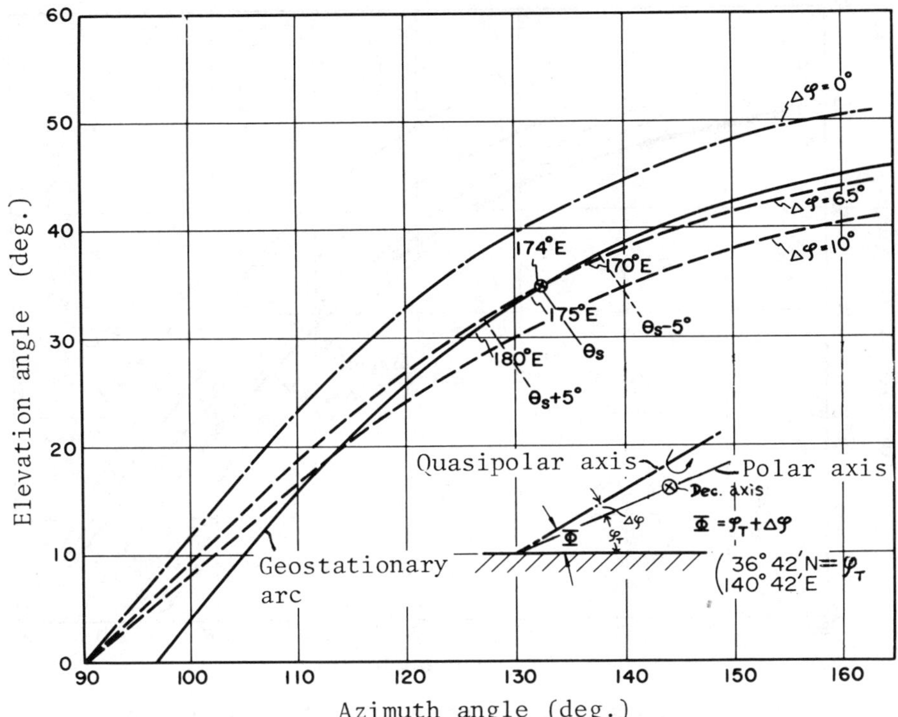

Fig. 3 Geostationary arc and pointing direction of polar-mount antenna at Ibaraki.

polar-mount antenna located at Ibaraki Earth station (36.7°N, 140.7°E).

Artificial satellites, however, can not be tracked by Ha axis rotation only, because they are at a finite distance. The solid curve of Fig. 3 shows the geostationary arc seen at Ibaraki Earth station, which is given by

$$A_S = \tan^{-1}\left(\frac{\sqrt{\cos^2\varphi_T - \cos^2\theta}}{-\sin\varphi_T \cdot \cos\theta}\right) \tag{2}$$

where

$\cos\theta = (R/a)\cdot\cos^2 E_S + \sin E_S \cdot \sqrt{1-(R/a)^2} \cdot \cos^2 E_S$,

A_S, E_S = azimuth angle and elevation angle of a satellite observed from the earth station,

θ = central angle between the Earth station and the subsatellite point,

R = geocentric distance of the Earth station,

a = geocentric altitude of the geostationary satellite.

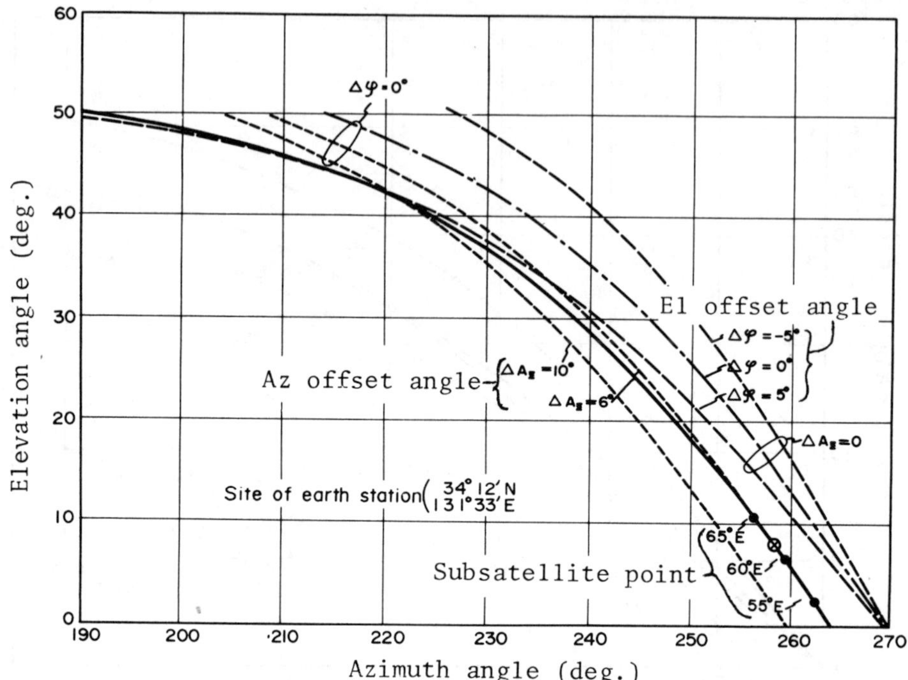

Fig. 4 Geostationary arc and pointing direction of polar-mount antenna at Yamaguchi.

As shown in Fig. 3, the geostationary arc is different from the pointing direction of the polar-mount antenna. However, it is interesting to see that the pointing direction can be kept very close to the geostationary arc, if the angle φ_T of Eq. (1) is slightly changed to $\varphi_T + \Delta\varphi$. For example, if a satellite is located at 174°E, the optimum angle $\Delta\varphi$ of the polar axis is found to be 6.5°. The broken curve in Fig. 3 shows this situation. The antenna can access the satellites lying within 174°E ±5° in longitude only by a rotation of the antenna around the inclined Ha axis. It should be noted that a small daily movement is superimposed to the motion in longitude, as is seen in Fig. 2a, and that this movement should be tracked by driving small reflectors automatically.

Figure 4 shows another example of geostationary arc and antenna pointing direction at Yamaguchi Earth station in Japan. The satellite is located at 62°E over the Indian Ocean, and the elevation angle seen from the Earth station is as low as 8°. The locus of the antenna pointing is again different from the geostationary arc. For the low elevation angle satellites, it can be seen from the figure that the azimuth angle of the

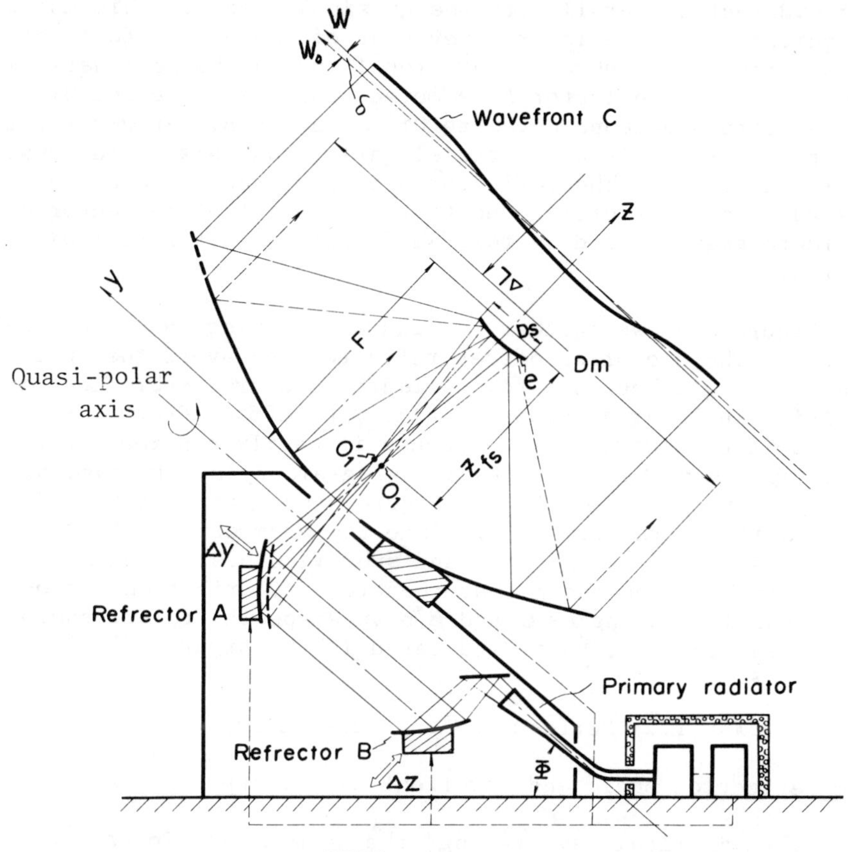

Fig. 5 Cassegrain antenna fed by beamguide reflectors.

rotation axis should be changed slightly keeping the inclination of the polar axis at the exact latitude. In this manner, if the longitudinal position of a satellite is determined, the optimum offset angle can be selected either in Az or El. This axis is named the quasipolar axis.

Figure 5 shows the proposed configuration of the limited steerable antenna applied to a Cassegrain type. It consists of a semifixed main reflector with a subreflector, a primary radiator on the ground, and a pair of movable beamguide reflectors. The longitudinal motion of a satellite can be tracked by rotating main and subreflector around the quasi-polar axis to a certain degree, and the perpendicular daily excursion of the satellite is tracked by controlling small beamguide reflectors. The beamguide reflector A is movable in the direction parallel to the quasipolar axis. This motion is equivalent to the lateral movement of one of the foci of the hyperboloidal subreflector, resulting in the beam deflection. When the reflector A is moved, the feeding point O_1 of the Cassegrain antenna moves to O_1'. The resultant wavefront C at the aperture is distorted slightly, and this would cause a gain reduction. The reflector B is provided to compensate this wavefront distortion due to the movement of reflector A.[5] The improvement gained by this is described in the following section.

Figure 6 shows another configuration using an offset reflector. The use of an offset reflector can avoid the aperture-blocking effect due to a primary feed or a subreflector, resulting in lower sidelobe levels. The offset Gregorian antenna is considered here, because it easily can realize a larger aperture than the offset Cassegrain type. In case of the configuration shown in the figure, the antenna has good polarization characteristics, since the asymmetrical field components caused by the main and subreflectors cancel each other. It has been shown that the cross-polarization components cancel when angles α and β have a special relationship[6-8] with eccentricity e, as is indicated in the figure.

IV. Performance of the Antenna

Beam Deflection Angle and Wavefront Distortion

By geometrical ray-tracing, the beam deflection angle and the wavefront distortion are calculated. The deflected wavefront W in Fig. 5 is obtained by averaging the distorted wavefront C. The beam deflection angle δ, then, is the angle between W and the original wavefront W_0. Table 1 gives the parameters that are used in the calculation.

Figure 7 shows the relation between the displacement Δy of the reflector A and the beam deflection angle δ. The horizontal unit and the vertical unit are normalized by the aperture diameter D_m of the main reflector and the half-power beamwidth HPBW, respectively. As shown in this figure, if the reflector displacement Δy is 20 by the horizontal unit (that is about 6.4λ), the beam deflection angle δ becomes about 2.5 HPBW (that is 0.55°). The angle δ is given by using the eccentricity e of the subreflector

$$\delta = K \cdot [(e-1)/(e+1)] \cdot (\Delta y/F_m) \tag{3}$$

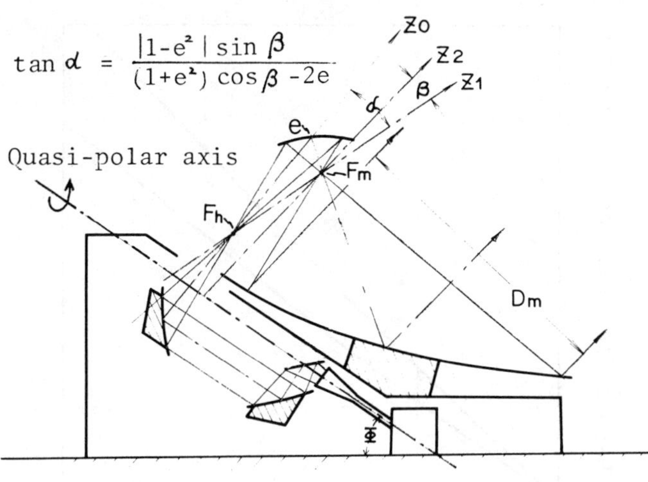

Fig. 6 Offset Gregorian antenna fed by beamguide reflectors.

Table 1 Antenna parameters for calculation (length normalized by wavelength)

Parameters	Case I	Case II
Main reflector (paraboloid)		
Aperture diameter D_m	320	320
Focal length F_m	82.5	82.5
Half aperture angle θ_m	88.5°	88.5°
Sub-reflector (hyperboloid)		
Eccentricity e	1.336	1.243
Distance between the foci of hyperboloid Z_{fs}	82.5	82.5
Aperture diameter D_s	44.8	32.0
Diameter ratio D_s/D_m	0.14	0.10
Primary radiator		
Half flare angle	15.5°	11.5°
Beamguide reflectors A, B		
Focal length F_A, F_B	25.0	25.0
Reflector interval g	75.0	75.0

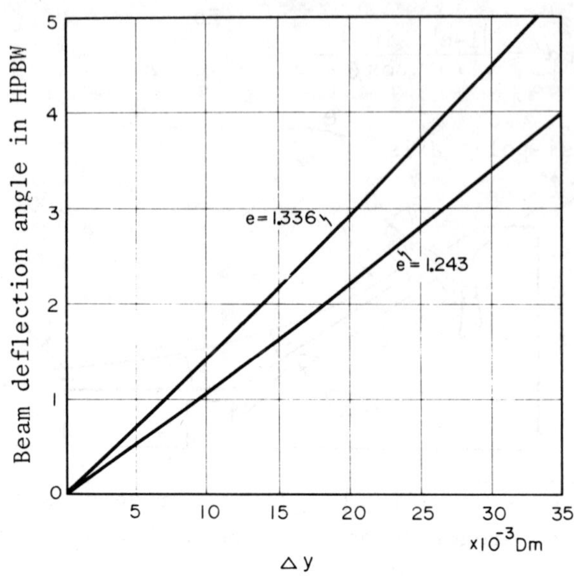

Reflector displacement

Fig. 7 Beam deflection angle vs. reflector displacement.

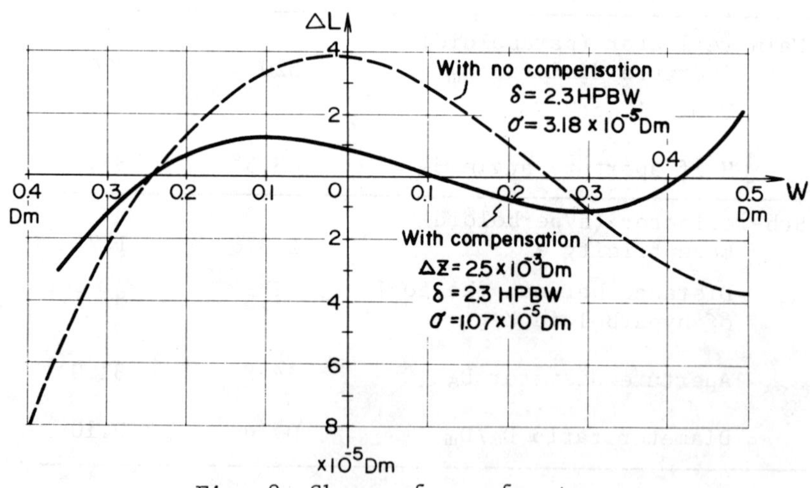

Fig. 8 Shape of wavefront.

where

K = beam deviation factor

F_m = focal length of main reflector

A larger eccentricity gives larger angle δ.

Compensation of Wavefront Distortion

Figure 8 shows the shape of equiphase plane over the antenna aperture for the cases with compensation and without it by the moving of small reflector B in the Z direction. By moving the reflector B in Z direction, the wavefront distortion due to beam deflection becomes fairly small.

Figure 9 shows the relation of beam deflection angle δ and the wavefront distortion σ for both cases. Large deflection causes larger distortion, and the gain reduction due to it could be very serious if the compensation is not made. The compensation is made without changing the beam direction.

The illumination areas of the main reflector and subreflector by the primary radiatory moves according to the movement of the reflector A, and the amount of spillover is increased. However, if the reflector B is moved to compensate the wavefront distortion, spillover loss also is reduced as well, although the conditions for both purposes do not match perfectly.

Gain Reduction

Figure 10 shows the approximate gain reduction obtained by calculating losses due to the wavefront distortion and spillover. From this figure, it can be understood that the

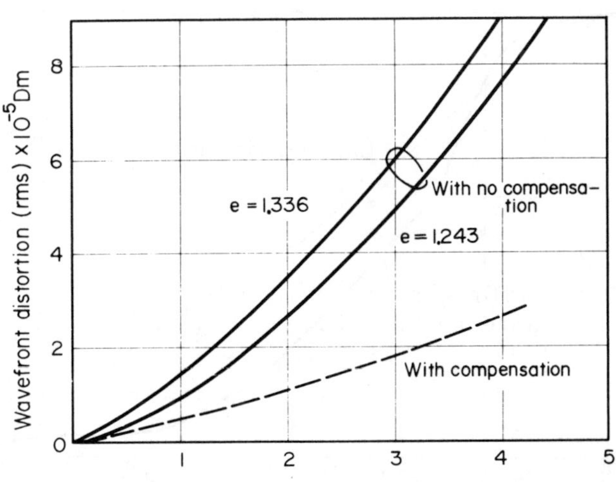

Fig. 9 Wavefront distortion vs. beam deflection angle.

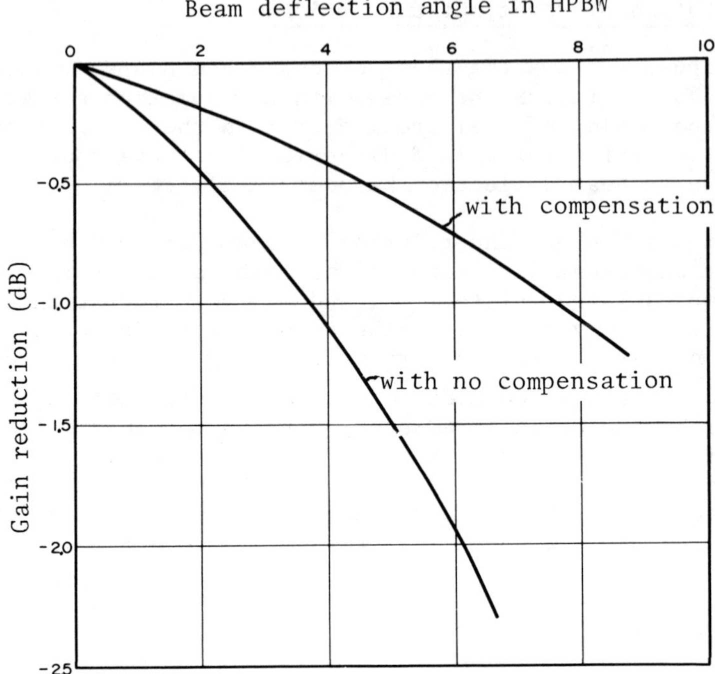

Fig. 10 Gain reduction due to beam deflection ($D_m = 320\lambda$).

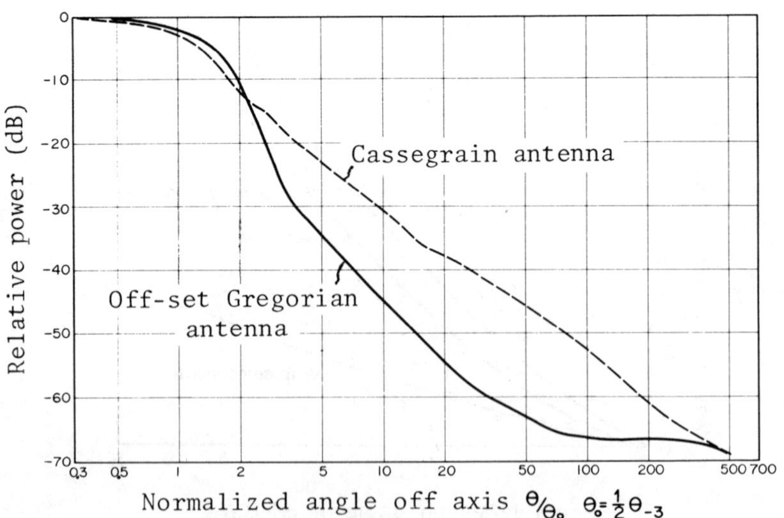

Fig. 11 Radiation pattern envelope.

compensation of the wavefront distortion is very effective. If the permissible gain reduction is assumed to be 1 dB, the beam deflection angle could be about 7 HPBW (half-power beam-width) in the compensated case. This should be compared to 3 or 4 HPBW in the noncompensated case. This deflection angle would be sufficient for the movement of Intelsat IV satellites, as shown in Fig. 2. In the offset-type antenna shown in Fig. 6, the side-lobe structure and gain reduction factor are different in the El plane because of the asymmetrical antenna structure. The deflection angles within 1-dB gain reduction are about 4 and 3 HPBW (noncompensation case) in the upward and downward directions in the El plane, respectively.

Antenna Noise Temperature

Figure 11 shows the normalized patterns of the Cassegrain antenna and the offset Gregorian antenna.[9,10] The angle and the gain are normalized by the half-angle of HPBW and the main beam gain, respectively.

If the radiation pattern is axially symmetric, the beam efficiency γ_i in the solid angle Ω_i is given by

$$\gamma_i = 1/4\pi \int_{\Omega_i} G(\Omega)\ d\Omega = \int_{\theta_i}^{\theta_{i+1}} p(\theta) \sin\theta\ d\theta \Big/ \int_0^{\pi} p(\theta) \sin\theta\ d\theta$$
$$\doteqdot p(\theta_i) \sin\theta_i \cdot \Delta\theta_i \Big/ \sum_i p(\theta_i) \sin\theta_i \cdot \Delta\theta_i \tag{4}$$

where Ω = solid angle,
$G(\Omega)$ = antenna gain function,
$p(\theta)$ = normalized antenna gain function.

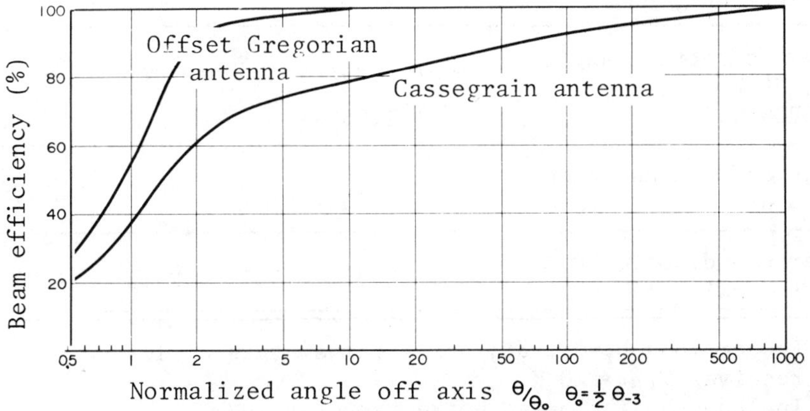

Fig. 12 Beam efficiency.

Figure 12 is the calculated beam efficiency obtained by Eq. (4). External antenna noise temperature T_a is given by

$$T_a = 1/4\pi \int_{4\pi} G(\Omega)\ T_b(\Omega) \doteqdot \sum_i \gamma_i \cdot T_i \qquad (5)$$

where $T_b(\Omega)$ = brightness temperature in solid angle Ω,
T_i = average brightness temperature in the i-th solid angle segment.

Brightness temperature $T_b(\Omega)$ can be found in literature.[11] Antenna noise temperature can be evaluated by Eq. (5) and the values are listed in Table 2. The antenna noise temperature of the offset Gregorian antenna is better by about 10°K at 5°

Table 2 Example of Earth station antenna parameters (G/T = 40.7 dB/°K)

Parameters		Cassegrain antenna		Offset Gregorian antenna	
Frequency, GHz		4	11	4	11
Antenna noise temperature due to external noise sources Ta, °K	Eℓ=5°	36	47	26	37
	30°	12	15	5	9
	90°	6	8	3	5
System noise temperature[a]	Eℓ=5°	67	78	58	68
Gain margin for beam deflection[b], dB		~ 1.5	~ 1.5	~ 1.5	~ 1.5
Beam deflection angle δ, HPBW (HPBW=0.15°)		~ 4(0.6°) ~ 7[c](1.05°)	~ 4	3 ~ 4	3 ~ 4
Required antenna gain G, dB		60.7	61.3	60.0	60.7
Aperture diameter D,m (η = 65%)		32.3	12.5	29.6	11.7

[a] Feeder loss L_f=0.2 dB; noise temperature of low-noise receiver T_{LNR} = 20°K.
[b] Include the increased noise temperature.
[c] With wavefront compensation.

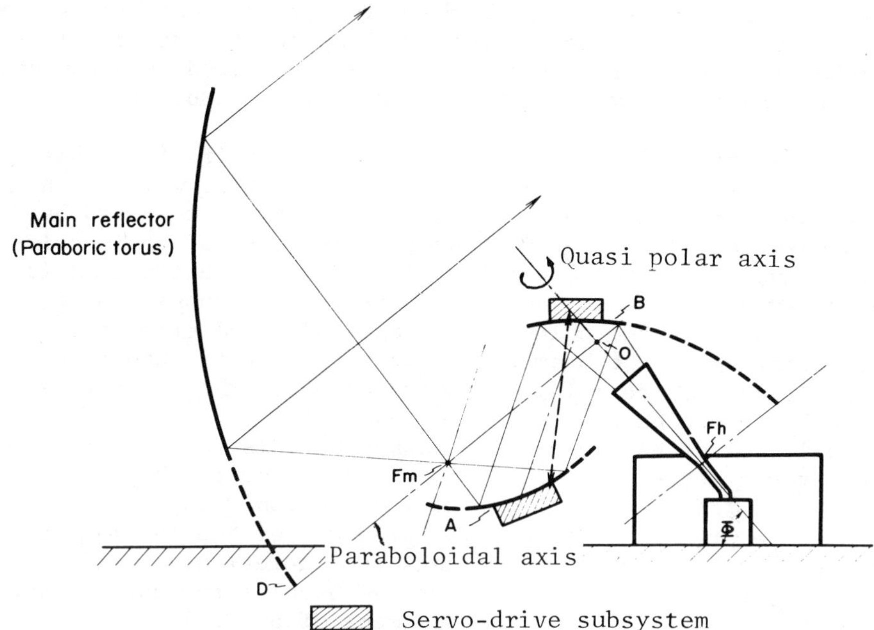

Fig. 13 Torus antenna fed by beamguide reflectors.

elevation than that of the Cassegrain antenna. Beam deflection will raise the sidelobe levels, but its contribution to noise temperature is estimated to be little.

Antenna System Parameters

Table 2 summarizes the calculated antenna characteristics to obtain the G/T of 40.7 dB/°K. The feeder loss L_f and the noise temperature T_{LNR} of the low-noise receiver are assumed to be 0.2 dB and 20°K, respectively. In this table, the gain margin for beam deflection is selected to be 1.5 dB, but this value may be reduced to about 1 dB in case of small angular movement. The required gain of the offset Gregorian antenna is smaller by about 0.7 dB than that of the Cassegrain antenna because of its low-noise performance. Therefore, the aperture diameter of offset Gregorian antenna can be made smaller than that of Cassegrain antenna.

V. Discussion

It has been shown in the previous sections that the geostationary satellite can be tracked by the proposed limited steerable structure, although the main reflector has to be rotated by a limited amount for the longitudinal drift. If completely fixed main reflector is to be required, the concept of torus antenna[12] may be introduced to this structure.

An example of the torus antenna with a pair of movable beamguide reflectors is shown in Fig. 13. It is well known that the mirror surface of the torus antenna is composed by palabolic and spherical curvatures. The curve of $\overset{\frown}{CD}$ is palaboloid and the curve perpendicular to $\overset{\frown}{CD}$ is spherical. O is the spherical center. The quasipolar axis is chosen to be perpendicular to the palaboloidal axis $\overline{DB}$. The effective feed point F_m of the beamguide system is at the midpoint between O and D. According to the rotation of the reflectors A and B, F_m can be pivoted around the point O. The drift direction of satellite can be tracked by this rotation and the movement crossed to this direction can be tracked by moving the reflector A in parallel direction with the rotation axis. The primary radiator, the transmitter and the receiver can be fixed on the ground in this configuration. On the contrary, the primary radiator of the usual torus antenna has to be moved together with the movement of a satellite.

VI. Conclusion

The new type of antenna described in this paper is applicable to the movement of Intelsat IV satellites and future geostationary satellites. This antenna may be applied to many Earth station antennas because of its inexpensive structure.

References

[1] Phillips, C.J.E. and Clarricoats, P.J.B., "Optimum Design of a Gregorian-Corrected Spherical-Reflector Antenna," Proceedings of the Institute of Electrical and Electronics Engineers, Vol. 117, April 1970, pp. 718-734.

[2] Byars, M., "An Approximate Method of Evaluating the Effects of Beam Steering by Subreflector Tilting in a Cassegrain System," IEE Conference on Earth Station Technology, Publ. 72, Oct. 1970.

[3] Karikomi, M., Kumazawa, H., and Kataoka, Y., "Space Communication Antennas Designed by the Ray Lattice Method," Rept. of Technical Group of the Institute of Electronics and Communication Engineers of Japan, A.P 70-73, 1971.

[4]Urasaki, S., and Mizusawa, M., "Radiation Characteristics of a Cassegrain Antenna with a Displaced Subreflector," Rept. of Technical Group of the Institute of Electronics and Communication Engineers of Japan, A.P 71-37, 1971.

[5]Akagawa, M., and Yokoi, H., "A Beam Scanning Method in the Antenna with Dual Reflector Antenna," National Conference of the Institute of Electronics and Communication Engineers of Japan, S6-12, July 1974.

[6]Graham, R., "The Polarization Characteristics of Offset Cassegrain Aerials," IEE International Conference on Radar-Present and Future, No. 105, Oct. 1973.

[7]Tanaka, H., and Mizusawa, M., "Reflector Design for Elimination of Cross-Polarization Components in the Antenna with Dual Offset Reflectors," Rept. of Technical Group of the Institute of Electronics and Communication Engineers of Japan, A.P 73-84, 1974.

[8]Mizuguchi, Y., "Radiation Characteristics of Offset Gregorian Antenna-Calculation by Means of Current Distribution Method," Rept. of Technical Group of the Institute of Electronics and Communication Engineers of Japan, A.P 74-76, 1975.

[9]Itohara, S., Endo, S., Hatakeyama, K., Mizusawa, M., and Betsudan, S., "Cassegrain Antenna Fed by Four-Reflector Beam-Waveguide," Rept. of Technical Group of the Institute of Electronics and Communication Engineers of Japan, A.P 71-39, 1971.

[10]Mizuguchi, Y., Akagawa, M., and Yokoi, H., "Radiation Characteristics of Offset Gregorian Antenna (Part II)," Rept. of Technical Group of the Institute of Electronics and Communication Engineers of Japan, A.P 75-45, 1975.

[11]Reed, H.H., "Noise Curves for High-Gain Antennas," Microwaves, Vol. 6, April 1967, pp. 46-49.

[12]Hyde, G., Kreutel, R.W., and Smith, L.V., "The Unattended Earth Terminal Multiple-Beam Torus Antenna," Comsat Technical Review, Vol. 4, Fall 1974, pp. 231-262.

AUTOMATIC COMPENSATION OF CROSS-POLARIZATION COUPLING IN COMMUNICATION SYSTEMS USING ORTHOGONAL POLARIZATIONS

Heinz Kannowade*
AEG-Telefunken, Backnang, Germany

Abstract

In frequency reuse satellite communication systems using orthogonally polarized waves, depolarization due mainly to rain and Faraday rotation reduces channel isolation. Several methods for compensating cross-polarization coupling have been proposed. The principles of the cancellation or cross-coupling method and of Chu's method are explained briefly. Possible circuit realizations for Chu's method with the application of either mechanically or electronically tuned components are given. The compensation can be performed in the dual polarized waveguide section following the antenna feed or in a system with two separate lines after the orthomode transducer. An automatic control system that avoids iterative methods or real-time computations is proposed. Control voltages are generated on the receive side by phase-sensitive demodulation of beacons contained in the communication channels. Slightly modified conventional tracking receivers may be used for the demodulation. The system can be used without modification for the reception of either dual linear or dual circular polarizations. How the control system reaches its steady state is demonstrated for several depolarization conditions.

Presented as Paper 76-304 at the AIAA/CASI 6th Communication Satellite Systems Conference, April 5-8, 1976, Montreal, Canada.

*Dipl.-Ing.

I. Introduction

The increasing requirements for communication system capacity can be satisfied either by accessing higher frequencies or by using the introduced frequencies more effectively. The capacity of a radio communication system with limited bandwidth can, in theory, be doubled by using orthogonally polarized waves. Such a frequency reuse system transmits two broadband channels either by two orthogonal linearly polarized waves or by two opposite-sensed circularly polarized waves. The degree of isolation between the two channels is related directly to the polarization purity of the signal waves. Reduction of polarization purity (cross-polarization coupling) is caused, in practice, by nonideal rf components, by misalignment of the antennas, by rain, and (in the case of dual linearly polarized satellite communication systems) by Faraday rotation. Some of these effects are time-varying in nature.

At present, the feasibility of introducing frequency reuse with orthogonal polarizations is being investigated, with principal interest in its application to satellite communication systems. In an operational system, some kind of compensation network must be provided in order to minimize cross-polarization coupling of the two broadband channels. Because of the time dependancy of the cross-polarization effects, this network must be incorporated in an automatic control system.

Figure 1 shows a possible structure of such a frequency reuse system. The two communication

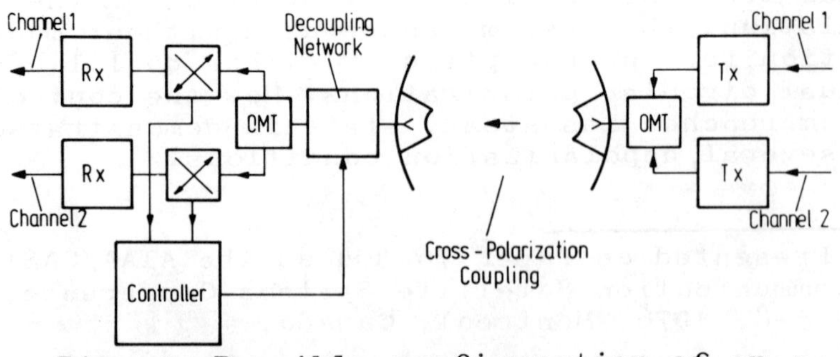

Fig. 1 Possible configuration of an adaptive compensation system.

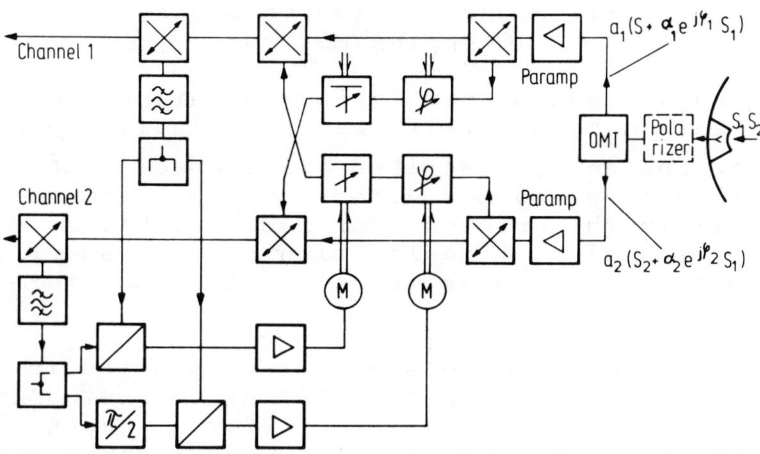

Fig. 2 Cancellation or cross-coupling method (with control system for channel 1).

channels are amplified separately and are combined by an orthomode transducer. Cross-polarization coupling occurs on the propagation path. In this example, the compensation network follows the receive antenna feed. After the decoupling network, the two channels are separated by an orthomode transducer. A controller that steers the compensation network decouples its input signals from the communication channels.

In this paper, a brief explanation is given of the principles of two compensation methods: The cancellation or cross-coupling method, and the method suggested by Chu. An automatic control system for the cancellation network is proposed. Possible circuit realizations for Chu's method are given. The compensation can be performed in a dual polarized waveguide section or in a system with two separate lines. An automatic control system that avoids iterative methods or real-time computations is proposed. Control voltages are generated on the receive side by phase-sensitive demodulation of beacons contained in the communication channels. Slightly modified conventional tracking receivers may be used for the demodulation. The system can be used for the reception of dual linear or dual circular polarizations without modification. How the control system reaches its steady state is demonstrated for several depolarization conditions.

II. Cancellation or Cross-Coupling Method

One obvious method for compensating cross-polarization coupling is the cancellation method as, for instance, described by DiFonzo.[1] The compensation is performed after the separation of the channels in the orthomode transducer. Figure 2 indicates a circuit realization of the cancellation method, including the control system for channel 1.

S_1 and S_2 are the signals transmitted in channels 1 and 2, respectively. At the network input, channel 1 contains (except for a common attenuation factor) the signal S_1 and a contribution from signal S_2 due to cross-polarization coupling. In channel 2, the reverse is true; i.e., the indices of S are reversed. Obviously the S_2 contribution in channel 1 can be cancelled by coupling a signal of equal amplitude and opposite phase from channel 2 to channel 1. The amplitude and phase of the cancellation signal are adjusted by a variable attenuator and a variable phase shifter.

The control circuit for automatic adjustment of the variable elements also is given in Fig. 2. For the generation of the control voltages, a beacon is required. The beacon is contained only in the transmitted channel 2 and is coupled out of both receive channels in a narrow band. The residual beacon in channel 1 is demodulated coherently twice: once with the beacon from channel 2, and once with the same signal shifted by 90°. These functions can be performed by a slightly modified conventional tracking receiver with one sum and one difference channel. The two generated dc voltages are amplified and feed two servomotors, which move the attenuator and the phase shifter. If electronic instead of mechanical attenuators and phase shifters are to be used, drive motors must be replaced by integrators.

It can be shown that this system balances automatically. The balancing of the phase shifter is independent of the position of the attenuator, but the control voltage for the attenuator is not independent of the position of the phase shifter. Therefore, after a perturbation of the system, the attenuator will reach its steady state only after the phase shifter.

The generation of control voltages for cross-coupling compensation in receive channel 2 can be performed in the same way as described previously with a beacon contained in the transmitted channel 1. The compensation network can be inserted only after the low-noise pre-amplifiers; otherwise, the increase of the system noise temperature due to the couplers would be too large.

Perfect compensation can be achieved only for the beacon frequencies. Cross-polarization coupling may remain at other frequencies in the band, because of nonuniform frequency characteristics of the network components and of the disturbance phenomena. With the compensation network placed after the LNA's, additional coupling remains because of different frequency responses of the low-noise amplifiers. Therefore, some consideration is given to the application of this compensation method to separate subchannels in the band, with the actual compensation performed in the intermediate frequency range. Such a system requires a beacon in each subchannel in order to derive the control voltages.

III. Compensation Achieved by a Transformation of Signal Wave Polarizations

Other compensation methods have been proposed which influence the polarization of the signal

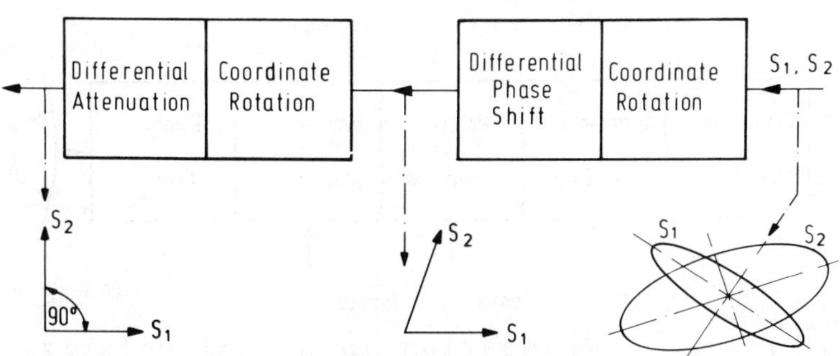

Fig. 3 Chu's compensation method.

waves. Compared with the cancellation method, they have the advantage that they work without or with reduced deterioration of the system noise temperature. However, they can be applied only to the entire communication band, and it is not yet clear how well this broadband performance can be realized.

A publication of Kreutel [2] should be mentioned. He has proposed a compensation network that is distributed between the transmitter and the receiver. The network works with a minimum of adaptive elements, however, and is restricted to certain types of depolarization, not including all types of depolarization that appear in practice.

Consideration in some detail now will be given to a compensation method suggested by Chu.[3] The compensation device may be placed before or behind the propagation path. We shall assume in the following that it is located in the receive station.

The polarization of the signal waves is, in general, elliptic and nonorthogonal. The ellipses are fat when the transmitted signals are dual circularly polarized, and they are slender in the case of dual linear polarization. The transformation is performed in two steps, as shown in Fig. 3. First the polarizations are made linear (but not, in general, orthogonal) by means of a coordinate rotation and differential phase shift. Second, the polarizations are orthogonalized by a further, independent coordinate rotation combined with a differential attenuation. Hence, four parameters must, in general, be adjusted to compensate for a given depolarization condition.

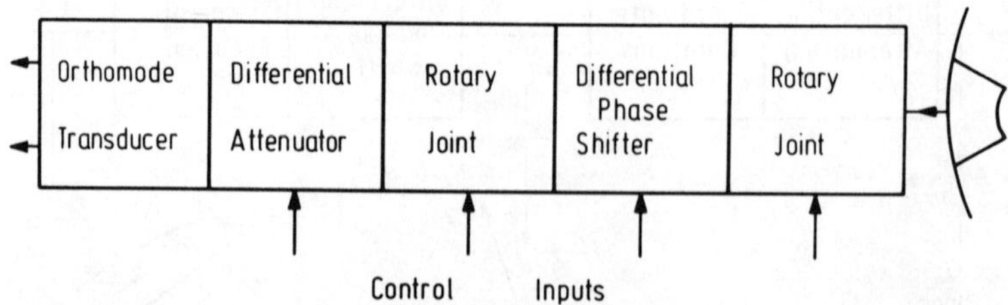

Fig. 4 Compensation in a dual polarized waveguide section.

In the case of dual circular polarizations, i.e., when fat polarization ellipses appear at the input of the network, the required differential phase shift is in the range of 90°. In order to reduce the variation range of the differential phase shifter, a polarizer may be inserted between the antenna and the network.

A. Circuit Realization of the Compensation Method Suggested by Chu

Consider first circuits that realize the different functions required by the Chu-type compensation network: coordinate rotation, differential phase shift, and differential attenuation. Two fundamental types of orthogonalizing networks can be distinguished, dependent on whether the orthomode transducer that separates the two channels is placed before or behind the network.

With the orthomode transducer placed after the network, the compensation elements must operate on the two received signal waves (in the H_{11}-mode), both contained in a dual polarized waveguide (see Fig. 4). A differential phase shifter is connected to the feedhorn of the receive antenna via a rotary joint. An independent angular adjustment of the subsequent differential attenuator is made possible by a second rotary joint. The two channels are separated by an orthomode transducer following the network.

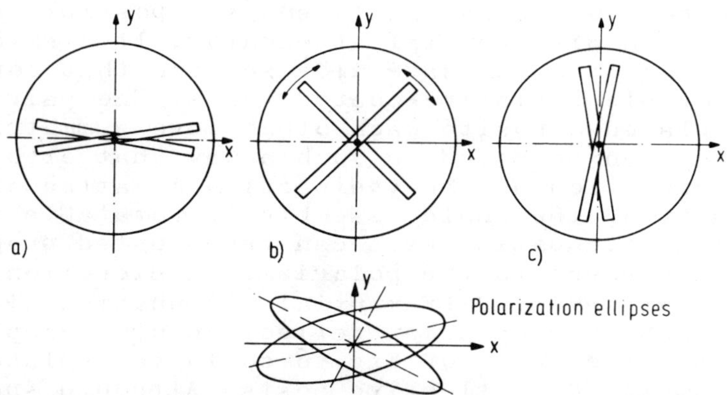

Fig. 5 Differential phase shifter (schematic cross section).

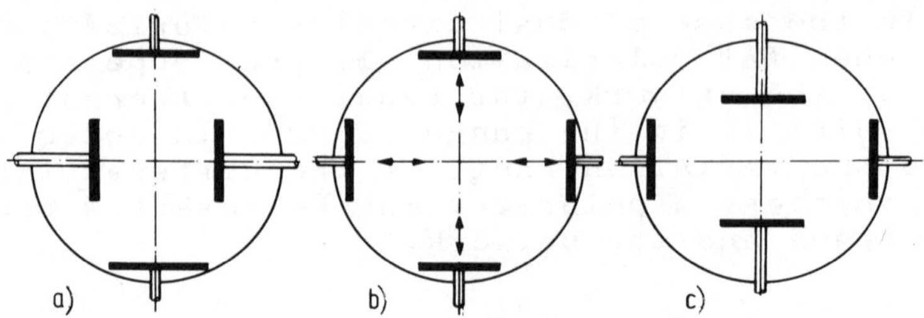

Fig. 6 Differential attenuator (schematic cross section).

The realization of a differential phase shifter is shown schematically in Fig. 5. The phase shift elements are installed in a circular waveguide section. The elements consist of rotatable dielectric vanes. Their axis of rotation is identical with the waveguide axis. The vanes rotate only in opposite directions and with equal angle increments, so that the assembly remains symmetric to the x and y axes. In position a, maximum delay is achieved for the x components with respect to the y components of the signal wave field strength vectors. With the vanes in position b, both components are delayed equally, and in position c maximum delay is achieved for the y components. By rotation of the whole assembly, the correct orientation of the differential phase shifter main axes with respect to the polarization ellipses of the received signal waves can be obtained.

Figure 6 schematically shows a possible realization of a differential attenuator. It also consists of a circular waveguide section that contains four movable resistive-coated vanes. The pairs of vanes placed opposite each other move symmetrically. They can be moved in such a way that either the x direction or the y direction is attenuated. In addition, the whole assembly is rotatable, so that the attenuation axis can be adjusted properly with respect to the polarization directions of the signal waves. Differential attenuation also can be achieved by power reflection or decoupling. The resistive vanes of Fig. 6 could be replaced, for example, by reflective posts. Although such nondissipative attenuation is favorable in terms of system noise temperature, it may complicate the

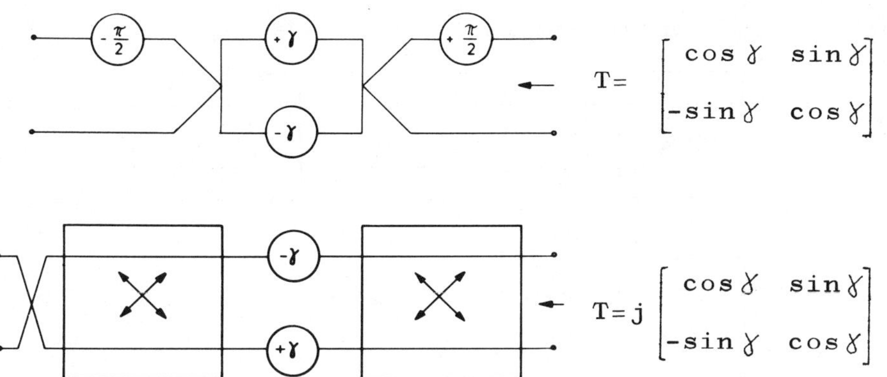

Fig. 7 Coordinate rotation.

realization of a differential attenuator. Reflective attenuation could result in multiple reflections and a frequency-sensitive response of the assembly.

With the orthomode transducer preceding the compensation network, the different functions (coordinate rotation, differential phase shift, and differential attenuation) must be realized in a system where the two channels are physically separate. Differential phase shift and differential attenuation obviously can be obtained by imposing different phase shifts and different attenuations on the signals in the two lines.

Two possible means of performing a coordinate rotation are shown in Fig. 7. In Fig. 7a, two hybrids, a differential phase shifter, and two 90° fixed phase shifters are used. The function of the circuit can be determined by investigating its transmission matrix. The matrix is given in Fig. 7a and, as can be seen, represents a coordinate rotation. (A differential phase shift of $2\gamma^{o}$ produces a coordinate rotation of γ^{o}.) As shown in Fig. 7b, the two 90° fixed phase shifters can be omitted if 3-dB couplers are used instead of hybrids.

Figure 8 shows the combination of a polarizer and a coordinate rotator. The polarizer can be used in the case of dual circular polarization in order to transform the fat polarization ellipses into slender ones. The polarizer function is per-

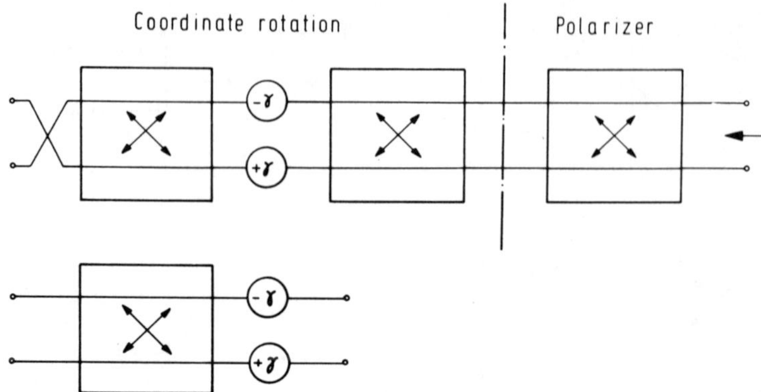

Fig. 8 Combined coordinate rotator and polarizer.

formed by an additional 3-dB coupler. An equivalent circuit is obtained by dropping the two 3-dB couplers in series, as shown in Fig. 8b. Thus, with the orthomode transducer following the antenna, the compensation network can be realized by the circuit in Fig. 9.

The following functions of the network are distinguishable: coordinate rotation, differential phase shift, differential attenuation, and fixed 45° coordinate rotation performed by a hybrid. The input 3-dB coupler can be dropped in the case of dual circular polarizations. The coordinate rotation at the input can be dropped if the complete feed system is rotatable. In this type of compensation network, mechanically or electronically tuned elements may be used. In an electronically tuned network, integrators are used instead of motors.

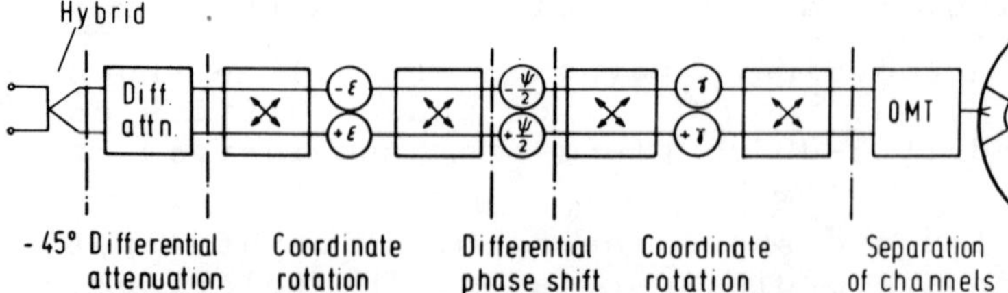

Fig. 9 Compensation network in a system with two separate lines.

B. Automatic Control Circuits Incorporating Chu's Network

In Fig. 10, a block diagram of an automatic control system incorporating Chu's network is shown. In this example, the different functions of the compensation network are realized in a dual polarized waveguide section.

In order to generate the voltages V_1 to V_4, which control the four actuators, a beacon must be transmitted in each of the two broadband channels (see Fig. 11). Let f_1 and f_2 be the beacon frequencies. Because of cross-polarization coupling, both beacons appear in both receiver channels. Four beacon signals therefore can be distinguished: b_{11}, b_{12}, b_{21}, and b_{22}. (The first index refers to the receive channel number and the second to the transmit channel number or frequency index.) For convenience, the beacon pairs are extracted from the communication channels after the low-noise preamplifiers and are converted to an intermediate frequency, by a common local oscilator. After amplification and separation of the beacon pairs with bandpass filters BP_1 and BP_2, the control voltages are generated by four coherent demodulations, as shown in Fig. 10. The fre-

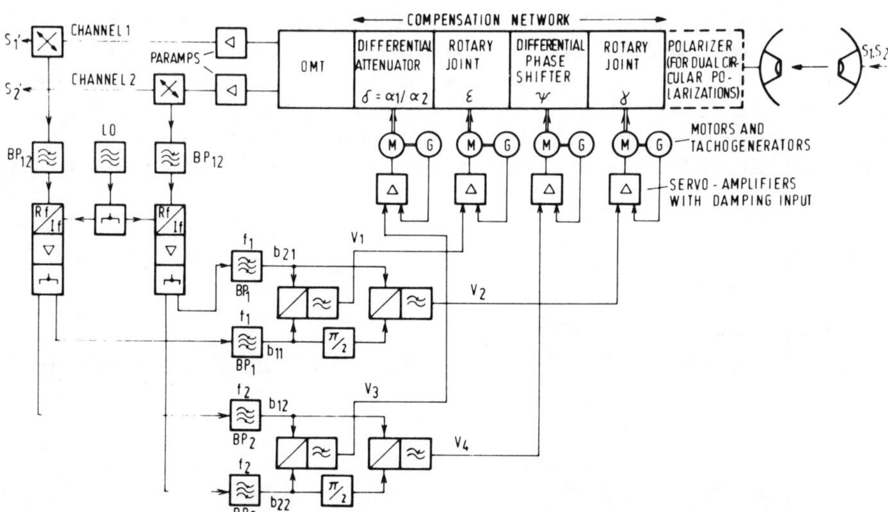

Fig. 10 Automatic control system for compensating cross-polarization coupling.

quency difference f_1-f_2 should be made small in order to keep the passband of filter BP_{12} narrow, but it must be great enough to enable the separation of f_1 and f_2 by the bandpass filters BP_1 and BP_2. As is common practice in tracking receivers, the beacons b_{11} and b_{22} also may be regenerated by phase-locked loops instead of narrow-band filtering.

In Fig. 10, functions from the IF amplifiers to the output of the demodulators also can be performed by two conventional tracking receivers, each having one sum and one difference channel. Only a second coherent demodulator with the sum signal shifted by 90° must be added to each receiver.

The four control voltages feed motors that drive the four actuators. A damping feedback for the stabilization of the system is indicated. In the case that electronic network components are used, the motors are replaced by integrators. When the compensation network is in the state of balance, no cross-coupling occurs, so that the beacon signals b_{12} and b_{21} and the control voltages V_1 to V_4 vanish, and the motors stop.

When perfect balance has been achieved for the beacon frequencies, cross-polarization coupling may remain at other frequencies in the band because of nonuniform frequency characteristics of the network components and of the disturbance phenomena. One important requirement of the rf component design will be to keep these residual errors small. With an automatic control system in

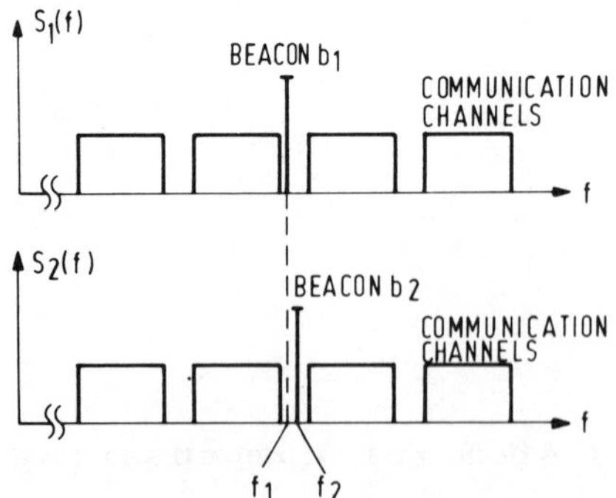

Fig. 11 Beacon concept.

operation, statistical data on cross-polarization coupling can be gathered by recording the values of the four parameters of the compensation network.

In the following, we shall demonstrate for some examples that balance is reached automatically by this system. We shall show separately for the differential attenuator and for the differential phase shifter how they reach their steady state when they are put out of balance.

To demonstrate the transient behavior of the differential attenuator, we plot the field of vectors formed by components V_1 and V_3 as a function of parameters ε and δ. The voltage V_1, which controls ε, is plotted in ε direction, and V_3, which controls δ, is plotted in δ direction. Assuming that the time rate of change of the parameters is tages, the vector field can be conceived as a velocity field of parameters ε and δ. The trajectory in the ε-δ space as the compensation network moves into the balanced state can be traced along the velocity vectors thus formed.

It can be shown that the balance of the differential phase shifter is unaffected by the dif-

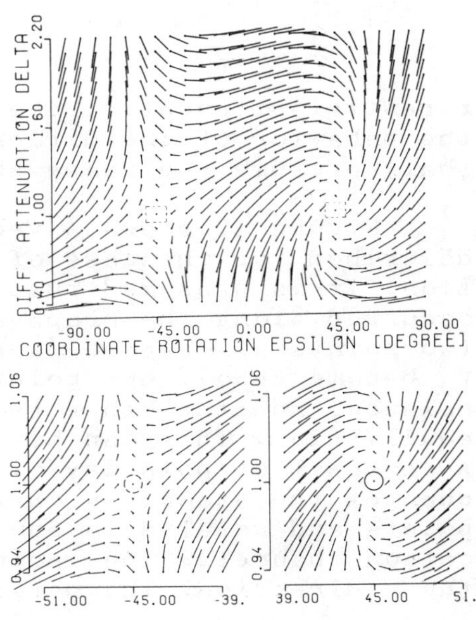

Fig. 12 Vector field $\{V_1, V_3\}$ (no depolarization).

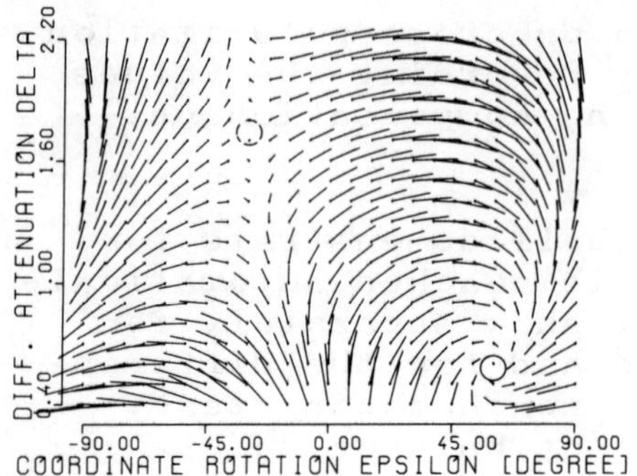

Fig. 13 Vector field $\{V_1, V_3\}$.

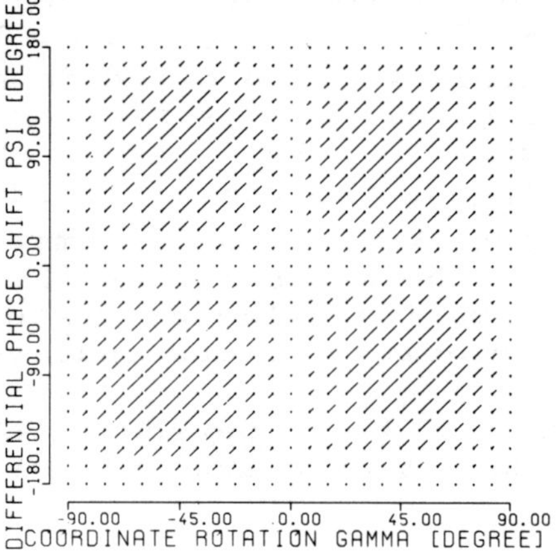

Fig. 14 Vector field $\{V_2, V_4\}$ (dual linear polarization without cross-coupling).

ferential attenuator. Therefore, in demonstrating the behavior of the differential attenuator, we shall presume the phase shifter to be balanced.

Figure 12 shows a vector field such as just described for the case of dual linear polarization. It is presumed that no depolarization has occurred along the propagation path. There exist two points of balance (where the voltages V_1 and V_3 become zero). One point represents a stable balance, and the other represents an unstable equilibrium. This can be seen in the detail views. A reversal of polarity of the voltage V_1 interchanges the stability characteristics of the two points and results in an interchange of communication channels. The vector field is periodic in the ε direction. The period is 180°.

The vector field of Fig. 13 corresponds to an arbitrarily chosen depolarization condition. Compared to Fig. 12, the points of balance have changed their position in the ε, δ plane, but the character of the field has not changed.

Both Figs. 12 and 13 show that the automatic settling into the stable point is not insured from all points of the ε, δ plane if the ε range is limited to one period (180°). Consequently, the angular range of the differential attenuator must not be limited unless some reset capability is provided in the system. The range of δ can, however, be limited to that required for balancing an extreme depolarization condition. During operation, it can be assumed that shifting of the stable point of balance occurs continuously and is slow compared with the reaction time of the system; under these conditions, balancing never is lost.

To demonstrate the transient behavior of the differential phase shifter, we plot the field of vectors formed by components V_2 and V_4 as a function of the angular orientation γ and of the differential phase shift Ψ. The vector fields are periodic in γ and Ψ; a full period in γ and Ψ is plotted. Each full period contains four balancing points, one of which is stable. Inverting the polarity of one of the control voltages shifts the condition of stability to one of the other balancing points. With two possible polarities for each

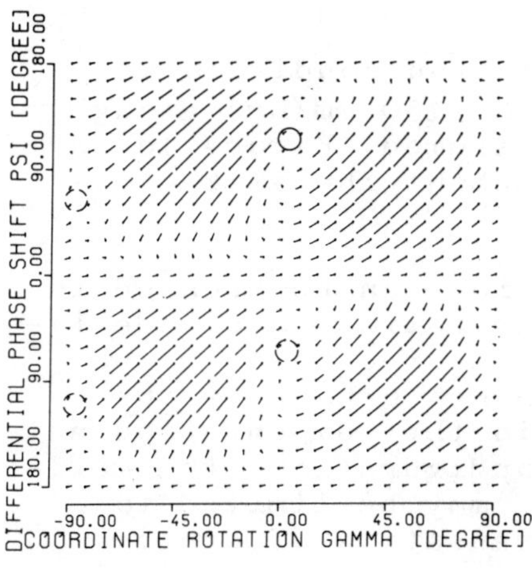

Fig. 15 Vector field $\{V_2, V_4\}$ (dual linear polarization).

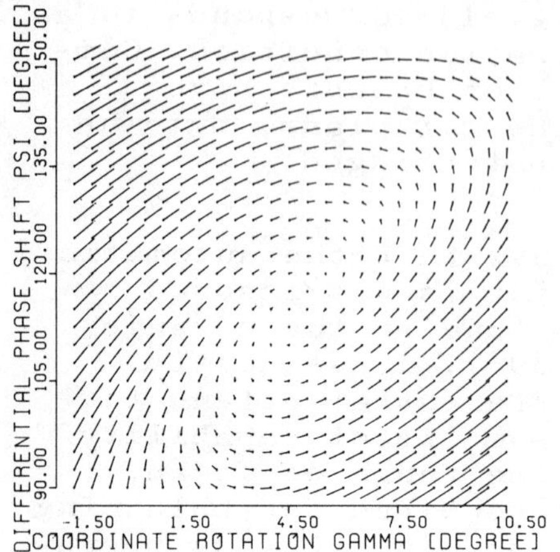

Fig. 16 Detail view.

of the voltages V_2 and V_4, there exist four combinations of polarities. Associated with each of these conditions is a different one of the four points as stable point. Points of stable and conditional balance are circled with solid and dotted lines, respectively.

Figure 14 refers to the case of dual linear polarization without cross-coupling. Since neither coordinate rotation alone nor differential phase shift alone causes cross-coupling (coordinate rotation is compensated for by a rotation of the following differential attenuator), points on the straight lines $\gamma = \pm n \cdot 90^o$ and $\Psi = \pm m \cdot 180^o$ (n, m integer) represent balance. Stable balance is, however, attained only on certain sections of these straight lines.

In Fig. 15, a depolarization condition at 11 GHz due to very heavy rain (150 mm/hr) along a 5-km propagation path is assumed. Figure 16 is a detail veiw of the region about the stable balance point.

Dual circular polarization can be treated in two different ways. The polarizations are made linear either by a polarizer inserted between the antenna and the compensation network or by the differential phase shifter in the compensation network itself. In the first condiguration, the maximum differential phase shift may be limited to values required for the compensation of cross-coupling. In the second, the range of the differential phase shifter must be increased in the order of 90°. The later configuration has the advantage that it is compatible to both circular and linear polarizations.

Figure 17 applies to the case of dual circular polarization with a polarizer in front of the network. The depolarization at 11 GHz again is assumed to be caused by very heavy rain (150 mm/hr) along a 5-km propagation path. A comparison of Figs. 14, 15, and 17 reveals that dual circular polarization is affected by rain much more than is dual linear polarization.

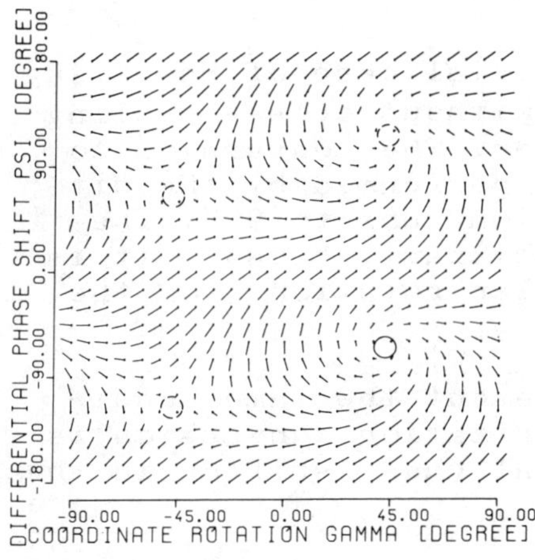

Fig. 17 Vector field $\{V_2, V_4\}$ (dual circular polarization).

Fig. 18 Vector field $\{V_2, V_4\}$ (dual circular polarization without cross-coupling, received without polarizer).

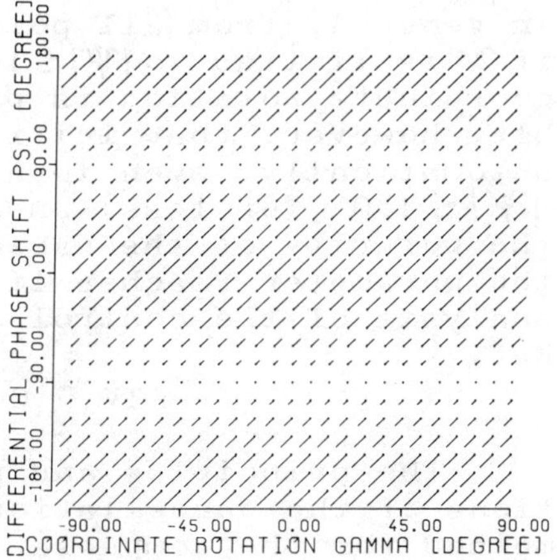

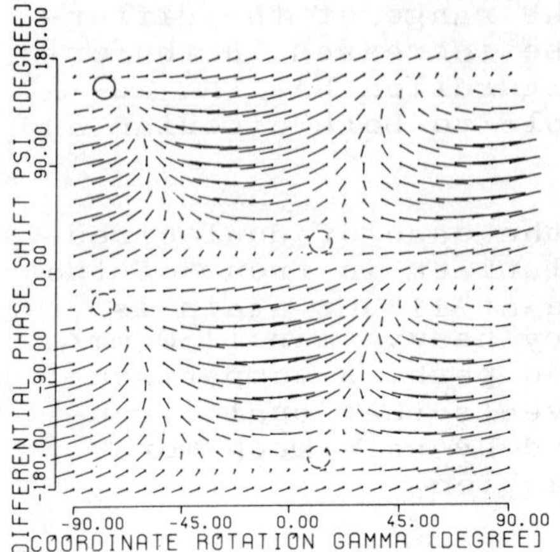

Fig. 19 Vector field $\{V_2, V_4\}$ (dual circular polarization, received without polarizer).

Figure 18 shows the trivial case of receiving dual circular polarization without cross-coupling in the absence of a polarizer. The conversion to dual linear polarization is performed by the differential phase shifter, which runs to 90°. The angular orientation γ is arbitrary, because it is compensated for by the angular rotation ε of the differential attenuator.

Figures 19 and 17 represent the same conditions, except that the polarization conversion is performed by the differential phase shifter in the former.

A comparison of Figs. 14 - 19 shows that initial acquisition of stable balance cannot be reached, in general, from all points of the γ, ψ plane if ψ is limited to $|\psi| \leq 180^o$, even when an infinite coordinate rotation is permitted. It seems possible, however, that initial balance can be obtained automatically, even in a system with limited $|\psi| \leq 180^o$ (or less) and $|\gamma| \leq 90^o$, by inverting the polarity of the parameter control voltage as the parameter reaches its limit. A more detailed analysis of the control system is published in Ref. 4.

IV. Conclusion

The principles and possible circuit realizations of the cancellation method and of Chu's method for the compensation of cross-polarization coupling have been discussed. The cancellation method is best suited for narrow-band compensation (single carriers or transponders). Because of its inherent losses, the compensation network must come after the low-noise preamplifiers, increasing the difficulties for realizing broadband performance. A beacon is required in each separately compensated band.

Chu's method has the advantage of minimum degradation of the system noise temperature. In addition, the method is capable of receiving either dual linear or dual circular polarization without modification, provided that the range of the differential phase shifter is sufficiently large. The compensation is applied to the whole 500-MHz satellite band.

For both methods, control voltages can be generated by coherent demodulation of beacon signals transmitted in each communication channel to be compensated separately. Slightly modified conventional tracking receivers with one sum and one difference channel can perform the demodulation. Each compensated communication channel requires one tracking receiver.

The transient behavior of the control system incorporating Chu's network was demonstrated for several depolarization conditions. In the examples chosen, the initial acquisition of balance can be achieved automatically, independent of the system initial state, even with the range of the differential phase shifter parameters limited, if the polarity of the voltages controlling each parameter is inverted when their limits are reached. The rotation of the differential attenuator is assumed to be unlimited.

References

[1]DiFonzo, D., "Interference Reduction Circuit," Memo. TCLT/72-2018, March 2, 1972, Comsat.

[2]Kreutel, R. W., "The Orthogonalization of Polarized Fields in Dual Polarized Radio Transmission Systems," Comsat Technical Review, Vol. 3, Fall 1973, pp. 375-386.

[3]Chu, T. S., "Restoring the Orthogonality of Two Polarizations in Radio Communication Systems," Part I, Bell System Technical Journal, Vol. 50, Nov. 1971, pp. 3063-3029; Part II, Vol. 52, March 1973, pp. 319-327.

[4]Kannowade, H., "An Automatic Control System for Compensating Cross-Polarization Coupling in Frequency Reuse Communication Systems," IEEE Transactions on Communications, Vol. COM-24, No. 9, September 1976, pp. 986-999.

Index to Contributors to Volume 55